Eberhard Breitmaier
Terpenes

Further of Interest

Fattorusso, E., Taglialatela-Scafati, O. (Eds.)
Modern Alkaloids
Structure, Isolation, Synthesis and Biology
2 Volumes
2007, ISBN 3-527-31521-7

Surburg, H., Panten, J.
Common Fragrance and Flavor Materials
Preparation, Properties and Uses
5., Completely Revised and Enlarged Edition
2006, ISBN 3-527-31315-X

Kraft, P., Swift, K. A. D. (Eds.)
Perspectives in Flavor and Fragrance Chemistry
2005, ISBN 3-906390-36-5

Ohloff, G.
Earthly Scents – Heavenly Pleasures
A Cultural History of Scents
2006, ISBN 3-906390-34-9

Eberhard Breitmaier

Terpenes

Flavors, Fragrances, Pharmaca, Pheromones

WILEY-VCH

WILEY-VCH Verlag GmbH & Co. KGaA

The Author

Prof. Dr. Eberhard Breitmaier
Engelfriedshalde 46
72076 Tübingen
Germany

1st Reprint 2008

All books published by Wiley-VCH are carefully produced. Nevertheless, authors, editors, and publisher do not warrant the information contained in these books, including this book, to be free of errors. Readers are advised to keep in mind that statements, data, illustrations, procedural details or other items may inadvertently be inaccurate.

Library of Congress Card No.:
applied for

British Library Cataloguing-in-Publication Data
A catalogue record for this book is available from the British Library.

Bibliographic information published by Die Deutsche Bibliothek
Die Deutsche Bibliothek lists this publication in the Deutsche Nationalbibliografie; detailed bibliographic data is available in the Internet at <http://dnb.ddb.de>.

© 2006 WILEY-VCH Verlag GmbH & Co. KGaA, Weinheim

All rights reserved (including those of translation into other languages). No part of this book may be reproduced in any form – by photoprinting, microfilm, or any other means – nor transmitted or translated into a machine language without written permission from the publishers. Registered names, trademarks, etc. used in this book, even when not specifically marked as such, are not to be considered unprotected by law.

Printing Strauss GmbH, Mörlenbach
Binding Litges & Dopf Buchbinderei GmbH, Heppenheim

Printed in the Federal Republic of Germany
Printed on acid-free paper

ISBN-13: 978-3-527-31786-8
ISBN-10: 3-527-31786-4

Contents

	Preface	IX
1	**Terpenes: Importance, General Structure, and Biosynthesis**	**1**
1.1	Term and Significance	1
1.2	General Structure: The Isoprene Rule	2
1.3	Biosynthesis	3
2	**Hemi- and Monoterpenes**	**10**
2.1	Hemiterpenes	10
2.2	Acyclic Monoterpenes	10
2.3	Monocyclic Monoterpenes	13
2.3.1	Cyclopropane and Cyclobutane Monoterpenes	13
2.3.2	Cyclopentane Monoterpenes	14
2.3.3	Cyclohexane Monoterpenes	15
2.3.4	Cymenes	18
2.4	Bicyclic Monoterpenes	19
2.4.1	Survey	19
2.4.2	Caranes and Thujanes	20
2.4.3	Pinanes	20
2.4.4	Camphanes and Fenchanes	21
2.5	Cannabinoids	22
3	**Sesquiterpenes**	**24**
3.1	Farnesanes	24
3.2	Monocyclic Farnesane Sesquiterpenes	25
3.2.1	Cyclofarnesanes and Bisabolanes	25
3.2.2	Germacranes and Elemanes	26
3.2.3	Humulanes	27
3.3	Polycyclic Farnesane Sesquiterpenes	28
3.3.1	Caryophyllanes	28
3.3.2	Eudesmanes and Furanoeudesmanes	29
3.3.3	Eremophilanes, Furanoeremonphilanes, Valeranes	31
3.3.4	Cadinanes	34
3.3.5	Drimanes	36
3.3.6	Guaianes and Cycloguaianes	37
3.3.7	Himachalanes, Longipinanes, Longifolanes	41
3.3.8	Picrotoxanes	42
3.3.9	Isodaucanes and Daucanes	42
3.3.10	Protoilludanes, Illudanes, Illudalanes	43
3.3.11	Marasmanes, Isolactaranes, Lactaranes, Sterpuranes	44
3.3.12	Acoranes	45

3.3.13	Chamigranes	45
3.3.14	Cedranes and Isocedranes	46
3.3.15	Zizaanes and Prezizaanes	47
3.3.16	Campherenanes and Santalanes	47
3.3.17	Thujopsanes	49
3.3.18	Hirsutanes	49
3.4	Other Polycyclic Sesquiterpenes	50
3.4.1	Pinguisanes	50
3.4.2	Presilphiperfolianes, Silphiperfolianes, Silphinanes, Isocomanes	50
4	**Diterpenes**	**52**
4.1	Phytanes	52
4.2	Cyclophytanes	52
4.3	Bicyclophytanes	54
4.3.1	Labdanes	54
4.3.2	Rearranged Labdanes	55
4.4	Tricyclophytanes	57
4.4.1	Pimaranes and Isopimaranes	57
4.4.2	Cassanes, Cleistanthanes, Isocopalanes	59
4.4.3	Abietanes and Totaranes	61
4.5	Tetracyclophytanes	63
4.5.1	Survey	63
4.5.2	Beyeranes	64
4.5.3	Kauranes and Villanovanes	64
4.5.4	Atisanes	66
4.5.5	Gibberellanes	66
4.5.6	Grayanatoxanes	67
4.6	Cembranes and Cyclocembranes	68
4.6.1	Survey	68
4.6.2	Cembranes	70
4.6.3	Casbanes	71
4.6.4	Lathyranes	71
4.6.5	Jatrophanes	72
4.6.6	Tiglianes	72
4.6.7	Rhamnofolanes and Daphnanes	73
4.6.8	Eunicellanes and Asbestinanes	73
4.6.9	Biaranes	74
4.6.10	Dolabellanes	74
4.6.11	Dolastanes	75
4.6.12	Fusicoccanes	75
4.6.13	Verticillanes and Taxanes	75
4.6.14	Trinervitanes and Kempanes	76
4.7	Prenylsesquiterpenes	77
4.7.1	Xenicanes and Xeniaphyllanes	78
4.7.2	Prenylgermacranes and Lobanes	78
4.7.3	Prenyleudesmanes and Bifloranes	79
4.7.4	Sacculatanes (Prenyldrimanes)	80

4.7.5	Prenylguaianes and Prenylaromadendranes	80
4.7.6	Sphenolobanes (Prenyldaucanes)	81
4.8	Ginkgolides	81

5 Sesterterpenes ... 82
5.1	Acyclic Sesterterpenes	82
5.2	Monocyclic Sesterterpenes	82
5.3	Polycyclic Sesterterpenes	83
5.3.1	Bicyclic Sesterterpenes	83
5.3.2	Tricyclic Sesterterpenes	84
5.3.3	Tetra- and Pentacyclic Sesterterpenes	85

6 Triterpenes ... 86
6.1	Linear Triterpenes	86
6.2	Tetracyclic Triterpenes, Gonane Type	88
6.2.1	Survey	88
6.2.2	Protostanes and Fusidanes	89
6.2.3	Dammaranes	89
6.2.4	Apotirucallanes	91
6.2.5	Tirucallanes and Euphanes	91
6.2.6	Lanostanes	92
6.2.7	Cycloartanes	93
6.2.8	Cucurbitanes	94
6.3	Pentacyclic Triterpenes, Baccharane Type	95
6.3.1	Survey	95
6.3.2	Baccharanes and Lupanes	97
6.3.3	Oleananes	97
6.3.4	Taraxeranes, Multifloranes, Baueranes	98
6.3.5	Glutinanes, Friedelanes, Pachysananes	99
6.3.6	Taraxastanes and Ursanes	100
6.4	Pentacyclic Triterpenes, Hopane Type	101
6.4.1	Survey	101
6.4.2	Hopanes and Neohopanes	102
6.4.3	Fernanes	103
6.4.4	Adiananes and Filicanes	104
6.4.5	Gammaceranes	104
6.5	Other Pentacyclic Triterpenes	105
6.5.1	Survey	105
6.5.2	Stictanes and Arboranes	106
6.5.3	Onoceranes and Serratanes	106
6.6	Iridals	107

7 Tetraterpenes ... 109
| 7.1 | Carotenoids | 109 |
| 7.2 | Apocarotenoids | 111 |

7.3	Diapocarotenoids	112
7.4	Megastigmanes	113
8	**Polyterpenes and Prenylquinones**	**115**
8.1	Polyterpenes	115
8.2	Prenylquinones	116
9	**Selected Syntheses of Terpenes**	**119**
9.1	Monoterpenes	119
9.1.1	Concept of Industrial Syntheses of Monoterpenoid Fragrances	119
9.1.2	(R)-(+)-Citronellal	121
9.1.3	Rose oxide	121
9.1.4	Chrysanthemic Acid Methyl Ester	122
9.1.5	α-Terpineol	123
9.1.6	(1R,3R,4S)-(−)-Menthol	124
9.1.7	Camphor from α-Pinene	124
9.1.8	α-Pinene and Derivatives for Stereospecific Syntheses of Chiral Monoterpenes	126
9.1.9	Hexahydrocannabinol	128
9.2	Sesquiterpenes	129
9.2.1	β-Selinene	129
9.2.2	Isocomene	130
9.2.3	Cedrene	132
9.2.4	Periplanone B	135
9.3	Diterpenes	138
9.3.1	Vitamin A (Retinol Acetate)	138
9.3.2	Cafestol	141
9.3.3	Baccatin III as the Precursor of Taxol	145
9.4	Triterpenes	152
9.4.1	Lupeol	152
10	**Isolation and Structure Elucidation**	**160**
10.1	Isolation from Plants	160
10.2	Spectroscopic Methods of Structure Elucidation	160
10.3	Structure Elucidation of a Sesquiterpene	161
10.3.1	Double Bond Equivalents	161
10.3.2	Functional Groups and Partial Structures detected by ^{13}C NMR	162
10.3.3	Skeletal Structure (Connectivities of Atoms)	163
10.3.4	Relative Configuration	169
10.3.5	Absolute Configuration	171
10.4	Determination of the Crystal Structure	173
10.5	Molecular Structure and Odor of Terpenes	176
	Bibliography	**180**
	Survey of Important Parent Skeletons of Terpenes	**185**
	Subject Index	**197**

Preface

Terpenes play an important role as fragrances in perfumery, as constituents of flavors for spicing foods, as environment-friendly luring compounds to trap damaging insects with the imitates of their own pheromones and, last but not least, as medicaments for the therapy of numerous diseases including tumors. Nevertheless, these natural compounds predominantly originating from plants are inadequately covered in the majority of the textbooks of organic and pharmaceutical chemistry. Voluminous encyclopediae and monographs dedicated to specialists on the field are rather confusing at first glance and unsuitable for the beginner. This justifies a concise and systematic introduction not available so far.

The introductory text outlines the significance of terpenes, the isoprene rule as the basic concept of their molecular structure, and their biogenesis. A review of the terpenes follows, arranged according to the number of isoprene units they contain and to their parent skeletons. The occurence of terpenes in plants and other organisms is scrutinized there, also considering their biological functions and pharmacological activities as well as their olfactoric properties. An additional section describes the total syntheses of some mono-, sesqui-, di-, and triterpenes, selected according to the originality of the preparative methods applied and to their didactic suitability, also including industrial processes, e.g. those for the production of mono- and sesquiterpenoid fragrances and of vitamin A acetate. Retrosynthetic disconnections easily reconstructable with the usual background of organic chemistry facilitate an understanding of the synthetic strategies. A final chapter deals with the isolation and structure elucidation of terpenes, drawing the path from the spectra to the molecular structure and sketching some relationships between the molecular shape of terpenes and their odor. Developed from lectures, this text is not comprehensive but rounded off, systematic, and as concise as possible. Consequently, it does not include the large field of steroids in spite of their biogenetic relation to terpenes.

The second German edition was translated to this English version, including some extensions concerning cymenes, cannabinoids, ginkgolides, taxine, and geohopanes. Many thanks are due to *Professor Dr. Gerhard Rücker*, Institute of pharmaceutical chemistry of the University of Bonn, for looking through the first German edition, to *Dr. Bill Down* for proof-reading this English version, and to some other collegues and reviewers for useful comments and corrections. Any suggestions for correction or improvement will be welcome for future electronic upkeeping and updating of this text.

Tübingen (Germany), Spring 2006   *Eberhard Breitmaier*

1 Terpenes: Importance, General Structure, and Biosynthesis

1.1 Term and Significance

The term terpenes originates from turpentine (*lat.* balsamum terebinthinae). Turpentine, the so-called "resin of pine trees", is the viscous pleasantly smelling balsam which flows upon cutting or carving the bark and the new wood of several pine tree species (Pinaceae). Turpentine contains the "resin acids" and some hydrocarbons, which were originally referred to as terpenes. Traditionally, all natural compounds built up from isoprene subunits and for the most part originating from plants are denoted as terpenes [1] (section 1.2).

Conifer wood, balm trees, citrus fruits, coriander, eucalyptus, lavender, lemon grass, lilies, carnation, caraway, peppermint species, roses, rosemary, sage, thyme, violet and many other plants or parts of those (roots, rhizomes, stems, leaves, blossoms, fruits, seed) are well known to smell pleasantly, to taste spicy, or to exhibit specific pharmacological activities. Terpenes predominantly shape these properties. In order to enrich terpenes, the plants are carved, e.g. for the production of incense or myrrh from balm trees; usually, however, terpenes are extracted or steam distilled, e.g. for the recovery of the precious oil of the blossoms of specific fragrant roses. These extracts and steam distillates, known as ethereal or essential oils ("essence absolue") are used to create fine perfumes, to refine the flavor and the aroma of food and drinks and to produce medicines of plant origin (phytopharmaca).

The biological and ecochemical functions of terpenes have not yet been fully investigated. Many plants produce volatile terpenes in order to attract specific insects for pollination or otherwise to expel certain animals using these plants as food. Less volatile but strongly bitter-tasting or toxic terpenes also protect some plants from being eaten by animals (antifeedants). Last, but not least, terpenes play an important role as signal compounds and growth regulators (phytohormones) of plants, as shown by preliminary investigations.

Many insects metabolize terpenes they have received with their plant food to growth hormones and pheromones. Pheromones are luring and signal compounds (sociohormones) that insects and other organisms excrete in order to communicate with others like them, e.g. to warn (alarm pheromones), to mark food resources and their location (trace pheromones), as well of assembly places (aggregation pheromones) and to attract sexual partners for copulation (sexual pheromones). Harmless to the environment, pheromones may replace conventional insecticides to trap harmful and damaging insects such as bark beetles.

1.2 General Structure: The Isoprene Rule

About 30 000 terpenes are known at present in the literature [2-7]. Their basic structure follows a general principle: *2-Methylbutane* residues, less precisely but usually also referred to as *isoprene* units, $(C_5)_n$, build up the carbon skeleton of terpenes; this is the isoprene rule [1] found by RUZICKA and WALLACH (Table 1). Therefore, terpenes are also denoted as *isoprenoids*. In nature, terpenes occur predominantly as hydrocarbons, alcohols and their glycosides, ethers, aldehydes, ketones, carboxylic acids and esters.

Table 1. Parent hydrocarbons of terpenes (isoprenoids).

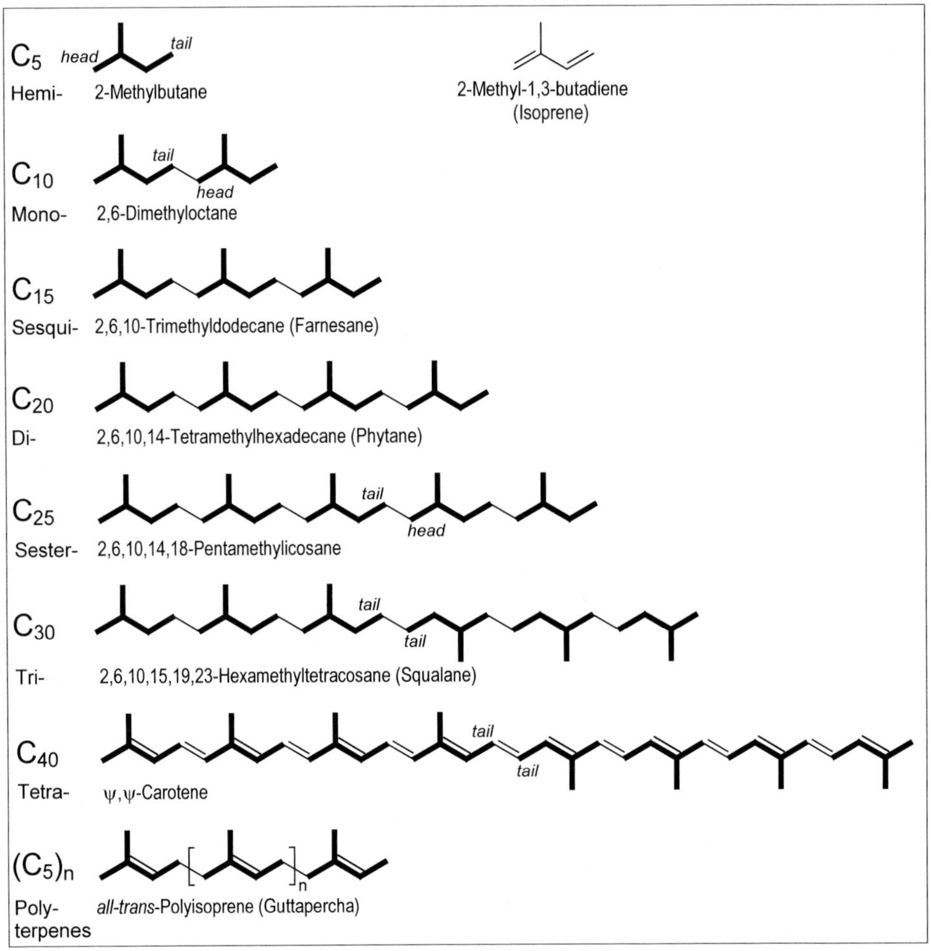

C_5 Hemi-	2-Methylbutane	2-Methyl-1,3-butadiene (Isoprene)
C_{10} Mono-	2,6-Dimethyloctane	
C_{15} Sesqui-	2,6,10-Trimethyldodecane (Farnesane)	
C_{20} Di-	2,6,10,14-Tetramethylhexadecane (Phytane)	
C_{25} Sester-	2,6,10,14,18-Pentamethylicosane	
C_{30} Tri-	2,6,10,15,19,23-Hexamethyltetracosane (Squalane)	
C_{40} Tetra-	ψ,ψ-Carotene	
$(C_5)_n$ Poly- terpenes	all-*trans*-Polyisoprene (Guttapercha)	

1.3 Biosynthesis

Depending on the number of 2-methylbutane (isoprene) subunits one differentiates between *hemi-* (C_5), *mono-* (C_{10}), *sesqui-* (C_{15}), *di-* (C_{20}), *sester-* (C_{25}), *tri-* (C_{30}), *tetraterpenes* (C_{40}) and *polyterpenes* (C_5)$_n$ with $n > 8$ according to Table 1.

The isopropyl part of 2-methylbutane is defined as the *head*, and the ethyl residue as the *tail* (Table 1). In mono-, sesqui-, di- and sesterterpenes the isoprene units are linked to each other from *head-to-tail*; tri- and tetraterpenes contain one *tail-to-tail* connection in the center.

1.3 Biosynthesis

Acetyl-coenzyme A, also known as *activated acetic acid*, is the biogenetic precursor of terpenes (Figure 1) [9-11]. Similar to the CLAISEN condensation, two equivalents of acetyl-CoA couple to acetoacetyl-CoA, which represents a biological analogue of acetoacetate. Following the pattern of an aldol reaction, acetoacetyl-CoA reacts with another equivalent of acetyl-CoA as a carbon nucleophile to give β-hydroxy-β-methylglutaryl-CoA, followed by an enzymatic reduction with dihydronicotinamide adenine dinucleotide (NADPH + H$^+$) in the presence of water, affording (*R*)-*mevalonic acid*. Phosphorylation of mevalonic acid by adenosine triphosphate (ATP) *via* the monophosphate provides the diphosphate of mevalonic acid which is decarboxylated and dehydrated to *isopentenylpyrophosphate* (isopentenyldiphosphate, IPP). The latter isomerizes in the presence of an isomerase containing SH groups to *γ,γ-dimethylallylpyrophosphate*. The electrophilic allylic CH$_2$ group of γ,γ-dimethylallylpyrophosphate and the nucleophilic methylene group of isopentenylpyrophosphate connect to *geranylpyrophosphate* as *monoterpene*. Subsequent reaction of geranyldiphosphate with one equivalent of isopentenyldiphosphate yields *farnesyldiphosphate* as a *sesquiterpene* (Fig. 1).

Dihydro nicotinamide adenine dinucleotide phosphate (NADPH + H$^+$)

Adenosine tri phosphate (ATP)

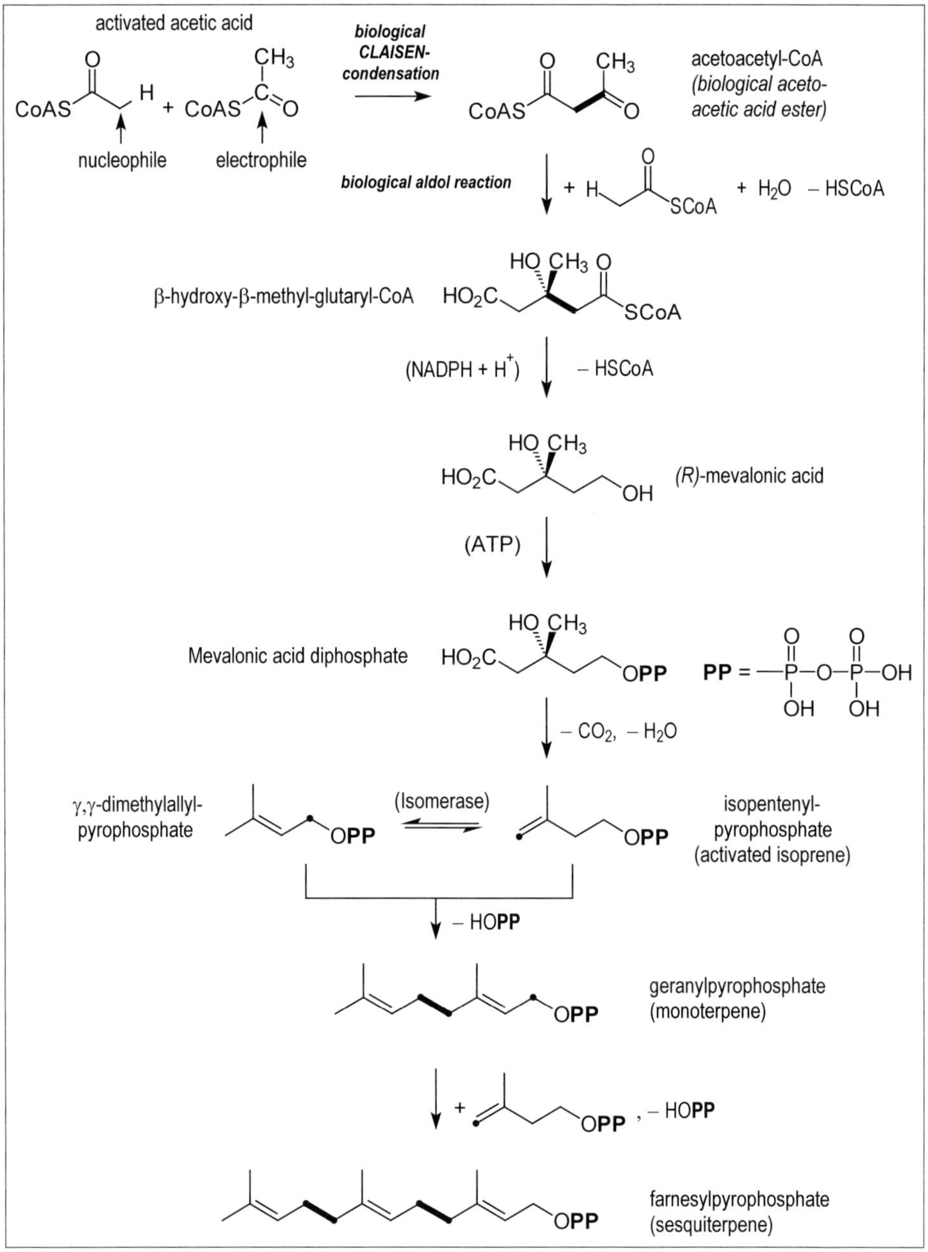

Figure 1. Scheme of the biogenesis of mono- and sesquiterpenes.

1.3 Biosynthesis

However, failing incoporations of ^{13}C-labeled acetate and successful ones of ^{13}C-labeled glycerol as well as pyruvate in hopanes and ubiquinones showed isopentenyldiphosphate (IPP) to originate not only from the acetate mevalonate pathway, but also from *activated acetaldehyde* (C_2, by reaction of pyruvate and thiamine diphosphate) and glyceraldehyde-3-phosphate (C_3) [12]. In this way, *1-deoxypentulose-5-phosphate* is generated as the first unbranched C_5 precursor of IPP.

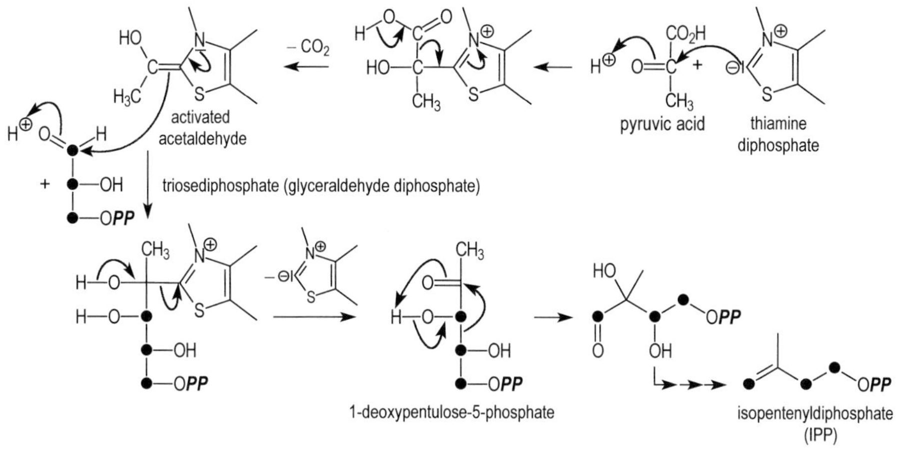

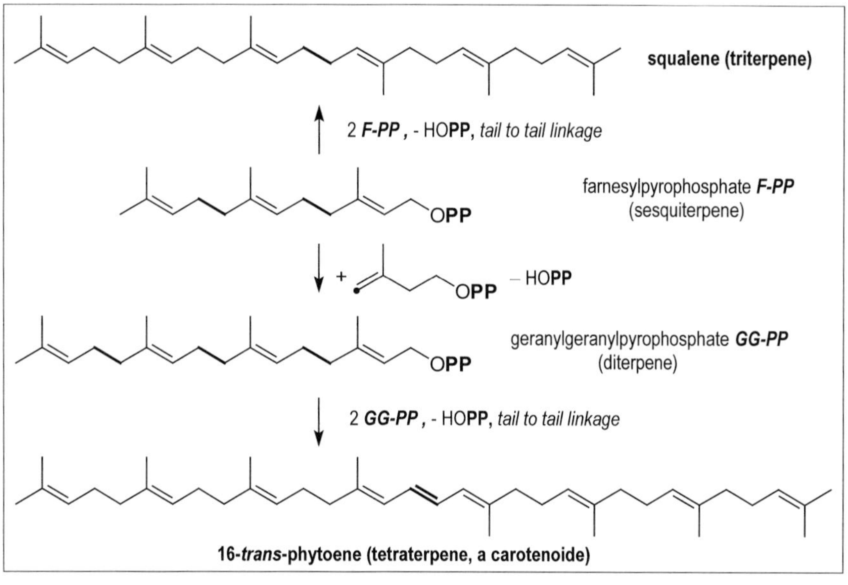

Figure 2. Scheme of the biogenesis of di-, tri- and tetraterpenes.

Geranylgeranylpyrophosphate as a diterpene (C_{20}) emerges from the attachment of isopentenylpyrophosphate with its nucleophilic head to farnesylpyrophosphate with its electrophilic tail (Fig. 2). The formation of sesterterpenes (C_{25}) involves an additional head-to-tail linkage of isopentenylpyrophosphate (C_5) with geranylgeranylpyrophosphate (C_{20}). A tail-to-tail connection of two equivalents of farnesylpyrophosphate leads to squalene as a triterpene (C_{30}, Fig. 2). Similarly, tetraterpenes such as the carotenoid 16-*trans*-phytoene originate from tail-to-tail dimerization of geranylgeranylpyrophosphate (Fig. 2).

The biogeneses of cyclic and polycyclic terpenes [9,10] are usually assumed to involve *intermediate carbenium ions*, but evidence for this *in vivo* was given only in some specific cases. In the simple case of monocyclic monoterpenes such as limonene the allylic cation remaining after separation of the pyrophosphate anion cyclizes to a cyclohexyl cation which is deprotonated to (*R*)- or (*S*)-limonene.

geranylpyrophosphate   (R+S)-limonene

The non-classical version of the intermediate carbenium ion (also referred to as a carbonium ion) resulting upon dissociation of the pyrophosphate anion from farnesylpyrophosphate explains the cyclization to several cyclic carbenium ions [8], as demonstrated for some sesquiterpenes (Fig. 3). Additional diversity arises from *1,2-hydride* and *1,2-alkyl shifts* (WAGNER-MEERWEIN rearrangements) and *sigmatropic reactions* (COPE rearrangements) on the one hand, and on the other hand from the formation of diastereomers and enantiomers provided that the cyclizations generate new asymmetric carbon atoms (Fig. 3) [8-10].

Thus, the non-classical carbenium ion arising from dissociation of the diphosphate anion from farnesylpyrophosphate permits formation of the monoyclic sesquiterpenes humulatriene and germacratriene after deprotonation (Fig.3). A COPE rearrangement of germacratriene leads to elematriene. Protonation of germacratriene following MARKOWNIKOW orientation initially provides the higher alkylated and therefore more stable carbenium ion which undergoes 1,2-hydride shifts resulting in bicyclic carbenium ions with an eudesmane or guaiane skeleton. Subsequent deprotonations yield diastereomeric eudesmadienes and guajadienes. Finally, eudesmanes may rearrange to eremophilanes involving 1,2-methyl shifts (Fig. 3).

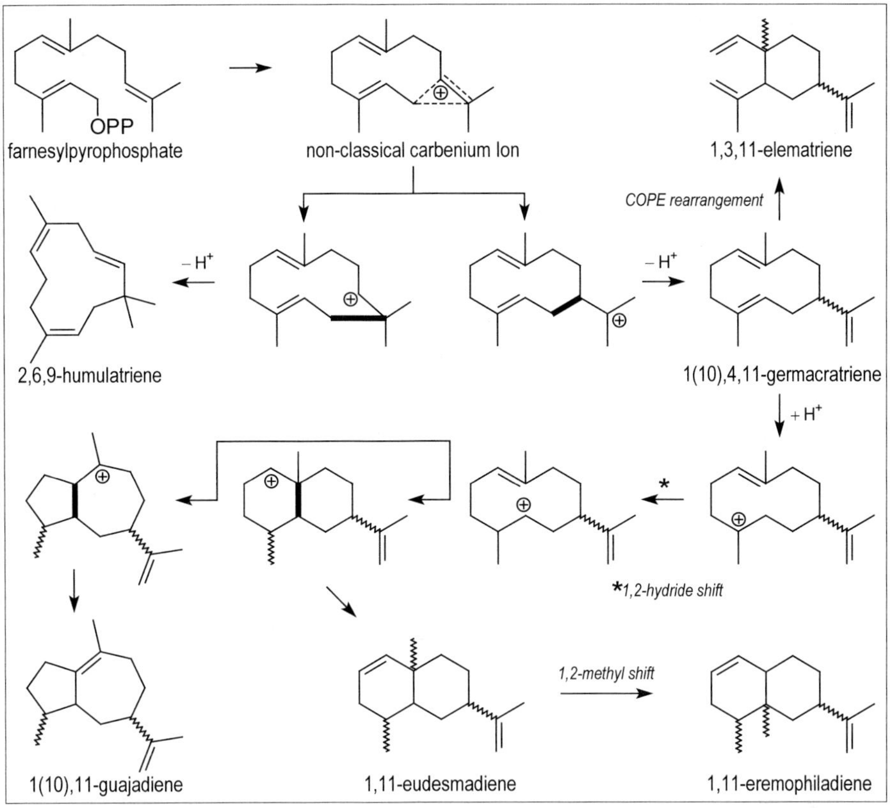

1.3 Biosynthesis

Figure 3. Biogenesis of some mono- and bicyclic sesquiterpenes from farnesylpyrophosphate.

A simliar cyclization generates the 14-membered skeleton of cembrane from which other polycyclic diterpenes are derived. 3,7,11,15-Cembratetraene, better known as cembrene A, emerges directly from geranylgeranylpyrophosphate (Fig. 2) involving the 1,14-cyclization of the resulting allylic cation [9,10].

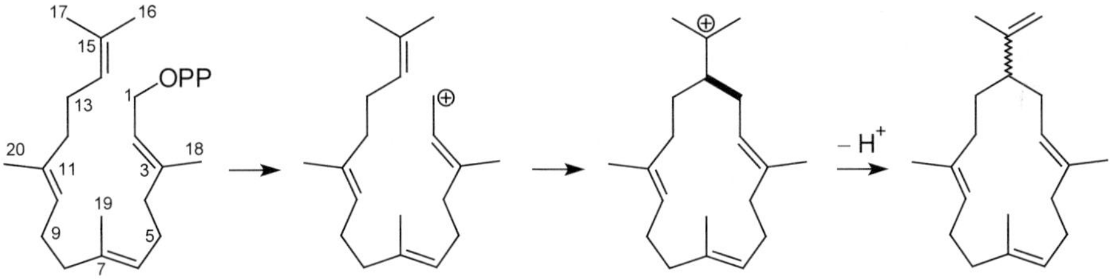

The biogenesis of pimarane, the parent compound of many polycyclic diterpenes, is assumed to arise from *iso*-geranylgeranylpyrophosphate [9,10]. After dissociation of the pyrophosphate anion, the remaining acyclic allylic cation undergoes a 1,3-sigmatropic hydrogen shift and thereby cyclizes to a monocyclic carbenium ion which, itself, isomerizes to the ionic precursor of the pimarane skeleton.

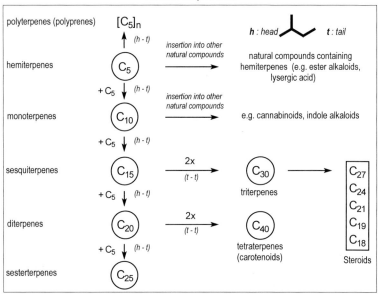

Table 2. Isoprenoids

1.3 Biosynthesis

2,3-Epoxysqualene has been shown by isotope labeling to be the biogenetic precursor of tetracyclic triterpenes with perhydrocyclopenta[*a*]phenanthrene as the basic skeleton (also referred to as *gonane* or *sterane*). Steroids [13] are derived from these tetracyclic triterpenes. These include cholestanes (C_{27}), pregnanes (C_{21}), androstanes (C_{19}) with *trans* fusion of the rings *A* and *B* (5α), estranes (C_{18}) with a benzenoid ring *A* (estra-1,3,5-triene; Fig. 4) [9,10] as well as cholic acid and its derivatives (C_{24}) with *cis* fusion of the rings *A* and *B* (5β). The biogenetic origins of tetracyclic triterpenes and steroids are summarized in Table 2.

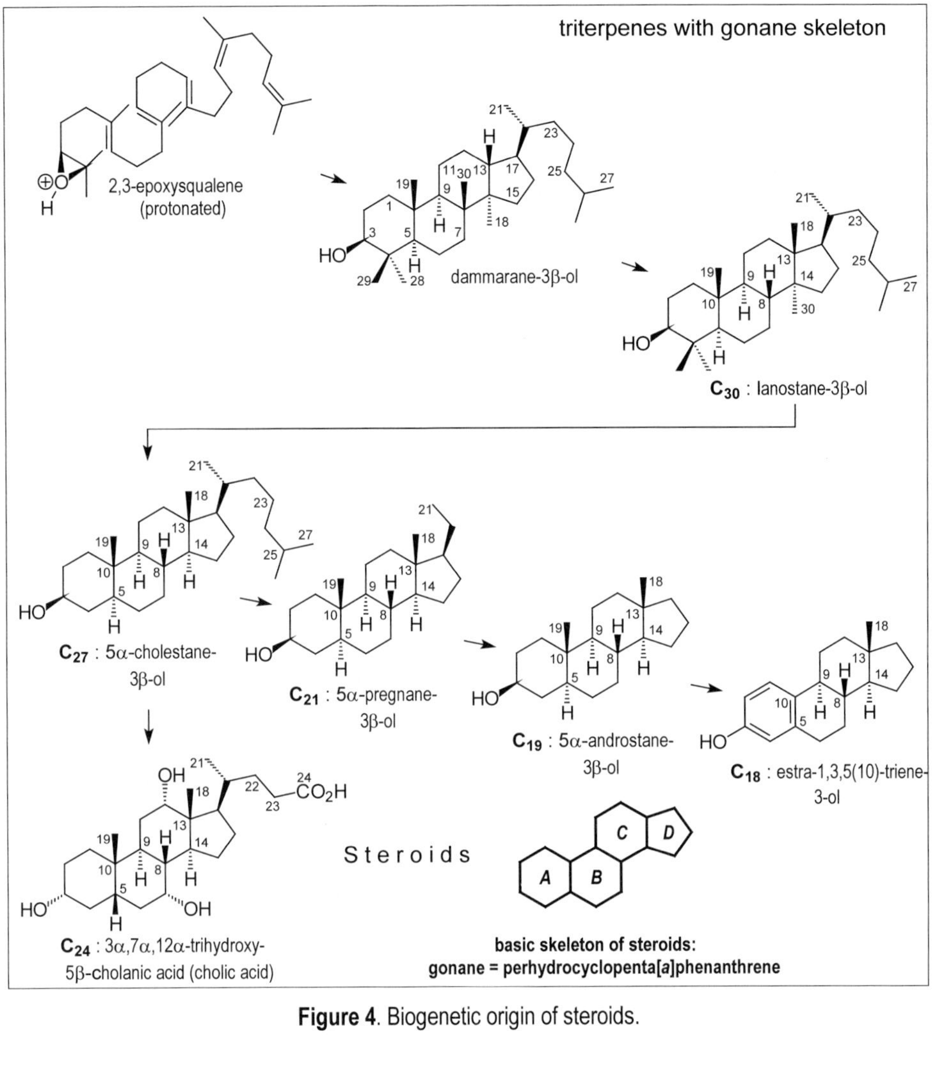

Figure 4. Biogenetic origin of steroids.

2 Hemi- and Monoterpenes

2.1 Hemiterpenes

About 50 hemiterpenes [2] are known. In contrast to non-natural 2-methyl-1,3-butadiene (isoprene), 3-methyl-2-buten-1-ol (prenol) occurs in ylang-ylang oil obtained from freshly picked flowers of the Cananga tree *Cananga odorata* (Annonaceae) and in the oil of hops fom *Humulus lupulus* (Cannabaceae). Terpenes biogenetically arise from isopentenyldiphosphate (isopentenylpyrophosphate) (section 1.3), hemiterpenoid lysergic acid from the amino acid tryptophane and the diphosphate of 2-carboxy-1-buten-4-ol. (*S*)-(–)-3-Methyl-3-buten-2-ol is found in the essential oils of oranges, grapefruit and hops. 4-Methoxy-2-methyl-2-butanthiol shapes the flavor of blackcurrant *Ribes nigrum* (Saxifragaceae). Tiglic acid, its regioisomers angelic and 3-methyl-2-butenoic acid as well as isovaleric acid, are the acid components of numerous natural esters (e.g. ester alkaloids).

2-methyl-1,3-butadiene (isoprene)

prenol (3-methyl-2-buten-1-ol)

isopentenyldiphosphate

2-carboxy-1-buten-4-ol

(*S*)-3-methyl-3-buten-2-ol

4-methoxy-2-methyl-2-butanthiol

tiglic acid

angelic acid

3-methylbutenoic acid

isovaleric acid

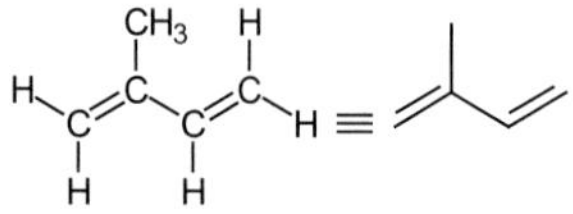

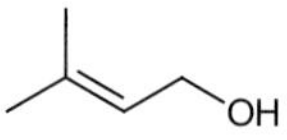

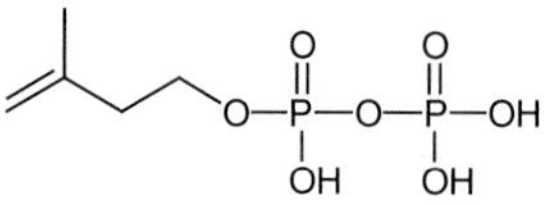

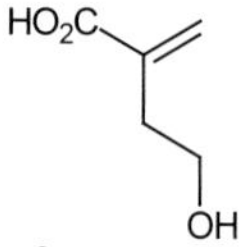

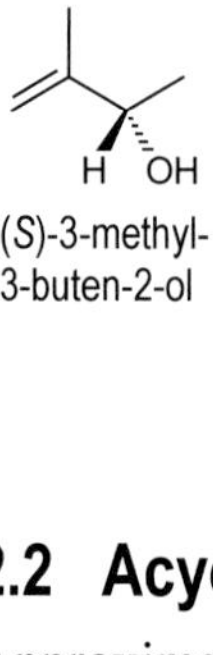

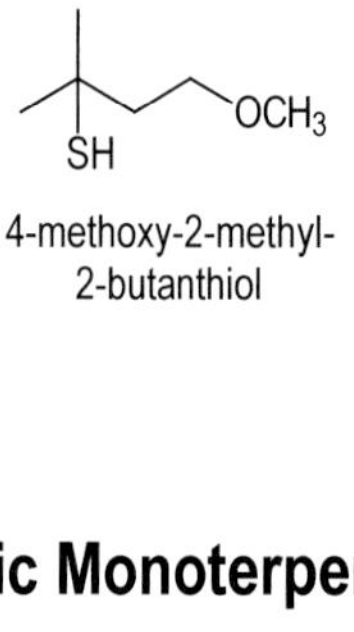

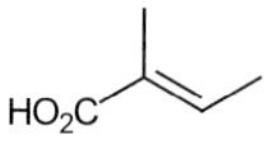

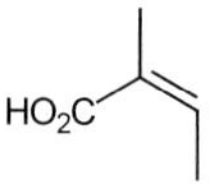

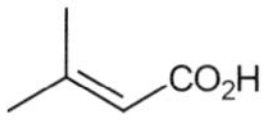

2.2 Acyclic Monoterpenes

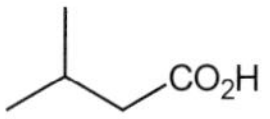

Approximately 1 500 monoterpenes are documented [2-7]. Most of these are linked in the head-to-tail manner and are derived from 2,6-dimethyloctane. (*R*)-3,7-Dimethyloctanol is a component of the nicely flowery to minty smelling geranium oil obtained from *Pelargonium graeveolens* (Geraniaceae). 2,6-Dimethyloctanoic acid occurs in the feather wax of several birds.

2,6-dimethyloctane

(*R*)-3,7-dimethyloctanol

2,6-dimethyloctanoic acid

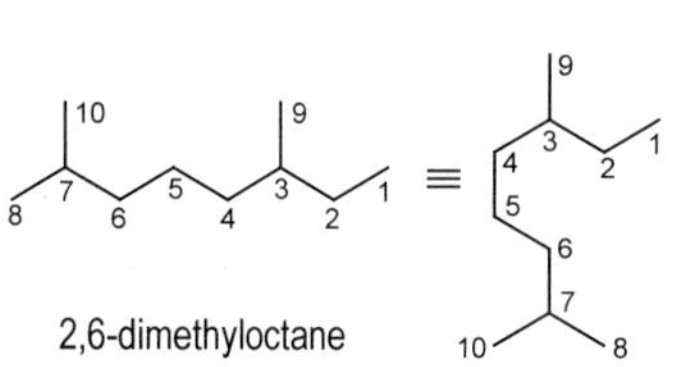

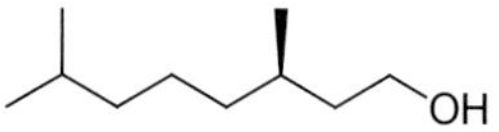

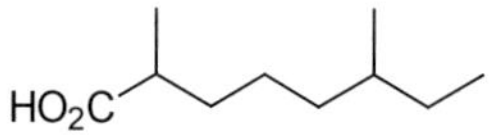

2.2 Acyclic Monoterpenes

Acyclic monoterpenoid trienes such as β-myrcene and configurational isomers of β-ocimene are found in the oils of basil (leaves of *Ocimum basilicum*, Labiatae), bay (leaves of *Pimenta acris*, Myrtaceae), hops (strobiles of *Humulus lupulus*, Cannabaceae), pettitgrain (leaves of *Citrus vulgaris*, Rutaceae) and several other essential oils.

α-myrcene   β-myrcene   (Z)-α-ocimene   (E)-α-ocimene   (Z)-β-ocimene   (E)-β-

Perillene, a monoterpenoid furan derived from β-myrcene, is a constituent of the essential oil obtained from *Perilla citridora* (Labiatae); among other monoterpenes, it also occurs in the pheromones of some mites and acts as a defense pheromone of the ant *Lasius fulginosus*. The isomeric rose furan is a fragrant component of the oil of rose obtained from fresh flowers of *Rosa damascena* (Rosaceae). 3-(4-Methyl-3-pentenyl)thiophene and derived cyclic tri- and tetrasulfides (1,2,3-trithiepine and 1,2,3,4-tetrathiocine) are found in the oil of hops.

perillene   3-methyl-2-(3-methyl-2-butenyl)furan (rose furan)   3-(4-methyl-3-pentenyl)-thiophene   5,8-dihydro-6-(4-methyl-3-pentenyl)-1,2,3,4-tetrathiocin   4,7-dihydro-5-(4-methyl-3-pentenyl)-1,2,3-trithiepine

Unsaturated monoterpene alcohols and aldehydes play an important role in perfumery. (R)-(−)-Linalool from the oils of rose, neroli (orange flowers) and spike (lavender) smells more woody, while the lavender fragrance of the (S)-(+)-enantiomer is more sweetish. The *cis-trans*-isomers of geraniol and nerol of the oil of palmarosa from the tropical grass *Cymbopogon martinii* var. *motia* (Poaceae), enantiomeric citronellols in the insect repellant oil of citronella from fresh grass of *Cymbopogon winterianus* and the lavandulols in the oil of spike from the flowering tops of *Lavandula angustifolia* (Labiatae) smell pleasantly flowery.

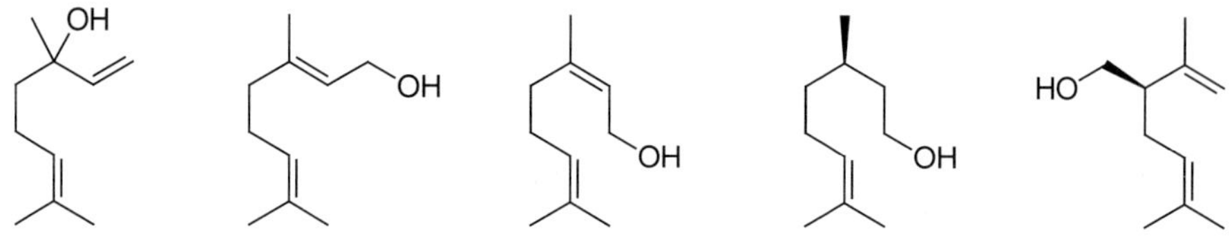

rac. linalool   geraniol (E)   nerol (Z)   (R)-(+)-citronellol   (R)-(–)-lavandulol

Oil of thyme contains β-myrcen-8-ol derived from β-myrcene, and (*R*)-(–)-ipsdienol as well as its non-chiral regioisomer is not only the aggregation pheromone [14-17] of the bark beetle *Ips confusus*, but also the fragrance of the blossoms of many orchids. Terpenoid pyran derivatives include diastereomeric rose oxides [(2*R*,4*R*)-*trans*- and (2*S*,4*R*)-*cis*-] as well as racemic nerol oxide, which essentially contribute to the pleasantly flowery green smell of the Bulgarian oil of rose [18].

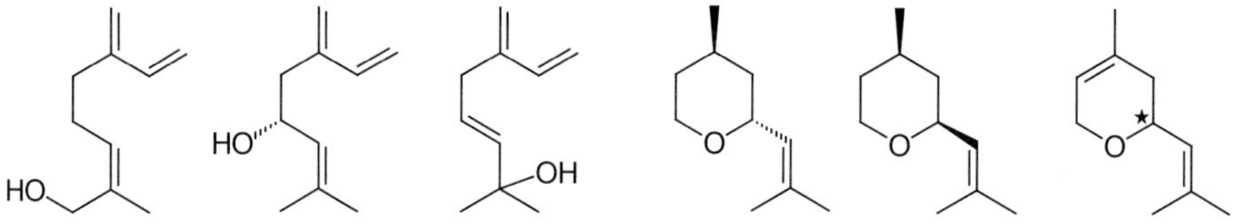

β-myrcen-8-ol   (R)-(–)-ipsdienol   2-methyl-6-methylene-3,7-octadien-2-ol   (2R,4R)-trans- rose oxide   (2S,4R)-cis-   rac. nerol oxide

Citral, widely used in perfumery, is a mixture of the (*E,Z*)-isomers geranial and neral. Both aldehydes occur in the oil of lemon grass from *Cymbopogon flexuosus* (Poaceae) growing in India; they smell pleasantly and strongly like lemon peel, similar to the insect repellant citronellal in the oil of citronella obtained from the fresh grass of *Cymbopogon nardus* (Poaceae). In contrast, the ketone (*E*)-tagetone and its dihydro derivative from *Tagetes glandulifera* (Asteraceae) emit an aromatic and bitter fruity odor.

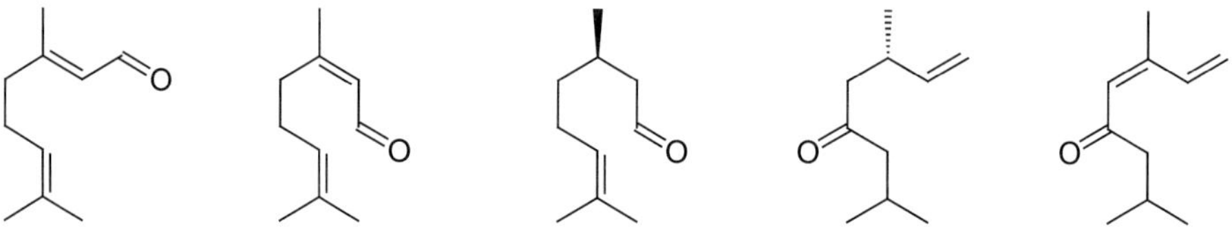

geranial (E)   neral (Z)   (R)-(+)-citronellal   (S)-(+)-dihydro-tagetone   (E)-tagetone

2.3 Monocyclic Monoterpenes

2.3.1 Cyclopropane and Cyclobutane Monoterpenes

(+)-Chrysanthemol from the leaves of *Artemisia ludiviciana* (Asteraceae) belongs to the cyclopropane monoterpenes [2]. Cinerins, jasmolins and pyrethrins (all including derivatives I and II) are esters of *trans*-chrysanthemic and pyrethric acid with terpenoid hydroxypentenones such as cinerolone, jasmolone and pyrethrolone. These are the active insecticidal constituents of pyrethrum recovered from dried flowers of several *Chrysanthemum* species (e.g. *Chrysanthemum cinerariaefolium*, Asteraceae). Some synthetic esters of chrysanthemic acid are also applied as insecticides.

$R^1 = CH_3$: (+)-*trans*-chrysanthemic acid
$R^1 = CO_2CH_3$: (+)-*trans*-pyrethric acid

(+)-chrysanthemol

$R^2 = CH_3$: (+)-cinerolone
$R^2 = C_2H_5$: (+)-jasmolone
$R^2 = CH=CH_2$: (+)-pyrethrolone

R^1	R^2	
CH_3	CH_3	: cinerin I
CO_2CH_3	CH_3	: cinerin II
CH_3	C_2H_5	: jasmolin I
CO_2CH_3	C_2H_5	: jasmolin II
CH_3	$CH=CH_2$	: pyrethrin I
CO_2CH_3	$CH=CH_2$	: pyrethrin II

Cyclobutane monoterpenes arise from the degradation of pinenes (section 2.3.3); some of them occur in several plants and as sexual pheromones of various beetles [14-16]. Examples include junionone in the fruits of the juniper tree *Juniperus communis* (Cupressaceae) and fragranol in the roots of *Artemisia fragrans* (Asteraceae). Grandisol, the 1-epimer of fragranol, is the major component in the sexual pheromone cocktail (grandlure) of the male boll weevil *Anthonomus grandis*, while the citrus flour beetle *Plenococcus citri* attracts its females with the regioisomeric 1-hydroxymethyl-2,2-dimethyl-3-(2-propenyl)cyclobutane.

junionone   (1S,2S)-fragranol   (1R,2S)-(+)-grandisol   (1R,3S)-(+)-1-hydroxymethyl-2,2-dimethyl-3-(2-propenyl)cyclobutane

2.3.2 Cyclopentane Monoterpenes

Apart from rare terpenoid monocyclic cyclopentane derivatives such as 1-acetyl-4-isopropenylcyclopentene from *Eucalyptus globulus* (Myrtaceae), about 200 cyclopentane monoterpenes [2] occur as *iridoides* and *seco-iridoides*. 4,7-Dimethylcyclopenta[*c*]pyran incorporates the basic skeleton of iridoids, while the C-6–C-7 bond of the cyclopentane rings opens in the seco-iridoids.

(+)-Iridomyrmecin, for instance, is an insecticidal and antibacterial pheromone of the Argentine ant *Iridomyrmex humilis*. The odor of stereoisomeric nepetalactones from the volatile oil of catnip obtained from *Nepeta cataria* (Labiatae) strongly attracts cats. On the other hand, nepetalactones belong to the pheromone cocktail of some leaf louses [14-17], isolated from the hind legs of the females.

1-acetyl-4-isopropenyl-cyclopentene   iridoid   seco-iridoid   (+)-iridomyrmecin   (4aS,7S,7aR)-(+)-nepetalactone

Hydroxylated iridoids occur as esters and glucosides, the latter being referred to as iridosides. In valepotriates such as the tranquilizing (+)-valtrate from valerian *Valeriana officinalis* (Valerianaceae) all three hydroxy groups are esterified, two of them with isovaleric acid as a hemiterpene. Antirrhinoside from the snapdragon *Antirrhinum tortuosum* and other *Antirrhinum* species (Scrophulariaceae) as well as asperuloside from herb of woodruff *Asperula odorata* (*Galium odoratum*, Rubiaceae) are found in many other plants and protect these as antifeedants.

(+)-valtrate   (−)-antirrhinoside   (−)-asperuloside

(−)-Loganin and (−)-secologanin as a seco-iridoid glucoside protecting an instable trialdehyde are the key intermediates of the biosynthesis of the *Strychnos* and other

2.3 Monocyclic Monoterpenes

monoterpenoid indole alkaloids; for evidence, both glucosides are isolated from the seeds and the pulp of the fruits of *Strychnos nux vomica* (Loganiaceae).

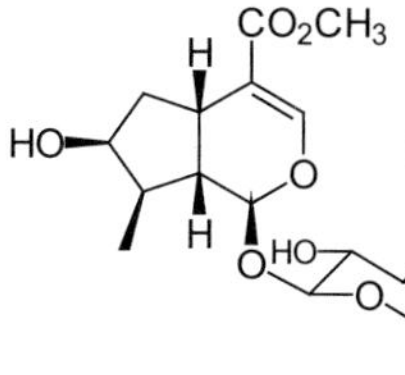

(−)-loganin

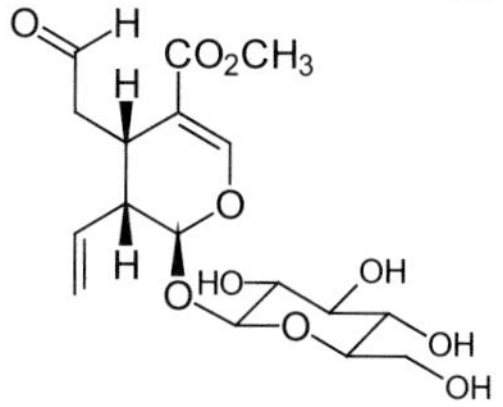

(−)-secologanin (stable glycoside) 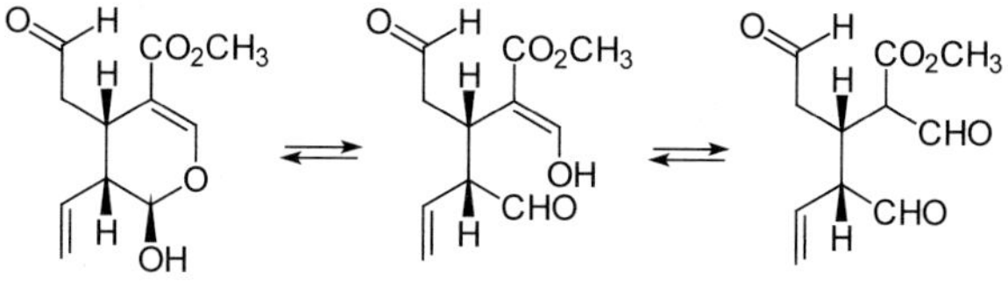   secologanin (unstable tautomers of the aglycon)

Yellowish (+)-jasmolactone A and other structural variants of this seco-iridoid are present in jasmine *Jasminum multiflorum* (Oleaceae). (−)-Oleuropein is the bitter-tasting and antihypertonic β-glucoside extracted from olives, the bark and the leaves of the olive tree *Olea europaea*, and also from ripe fruits of *Ligustrum lucidum* and *L. japonicum* (Oleaceae); it was the first seco-iridoid to be isolated.

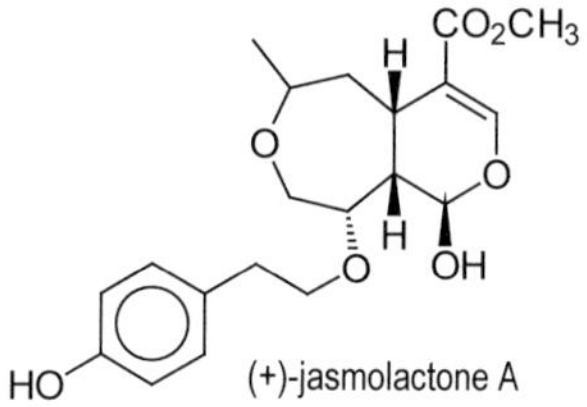

(+)-jasmolactone A 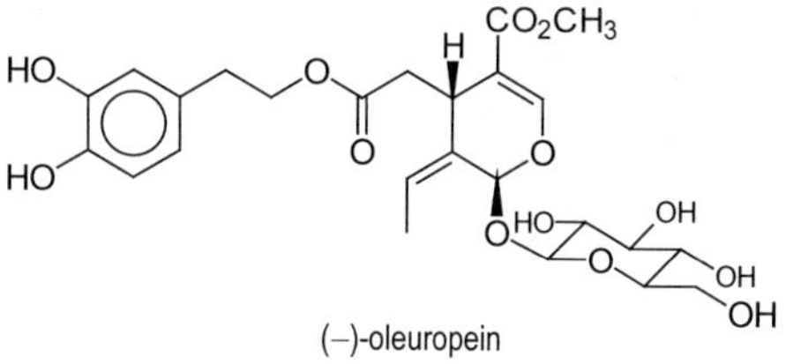   (−)-oleuropein

2.3.3 Cyclohexane Monoterpenes

Monocyclic terpenes [2-7] are derived, for the most part, from the *cis-trans*-isomers of *p*-menthane. *Trans-p*-menthane itself occurs in the oil of turpentine. Its *o*- and *m*-isomers are rarely occurring rearrangement products of *p*-menthane.

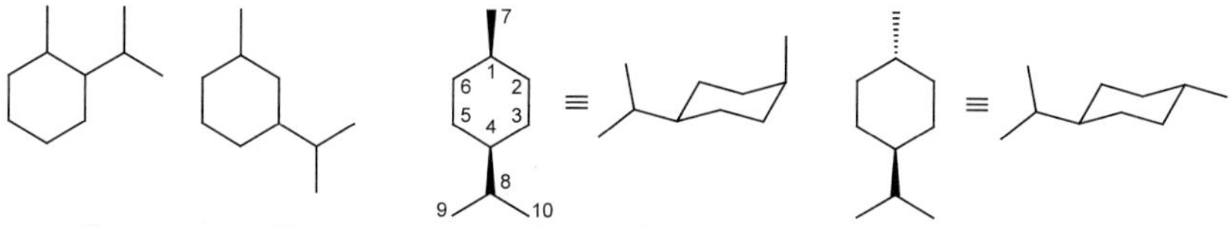

o-menthane   m-menthane   cis-p-menthane   trans-p-menthane

Limonene is an unsaturated monocyclic terpene hydrocarbon occurring in various ethereal oils; its (R)-(+)-enantiomer, smelling like oranges, is the dominant component of mandarin peel oil from *Citrus reticulata* and the oil of orange from *C. aurantium* (Rutaceae), respectively, while the (S)-(–)-enantiomer, concentrated in the oil of fir-cones obtained from young twigs and cones of *Abies alba* (Pinaceae), also smells like oranges, but is more balsamic with a terebinthinate touch [18]. Oils of eucalyptus as isolated from *Eucalyptus phellandra* (Myrtaceae), for example, predominantly consist of (S)-(–)-α-phellandrene. (R)-(+)-β-Phellandrene is found in the oil of water fennel from *Phellandrium aquaticum* (Umbelliferae), while the (S)-(–)-enantiomer occurs in Canada balsam oil from the balm fir *Abies balsamea* and in oils of pine needles (e.g. from *Pinus contorta*, Pinaceae). Menthadienes such as α- and β-terpinene as well as terpinolene are fragrant components of several ethereal oils originating from *Citrus*, *Mentha*-, *Juniperus*- and *Pinus* species; terpinolene additionally acts as an alarm pheromone [14-17] of termites.

Δ1,8(9) -
(R)-(+)-
limonene

Δ1(7),2 -
(S)-(+)-
β-phellandrene

Δ1,5 -
(R)-(–)-
α-phellandrene

Δ1,4(8) -
terpinolene

Δ1,4 -
γ-terpinene

Δ1,3 -
menthadiene
α-terpinene

p-Menthan-3-ol forms four pairs of enantiomers. The levorotatory (1*R*,3*R*,4*S*)-enantiomer referred to as (–)-menthol represents the most significant. It is the major component of peppermint oil obtained from the fresh-flowering plant *Mentha piperita* (Labiatae), and is widely used for flavoring and as fragrance in confectionery and perfumery.

(–)-menthol

(–)-isomenthol

(+)-neomenthol

(–)-neoisomenthol

(–)-Menthol has mildly anesthetic, antipruritic, antiseptic, carminative, cooling and gastric sedative actions [19], and is applied as an antipruritic and in nasal inhalers. It smells and tastes sweetish-minty, is fresh and strongly cooling, in contrast to the

2.3 Monocyclic Monoterpenes

(+)-enantiomer which smells and tastes herby-minty and weakly cooling [18]. (+)-Neomenthol occurs in the Japanese peppermint oil from *Mentha arvensis* (Labiatae); (–)-neoisomenthol is found in the oil of geranium from *Pelargonium roseum* and allied species (Geraniaceae).

Regioisomeric *p*-menthenols are represented by (–)-pulegol in several peppermint oils (*Mentha gentilis* and *spirata*, Labiatae), (–)-isopulegol from *Mentha rotundifolia* (Labiatae) and the fungus *Ceratocystis coerulescens*, (–)-piperitol from various *Mentha* and *Eucalyptus* species as well as α-terpineol, which has the very pleasant smell of lilac blossoms, and is an important raw material in perfumery obtained from different ethereal oils (*Artemisia, Eucalyptus, Juniperus, Mentha*). α-Thio-terpineol, *p*-menth-1-en-8-thiol, constitutes the impact compound of grapefruit juice (*Citrus paradisi,* Rutaceae). It is the flavor with the lowest threshold value known to date; concentrations down to 10^{-4} mg ton^{-1} of water can be smelled and tasted [18]. (4*S*,6*R*)-Mentha-1,8-dien-6-ol, also known as (–)-carveol, is a flavor in some ethereal oils of *Citrus*.

(1*R*,3*R*)-(–)-pulegol   (1*R*,3*R*,4*S*)-(–)-isopulegol   (3*S*,4*S*)-(–)-piperitol   (*S*)-(–)-α-terpineol   (*S*)-(–)-*p*-menth-1-en-8-thiol   (4*S*,6*R*)-(–)-carveol

Naturally abundant monoterpene aldehydes include (–)-perillaaldehyde in mandarin peel oil (*Citrus reticulata*, Rutaceae) and *Perilla arguta* (Labiatae) as well as phellandral in the oil of water fennel (*Phellandrium aquaticum*, Umbelliferae) [18].

(–)-perillaaldehyde   (–)-menthone   (–)-isomenthone   (+)-pulegone   (–)-isopulegone

(–)-phellandral   (–)-piperitone   (–)-dihydrocarvone   (–)-carvenone   (+)-carvone

Associated with structurally related menthols, saturated monoterpene ketones such as (–)-menthone with a slight peppermint odor, and (–)-isomenthone, as well as unsaturated ketones such as (+)-pulegone, smelling pleasantly like peppermint with a touch of camphor, and (–)-isopulegone, occur in peppermint oils of different origins (e.g. from *Mentha pulegium*, Labiatae) [18]. (–)-Piperitone from various oils of *Eucalyptus*, is used as masking odor in dentrifrices. (–)-Dihydrocarvone, (–)-carvenone and (+)-carvone are found in the oils of caraway and dill from *Carum carvi* (Umbelliferae) and *Anethum graveolens* (Umbelliferae), respectively, which are used for flavoring liqueurs and soaps and as carminatives. (S)-(+)-Carvone, with the typical odor of caraway, and its (R)-(–)-enantiomer in the oil of spearmint from *Mentha spicata* (Labiatae), smelling more like peppermint, exemplify the influence of absolute configuration on olfactory properties.

Oxygen-bridged derivatives of *p*-menthane such as the bicyclic ethers 1,4-cineol from *Juniperus* or *Artemisia* species and *Cannabis sativa* as well as 1,8-cineol (eucalyptol, the chief component of eucalyptus oil), stamp the spicy odor of the oils of cardamom, eucalyptus and lavender. The oil of cardamom obtained from *Elettaria cardamomum* and *E. major* (Zingiberaceae) is used to spice food and alcoholic drinks. Oils of eucalyptus and lavender are predominantly applied as fragrances and flavors in perfumery and pharmacy [18].

Steam distillation of the aboveground parts of the flowering and fruiting plant *Chenopodium ambrosioides* (Chenopodiaceae) yields the disagreeably smelling oil of chenopodium. This contains bicyclic monoterpene peroxides such as 1,4-epidioxy-*p*-menth-2-ene (ascaridole) and 3,6-epidioxy-*p*-menth-1-ene, which explode upon heating (100 °C). Oil of chenopodium, also known as the oil of American wormseed, was used as antihelmintic; however, as overdoses have caused intoxications, synthetic antihelmintics are preferred nowadays in human medicine.

1,4-epoxy-*p*-menthane (1,4-cineol)

1,8-epoxy-*p*-menthane (1,8-cineol, eucalyptol)

1,4-epidioxy-*p*-menth-2-ene (ascaridol)

3,6-epidioxy-*p*-menth-1-ene

2.3.4 Cymenes

Benzenoid menthanes are referred to as cymenes. The *o*-isomer has not yet been found in nature. *m*-Cymene is a constituent of the ethereal oil of blackcurrant (*Ribes nigrum*, Saxifragaceae); *p*-cymene occurs in the ethereal oils of cinnamon,

2.4 Bicyclic Monoterpenes

cypress, eucalyptus, thyme, turpentine and others; both are used as fragrances in perfumery. Carvacrol is isolated from the oils of marjoram, origanum, summer savoy and thyme, and applied as a disinfectant. Thymol exists in the oil of thyme (*Thymus vulgaris*, Labiatae) and is the predominant constituent of the ethereal oil obtained from the seeds of *Orthodon angustifolium* (Labiatae); it is applied as a topical antiseptic and antihelmintic. *p*-Cymen-8-ol was found in the frass of the woodworm *Hylotrupes bajulus* (Cerambycodae). Cuminaldehyde, a constituent of various essential oils as well as eucalyptus and myrrh, with its strong persistent odor, is used in perfumery.

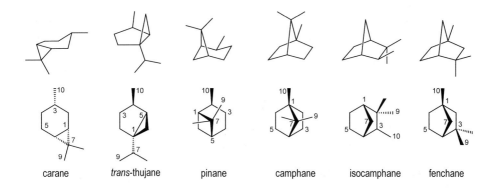

2.4 Bicyclic Monoterpenes

2.4.1 Survey

Bicyclic cyclopropanes carane and thujane, bicyclic cyclobutane pinane, and bicyclo[2.2.1]heptanes such as camphane, isocamphane and fenchane are the most important skeletons of naturally occurring bicyclic monoterpenes [2-7].

2.4.2 Caranes and Thujanes

(+)-3-Carene (3,7,7-trimethylbicyclo[4.1.0]hept-3-ene) is a component of the oil of turpentine from the tropical pine *Pinus longifolia*, also occurring in some species of fir (*Abies*), juniper (*Juniperus*) and *Citrus*. The ethereal oil from wood pine trees *Pinus silvestris* contains the enantiomer (–)-3-carene. Carboxylic acids derived from carane and carene such as (+)-chaminic acid are found in *Chamaecyparis nootkatensis* (Cupressaceae).

Derivatives of thujane are more abundant in plants. (+)-3-Thujanones (thujone and its 4-epimer isothujone) in the oil of thuja obtained from young twigs of the tree of life (*Thuja occidentalis*, Cupressaceae) and in the oils of other plant families (Pinaceae, Labiatae, Asteraceae), smell similar to menthol but cause convulsions upon ingestion. Thujol [(–)-thujan-3α-ol] and its 4-epimer (+)-isothujol are found in *Artemisia*, *Juniperus* and *Thuja* species. (+)-4(10)-Thujene, better known as (+)-sabinene, occurs in the oil of savin obtained from fresh tops of *Juniperus sabina* (Cupressaceae). Its regioisomer 3-thujene is found in the oils of coriander-, dill-, eucalyptus-, thuja-, juniper, and incense (the latter from *Boswellia serrata*, Burseraceae) [2,18].

(+)-3-carene   (+)-chaminic acid   (+)-4(10)-thujene [(+)-sabinene]   (–)-3-thujene   (–)-3-thujanol [(–)-thujol]   (+)-3-thujanone

2.4.3 Pinanes

The oil of turpentine obtained on large scale from the wood of various pine trees (*Pinus caribeae*, *P. palustris*, *P. pinaster*) and by way of cellulose production (sulfurated oil of turpentine) [18] contains more than 70 % of α- and up to 20 % of β-pinene. Enantiomers of both regioisomers are found in many other conifers (Pinaceae). (+)-*trans*-Verbenol, a pinenol occurring in the oil of turpentine, and (+)-verbenone, a pinenone, belong to the aggregation pheromones [14-17] of bark beetles *Ips confusus* and *Ips typograhicus* inducing the death of conifers. Moreover, (+)-verbenone is a constituent of Spanish verbena oil obtained from *Verbenia triphylla* (Verbenaceae); regioisomeric (–)-pinocarvone and (–)-pinocarveol occur in several oils of eucalyptus such as *Eucalyptus globulus* (Myrtaceae); both belong to the sexual pheromones of the pine moth. (+)-Myrtenol is found in the orange *Citrus sinensis* (Rutaceae), the corresponding aldehyde (+)-myrtenal in *Hernandia peltata*

(Hernandiaceae); the grass *Cyperus articulatus* (Cyperaceae) contains the levorotatory enantiomers.

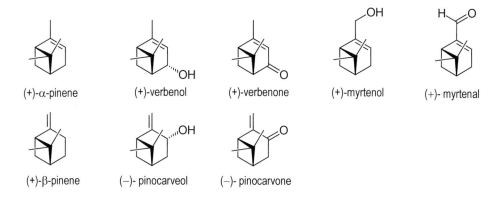

(+)-α-pinene   (+)-verbenol   (+)-verbenone   (+)-myrtenol   (+)- myrtenal

(+)-β-pinene   (−)- pinocarveol   (−)- pinocarvone

2.4.4 Camphanes and Fenchanes

Naturally occurring camphanes include the borneols with an *endo* hydroxy group, the isoborneols with an *exo* OH and 2-camphanone (2-bonanone) referred to as camphor. (+)-Borneol from the camphor tree *Cinnamomum camphora* (Lauraceae) and from the roots of ginger-like *Curcuma aromatica* (Zingiberaceae), both of which grow in Eastern Asia, is known as Borneo camphor. (−)-Isoborneol was isolated from *Achillea filipendulina* (Asteraceae).

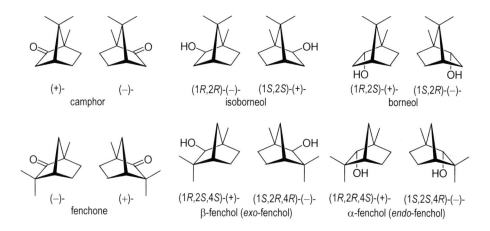

(+)-   (−)-   (1R,2R)-(−)-   (1S,2S)-(+)-   (1R,2S)-(+)-   (1S,2R)-(−)-
camphor   isoborneol   borneol

(−)-   (+)-   (1R,2S,4S)-(+)-   (1S,2R,4R)-(−)-   (1R,2R,4S)-(+)-   (1S,2S,4R)-(−)-
fenchone   β-fenchol (*exo*-fenchol)   α-fenchol (*endo*-fenchol)

(+)-Camphor, known as Japan camphor, is the main constituent of the camphor tree, but also occurs in other plant families, e.g. in the leaves of rosemary *Rosmarinus officinalis* and sage *Salvia officinalis* (Labiatae). It gives off the typical camphor-like odor of spherical molecules [18], acts as an analeptic, a topical analgesic, a topical antipruritic, antirheumatic, antiseptic, carminative, counterirritant and, correspondingly, finds versatile application [19]. For the preparation on a large scale, the crushed wood of adult camphor trees is steam-distilled whereupon (+)-camphor partly crystallizes from the distillate.

Fenchane derivatives occur as fenchones and fenchols in several ethereal oils. Oil of fennel, obtained from the dried fruit of *Foeniculum vulgare* (Umbelliferae), contains up to 20 % (+)-fenchone, and is associated with limonene, phellandrene and α-pinene. (–)-Fenchone is isolated from the tree of life *Thuja occidentalis* (Cupressaceae), which is cultivated as hedges. The dextrorotatory enantiomer of α-fenchol with an *endo* OH, requested in perfumery, as well as its stereoisomers are found in fresh lemon juice, in oil of turpentine obtained from *Pinus palustris* (Pinaceae), in ethereal oils originating from the Lawson white cedar *Chamaecyparis lawsoniana* (Cupressaceae) and other plant families such as *Ferula*, *Juniperus*, and *Clausena* species [18].

Camphene, with its slight camphoric odor, is used in perfumery and is derived from the isocamphane skeleton; its enantiomers readily undergo racemization and occur as such or as the racemate in bergamot oil, as well as in the oils of citronella and turpentine. (+)-α-Fenchene is found in the ethereal oils of the giant tree of life *Thuja plicata* (Cupressaceae) and valerian *Valeriana officinalis* (Valerianaceae), (+)-β-fenchene in the fruits of caraway *Carum carvi* (Umbellifereae).

(1S,4R)-(–)-camphene   (1R,4S)-(+)-camphene   (1S,4S)-(+)-β-fenchene   (1S,4R)-(+)-α-fenchene

2.5 Cannabinoids

About 70 among more than 400 constituents of the Indian hemp *Cannabis sativa* var. *indica* (Moraceae, Cannabaceae) belong to the cannabinoids. These are benzopyrans derived biogenetically from the monoterpene *p*-menth-1-ene and a phenol. More precisely, the carbon atoms C-3 and C-8 of *p*-menth-1-ene (section 2.3.3, bold partial structure in the formula) close a dihydrobenzopyran ring with 5-*n*-pentylresorcinol (olivetol) in (–)-Δ⁹-tetrahydrocannabinol.

2.5 Cannabinoids

Illegal drugs prepared from Indian hemp include *marihuana*, a tobacco-like fermented mixture of dried leaves and blossoms, coming from Central Africa, Central America, U.S., as well as Southeast Asia, and *hashisch*, the resin secreted by the glands in the flowering tops of female plants, coming from the Middle East and South Asia, and having higher content of active substances. Regioisomers Δ^8- and Δ^9-tetrahydrocannabinol (THCs), differing by the position of their alkene CC double bond, are the most important addictive, analgesic, euphorizing, hallucinatory, laxative, and sedative constituents [19]. The THCs are formed upon ageing and drying (smoking) of the drugs; this involves decarboxylation of the genuine carboxylic acids such as Δ^9-tetrahydrocannabinol-2-carboxylic acid. Additional sedative constituents [19] include cannabinol (CBN) with two benzenoid rings and (−)-cannabidiol (CBD) with an opened heterocycle.

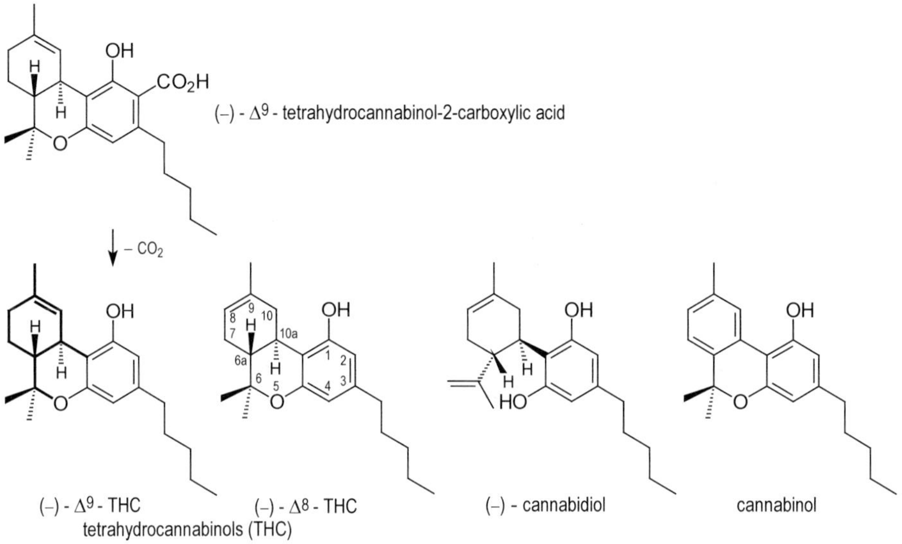

(−) - Δ^9 - tetrahydrocannabinol-2-carboxylic acid

− CO_2

(−) - Δ^9 - THC
tetrahydrocannabinols (THC)

(−) - Δ^8 - THC

(−) - cannabidiol

cannabinol

In former times, cannabis preparations were legally applied in the U.S. for the therapeutic treatment of several diseases such as asthma, constipation, epilepsy, hysteria, insomnia, and rheumatism.

3 Sesquiterpenes

3.1 Farnesanes

2,6,10-Trimethyldodecane or farnesane, the parent compound of about 10 000 sesquiterpenes known to date [2,8], is found in the oil slate. The *(E,E)*-isomer of α-farnesene is a component of the flavors and natural coatings of apples, pears and other fruits. Associated with *(E)*-β-farnesene, it also occurs in several ethereal oils, for example those of camomile, citrus, and hops. Aldehydes such as α- and β-sinensal derived from α- and β-farnesene contribute to the flavor of the oil of orange expressed from the fresh peel of ripe fruits of *Citrus sinensis* (Rutaceae); mandarin peel oil from *Citrus reticulata* and *C. aurantium* (Rutaceae) contains 0.2% of α-sinensal with the smell of oranges. (*S*)-(+)-Nerolidol in the oil of neroli obtained from orange flowers and found in many other flowers is used in perfumery [18], similar to farnesol from *Acacia farnensiana* (Mimosaceae) and the oils of bergamot, hibiscus, jasmine and rose, and pleasantly smelling blossoms such as lily of the valley. (*S*)-2,3-Dihydrofarnesol, known as terrestrol, is the marking pheromone of the male bumble bee *Bombus terrestris*.

farnesane

α-farnesene *(E,E)*

β-farnesene *(E)*

α-sinensal *(E,E,E)*

β-sinensal *(E,E)*

(*S*)-(+)-nerolidol *(E)*

farnesol *(E,E)*

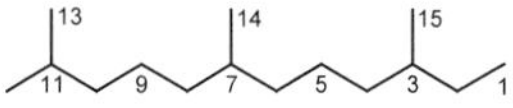

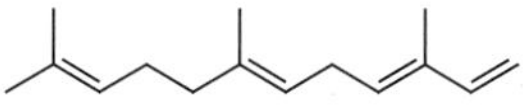

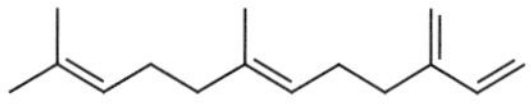

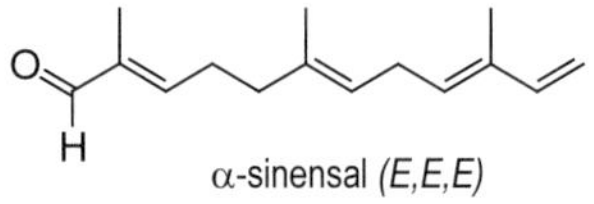

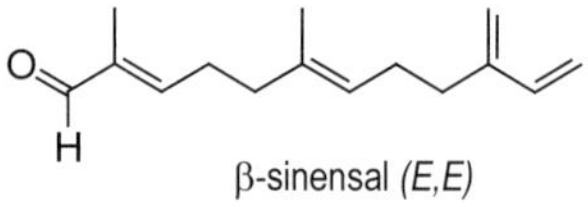

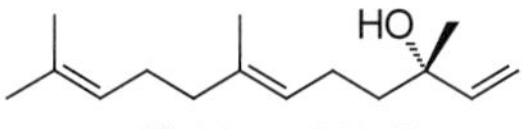

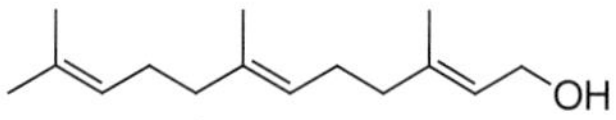

The chains of some furanoid farnesane derivatives are terminated by furan rings. Dendrolasin from sweet potatoes, also isolated from some marine snails, for example, is an alarm and defense pheromone of the ant *Dendrolasius fulginosus*. Sesquirosefuran and longifolin occur in the leaves of *Actinodaphne longifolia*.

dendrolasin

sesquirosefuran

longifolin

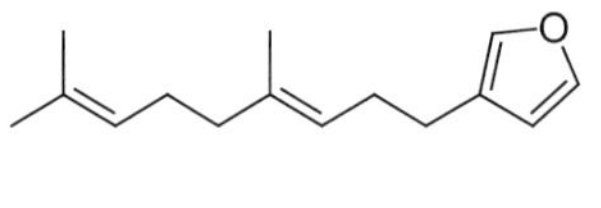

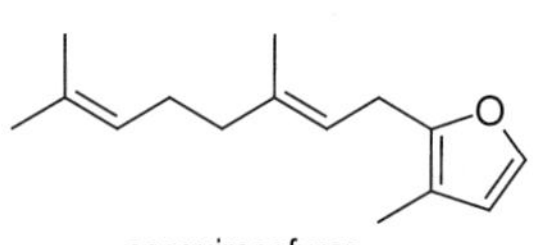

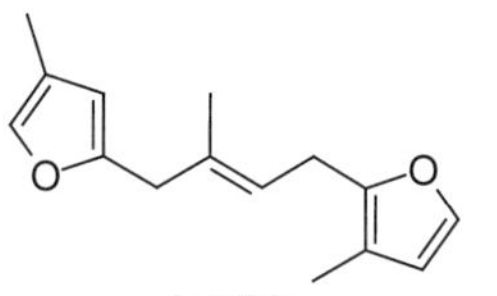

3.2 Monocyclic Farnesane Sesquiterpenes

3.2.1 Cyclofarnesanes and Bisabolanes

Cyclofarnesanes formally arise when carbon atoms C-6 and C-7 of farnesane close a ring. Abscisic acid, occurring in the leaves of cabbage, potatoes, roses, and young fruits of cotton, and (+)-dihydroxy-γ-ionylidene acetic acid produced by the fungus *Cercospora cruenta* which is antibacterially active, are examples. Abscisic acid acts as an antagonist of plant growth hormones and controls flowering, falling of fruits and shedding of leaves.

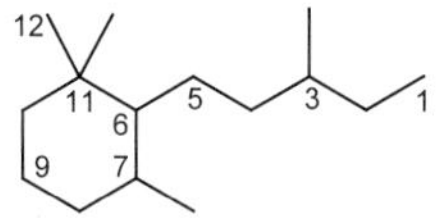

cyclofarnesane 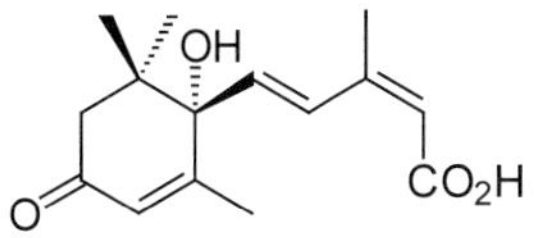   (S)-(+)-abscisic acid 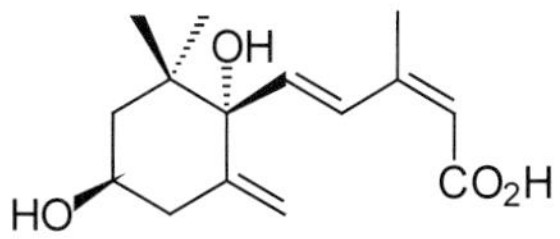   (+)-dihydroxy-γ-ionylidene acidic acid

Formally, C-1 and C-6 of farnesane close a cyclohexane ring in the bisabolanes, which represent a more prominent class of monocyclic sesquiterpenes. Additional cyclizations increase the diversity. More than 100 bisabolane derivatives of plant origin are known to date [2].

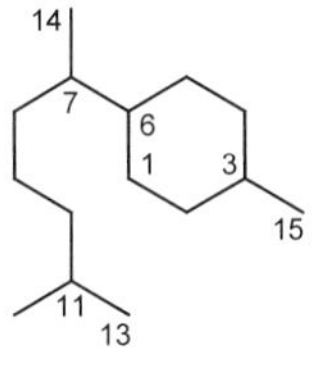

bisabolane 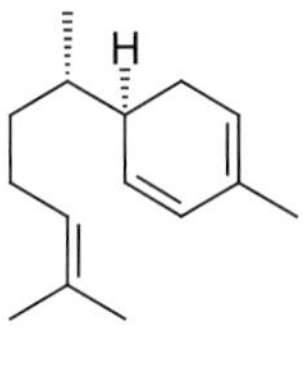   (−)-zingiberene 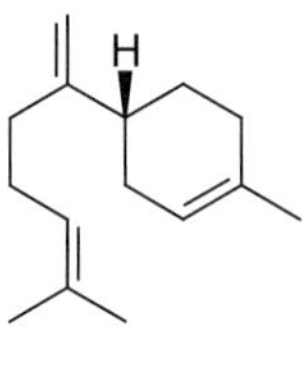   β-bisabolene 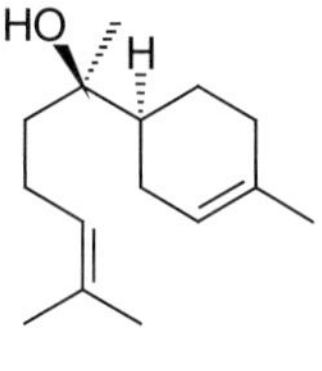   (+)-α-bisabolol 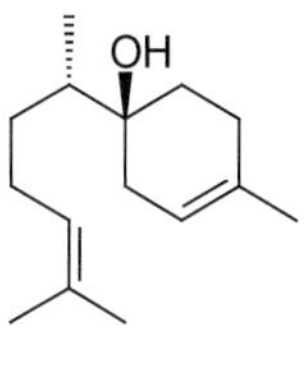   (+)-β-bisabolol

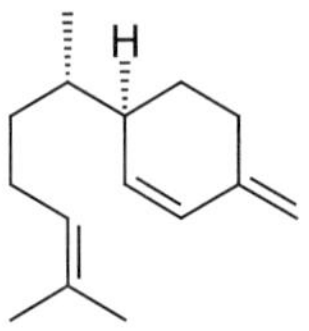

β-sesquiphellandrene 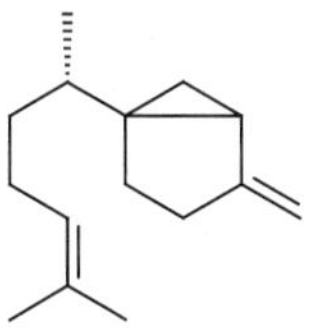   sesquisabinene 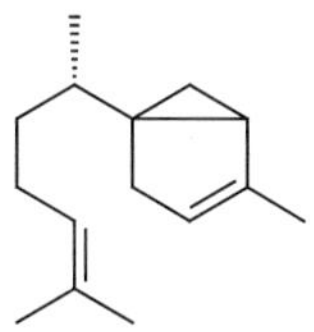   sesquithujene 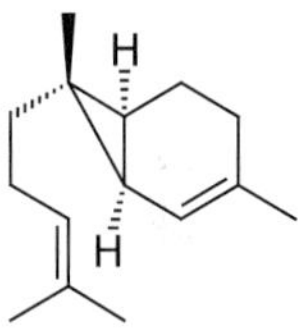   (−)-sesquicarene

Oil of ginger obtained from the rhizome of *Zingiber officinalis* (Zingiberaceae) consists predominantly of (−)-zingiberene (20-40%), β-sesquiphellandrene and (+)-

β-bisabolene. The latter also occurs in *Chamaecyparis nootkatensis* (Cupressaceae) and in the Sibirian pine tree *Pinus sibirica* (Pinaceae). (+)-α- and (+)-β-bisabolol are fragrant sesquiterpenes found in the essential oils of various plants; they also contribute to the odors of camomile and of bergamot oil from unripe fruits of *Citrus aurantium* var. *bergamia* (Rutaceae) [18] growing in southern Italy.

Sesquisabinene from pepper *Piper nigrum* (Piperaceae), sesquithujene from ginger *Zingiber officinalis* (Zingiberaceae) and sesquicarene from *Schisandra chinensis* represent bicyclic bisabolanes.

3.2.2 Germacranes and Elemanes

Germacranes formally result from ring closure of C-1 and C-10 of farnesane. 1(10),4-Germacradienes such as 1(10),4-germacradien-6-ol present as a glycoside in *Pittosporum tobira* may undergo COPE rearrangements to elemadienes, exemplified by shyobunol from the oils of galbanum and kalmus, so that some isolated elemane derivatives are supposed to be artifacts arising from germacranes.

More than 300 naturally occurring germacranes are reported [2]. Among these are germacrene B [1(10)-*E*,4-*E*,7(11)-germacratriene] from the peel of *Citrus junos*, germacrene D from bergamot oil (*Citrus bergamia*, Rutaceae), and germacrone [1(10)-*E*,4-*E*,7(11)-germacratrien-8-one], derived from germacrene B, a pleasantly flowery to herby-smelling component isolated from the essential oil of myrrh (*Commiphora abyssinica*, Burseraceae) [18] as well as from ethereal oils of *Geranium macrorhyzum* (Geraniaceae) and *Rhododendron adamsii* (Ericaceae).

3.2 Monocyclic Farnesane Sesquiterpenes

Periplanones A-D act, in contrast to long-range pheromones, as close proximity sex-excitants [16,17]. They are found in the alimentary tract and excreta of the female American cockroach *Periplaneta americana*, and cause the males to run and to perform the courtship display. Periplanones A and B, occurring in a ratio of 10:1, are 100 times more active than C and D.

periplanones

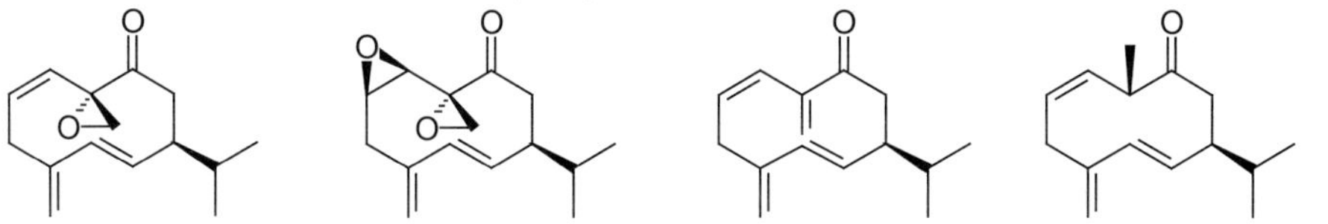

A B C D
     1,4(15),5-germacratrien-9-one

About 50 elemanes known to date [2] comprise β-elemenone from the oil of myrrh, representing the COPE rearrangement product of germacrone, (−)-bicycloelemene from peppermint oils of various provenance (e.g. *Mentha piperita* or *Mentha arvensis*), and β-elemol which is not only a minor component of Javanese oil of citronella but is also found in the elemi oil with an odor like pepper and lemon, expressed from the Manila elemi resin of the tree *Canarum luzonicum* (Burseraceae).

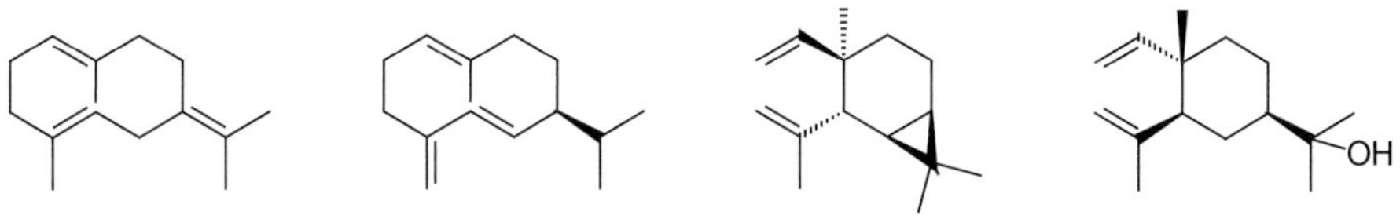

germacrene B   (−)-germacrene D   (−)-bicycloelemene   (−)-β-elemol

3.2.3 Humulanes

Ring closure of C-1 and C-11 of farnesane, not only formally but also in biogenesis via farnesyldiphosphate, produces the sesquiterpene skeleton of more than 30 naturally occurring humulanes [2]. Regioisomeric α- and β-humulene occur in the leaves of *Lindera strychnifolia* (Lauraceae).

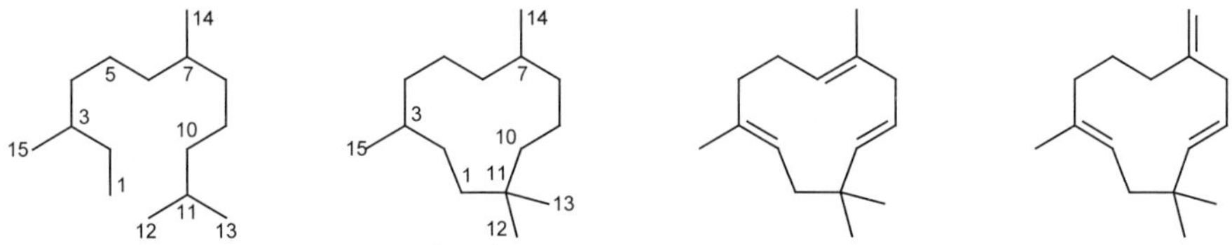

farnesane   humulane   α-humulene   β-humulene

Humulenes are, associated with epoxyhumuladienes derived from α-humulene as well as (−)-humulol and (+)-humuladienone, prominent constituents of the essential oils of hops (*Humulus lupulus*, Cannabaceae), cloves (*Caryophylli flos*, Caryophyllaceae) and ginger (*Zingiber zerumbeticum*, Zingiberaceae).

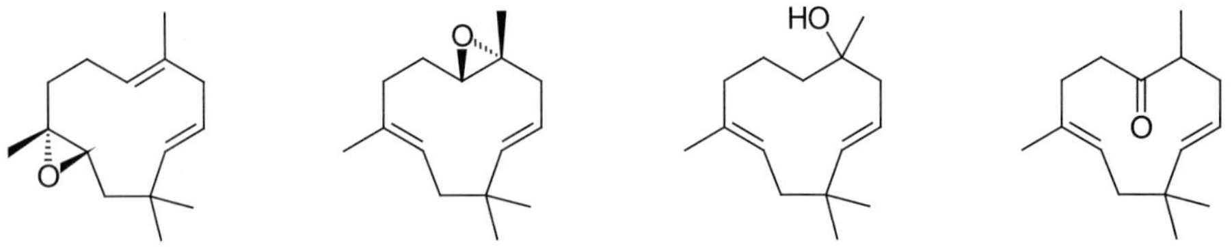

(−)-2,3-epoxy-6,9-humuladiene   (−)-6,7-epoxy-2,9-humuladiene   (−)-humulol   (+)-humula-2,9-dien-6-one

3.3 Polycyclic Farnesane Sesquiterpenes

3.3.1 Caryophyllanes

Approximately 30 naturally abundant caryophyllanes [2] are derived from humulanes in which C-2 and C-10 close a cyclobutane ring.

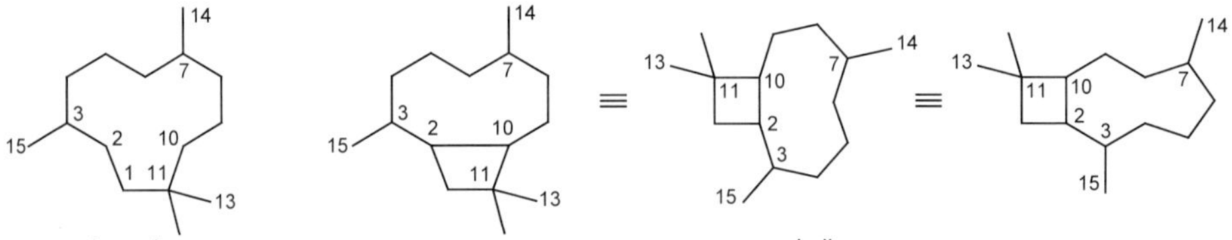

humulane   caryophyllane

(−)-β-Caryophyllene occurs as a mixture with its *cis* isomer isocaryophyllene in the clove oil (up to 10 %) from dried flower buds of cloves (*Caryophylli flos*, Caryophyllaceae), in the oil obtained from stems and flowers of *Szygium aromaticum* (Myrtaceae), as well as in the oils of cinnamon, citrus, eucalyptus, sage, and thyme [18]. Clove oil, with its pleasantly sweet, spicy and fruity odor, is used not only in perfumery and for flavoring chewing gums, but also as a dental analgesic, carminative and counterirritant.

3.3 Polycyclic Farnesane Sesquiterpenes

Other representatives include (–)-6,7-epoxy-3(15)-caryophyllene from the leaves, flowers and stems of cloves, (+)-6-caryophyllen-15-al from the oil of sage (*Salvia sclarea*, Labiatae), and (–)-3(15),7-caryophylladien-6-ol from Indian hemp *Cannabis sativa* var. *indica* (Cannabaceae).

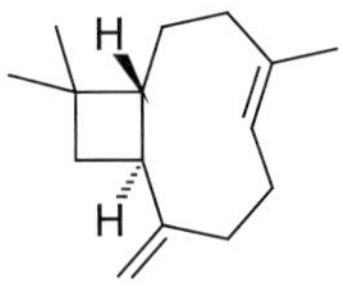

(–)-β-caryophyllene (*trans* isomer)

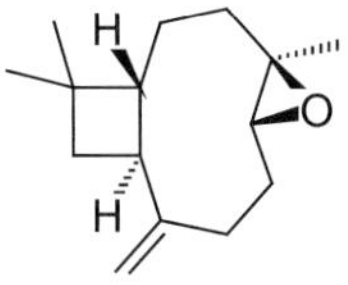

(–)-6,7-epoxy-3(15)-caryophyllene

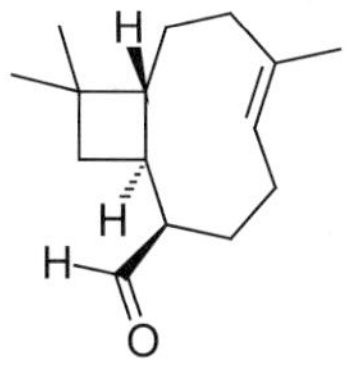

(+)-6-caryophyllen-15-al

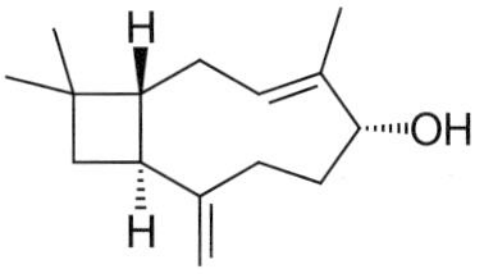

(–)-3(15),7-caryophylladien-6-ol

3.3.2 Eudesmanes and Furanoeudesmanes

Carbon atoms C-1 and C-10 in addition to C-2 and C-7 of farnesane link up to close the eudesmane bicyclic skeleton of sesquiterpenes with *trans*-decalin as core structure with corresponding numbering of the ring positions. To date, about 500 eudesmanes, formerly referred to as selinanes, have been documented in the literature [2,8].

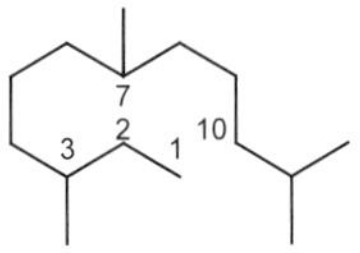

farnesane

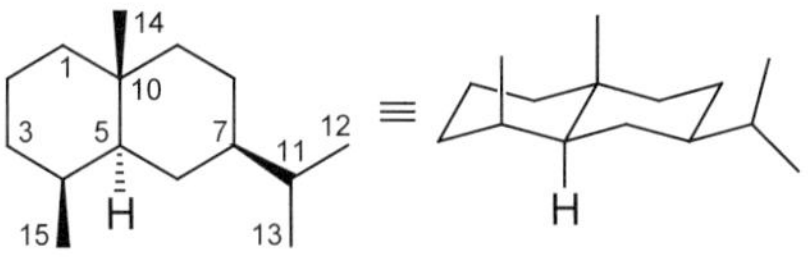

eudesmane (selinane)

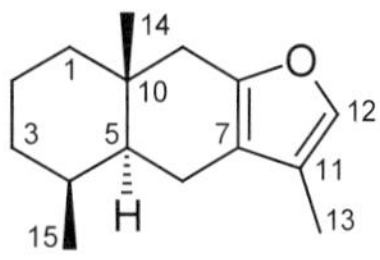

furanoeudesmane

Well-known eudesmane derivatives in flavors and fragrances include α- and β-selinene from the oils of *Cannabis sativa* var. *indica* (Moraceae), celery (*Apium graveolens*, Umbelliferae) and hops (*Humulus lupulus*, Moraceae), (+)-α- and (+)-β-eudesmol from some oils of eucalyptus (*Eucalyptus macarthuri*), (–)-epi-γ-eudesmol with its woody odor from the north African oil of geranium (*Pelargonium odoratissimum* and allied species), and the almost odorless diastereomeric (+)-γ-eudesmol from various ethereal oils [18]. (+)-β-Costus acid and (+)-β-costol belong to the constituents of the essential oil obtained from the roots of *Saussurea lappa* (Asteraceae) which is used to treat stomach ailments in Chinese and Japanese popular medicine.

(−)-α-eudesmene
(α-selinene)

(−)-β-eudesmene
(β-selinene)

(+)-α-eudesmol
(3-selinen-11-ol)

(+)-β-eudesmol
(4(15)-selinen-11-ol)

(+)-γ-eudesmol
(4-selinen-11-ol)

(−)-*epi*-γ-eudesmol
(4-selinen-11-ol)

(+)-β-costus acid

(+)-β-costol

Several structural variants of lactones derived from eudesmane occur in *Artemisia*-species (Asteraceae). These include various 3-oxo-12,6-eudesmanolides such as (+)-santonane from the flowers of *A. pauciflora*, (−)-taurin from *A. taurica* (not to be confused with taurine = 2-aminoethanesulfonic acid) and the antihelmintic but toxic santonines [19] isolated from the dried unexpanded flowerheads of *A. maritima* (contents up to 1.5% of α-santonin) and allied species.

(−)-4α,5α,6α,11α-3-oxo-
12,6-eudesmanolid
(santonan)

(−)-6α,11α-3-oxo-
4-eudesmen-12,6-olid
(taurin)

(−)-6α,11βH-3-oxo-
1,4-eudesmadien-
12,6-olid (α-santonin)

(−)-6α,11αH-3-oxo-
1,4-eudesmadien-
12,6-olid (β-santonin)

Furanoeudesmanes such as (−)-furanoeudesma-1,3-diene, furanoeudesma-1,4-dien-6-one and furanoeudesma-1,4(15)-diene, known as (−)-lindestrene, belong to the sweetish balsamic-smelling constituents of the yellowish red gum-resin myrrh, used as a carminative and astringent and obtained from *Commiphora* species (e.g. *Commiphora abyssinica, C. molmol*, Burseraceae) [18]. Tubipofuran, a diastereomer of

3.3 Polycyclic Farnesane Sesquiterpenes

(−)-furanoeudesma-1,3-diene isolated from *Tubipora musica* exhibits cyto- and ichthyotoxic activity (killing cells and fish, respectively).

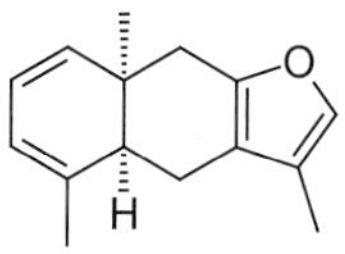

(+)-tubipofuran

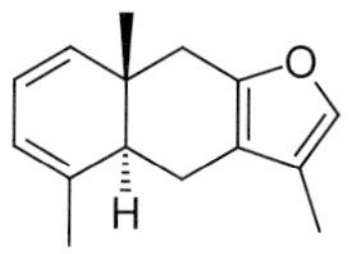

(−)-furanoeudesma-1,3-diene

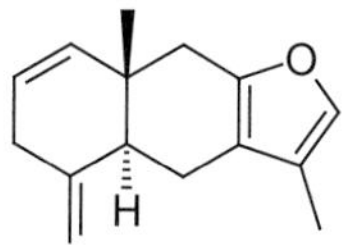

(−)-furanoeudesma-1,4(15)-diene

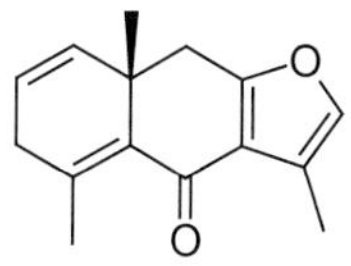

(−)-furanoeudesma-1,4-dien-6-one

3.3.3 Eremophilanes, Furanoeremophilanes, Valeranes

A methyl shift from C-10 to C-5 in eudesmane leads to the basic skeleton of more than 150 eremophilane and furanoeremophilane derivatives isolated so far from higher plants [2].

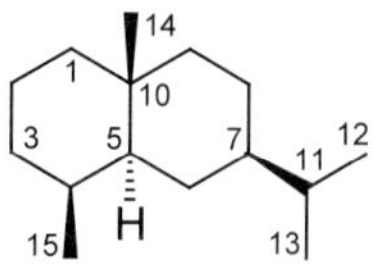

eudesmane

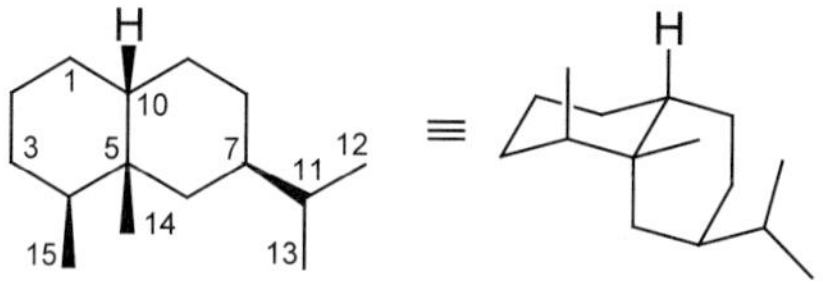

eremophilane

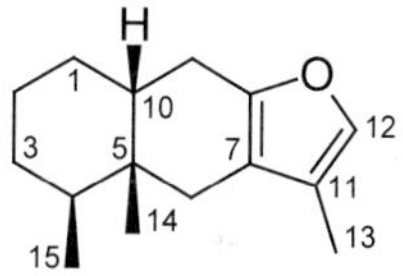

furanoeremophilane

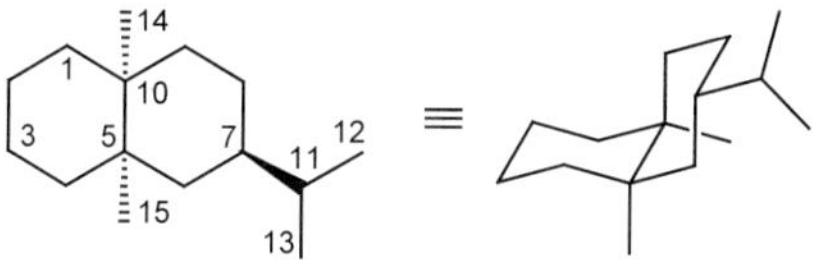

valerane

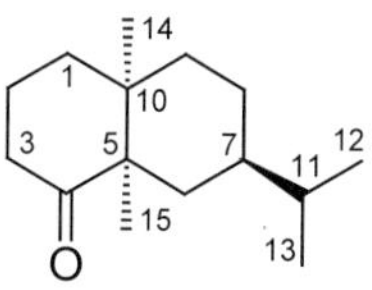

(−)-valeranone

In contrast, valeranes arising from migration of the methyl group C-15 in eudesmane from C-4 to C-5, in contrast, very rarely occur. Examples include the valerenones from the roots of valerian *Valeriana officinalis* and from *Nardostachys jatamansi* (Valerianaceae).

The Australian tree *Eremophila mitchelli* gave its name to the eremophilanes with both methyl groups in β-positions of the decalin bicycle; the wood of *Eremophila mitchelli* contains various eremophiladienones. Non-toxic metabolites of the fungus

Penicillium roqueforti growing in some kinds of cheese and referred to as eremofortins are more prominent representatives of the eremophilanes.

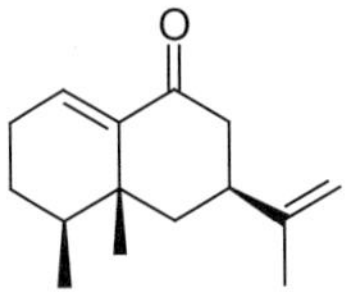

1(10),11-eremo-
philadien-9-one

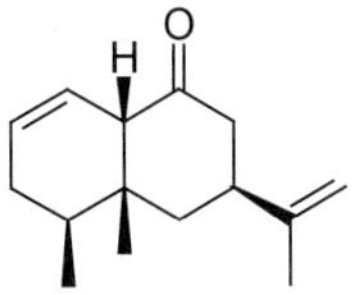

1,11-eremophila-
dien-9-one

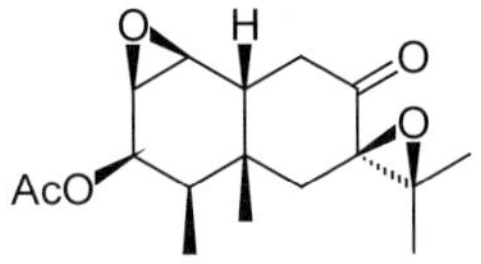

(+)-eremofortin A

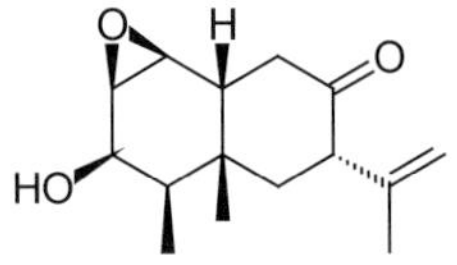

(+)-eremofortin B

Eremophilanes with both methyl groups in α-positions of the decalin core structure are referred to as valencanes, exemplified by (−)-nootkatene from the nootka cypress *Chamaecyparis nootkatensis* (Cupressaceae) and (+)-11-eremophilen-2,9-dione from the oil of grapefruit. Additional examples include (+)-valerianol [1(10)-eremophilen-11-ol] from valerian (*Valeriana officinalis*) as well as nootkatone [1(10),11-eremophiladien-2-one] from the oil of grapefruit (*Citrus paradisii*) and the nootka cypress which is added as a flavor to drinks. The regioisomeric (+)-isonootkatene [α-vetivone, 1(10),7(11)-eremophiladien-2-one] is a main constituent of the oil of vetiver distilled from the roots of tropical vetiver grass *Vetiveria zizanoides* (Poaceae) used in soap formulations and perfumery.

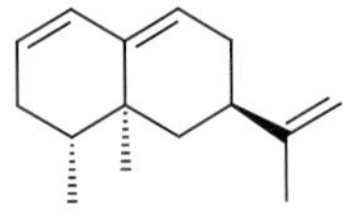

(−)-1,9,11-eremophilatriene
(nootkatene)

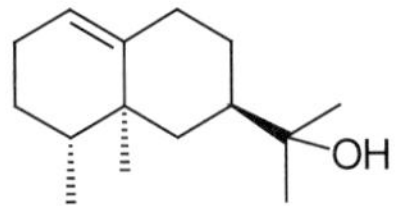

(+)-1(10)-eremophilen-11-ol
(valerianol)

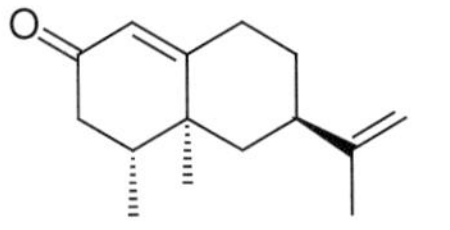

(+)-1(10),11-eremophiladien-
2-one (nootkatone)

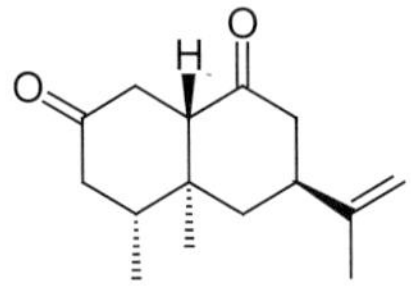

(+)-11-eremophilen-
2,9-dione

More than 100 furanoeremophilanes are described as constituents in higher plants [2], chiefly in *Senecio* species (Asteraceae, formerly Compositae).

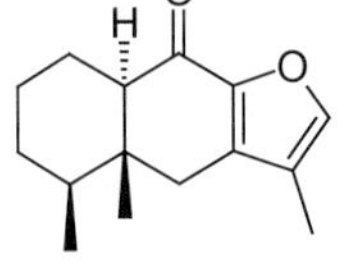

furanoeremo-
philan-9-one

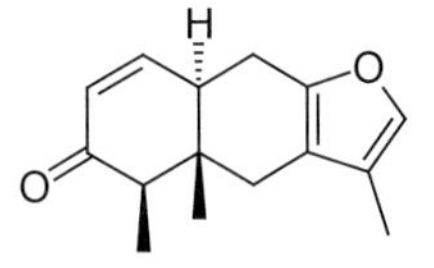

(+)-furanoeremophil-
1-en-3-one

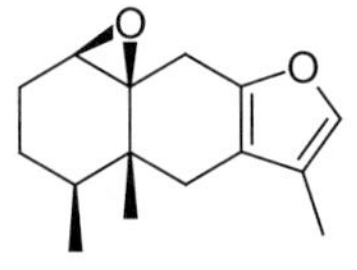

(−)-1,10-epoxy-
furanoeremophilane

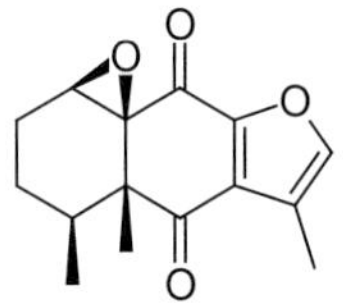

(−)-1,10-epoxyfuranoeremo-
philan-6,9-dione

3.3 Polycyclic Farnesane Sesquiterpenes

Furanoeremophilan-9-one, a constituent of golden ragwort (squaw weed, life root) from the dried plant of *Senecio aureus*, (+)-furanoeremophilen-3-one from *Senecio nemorensis*, (−)-1,10-epoxyfuranoeremophilane from *Senecio glastifolius*, and (−)-1,10-epoxyfuranoeremophilan-6,9-dione from *Senecio smithii* represent typical examples.

Nardosinanes such as (−)-kanshone A, including some structural variants, and nardosinone, a 1,2-dioxolane (cyclic peroxide) isolated from *Nardostachys chinensis* (Valerianaceae) formally emerge from eremophilane by migration of the isopropyl-group from C-7 to C-6.

Aristolanes are 6,11-cycloeremophilanes. Some representatives are found in Aristolochiaceae, for example 9-aristolen-8-one and 1(10)-aristolen-12-al in *Aristolochia debilis*. 1(10)-Aristolene is a constituent of the Gurjun balm flowing from the caves cut into the giant Dipterocarpaceae growing in Bengal upon setting afire these trees.

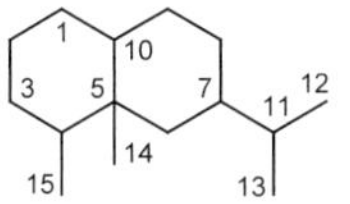

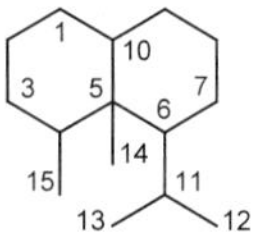

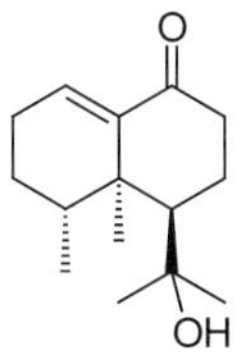

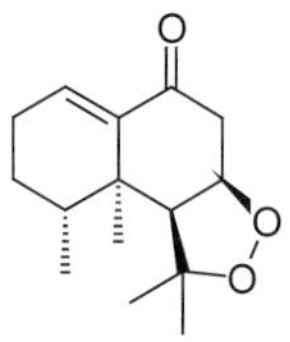

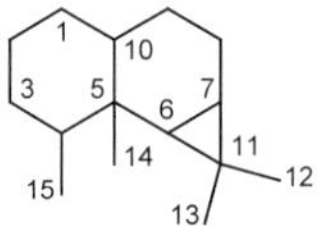

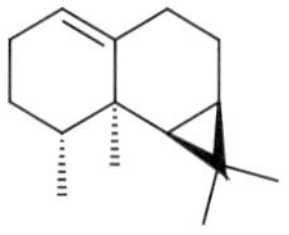

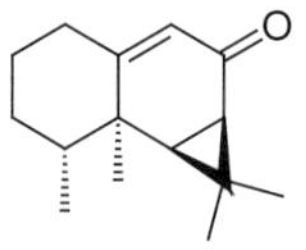

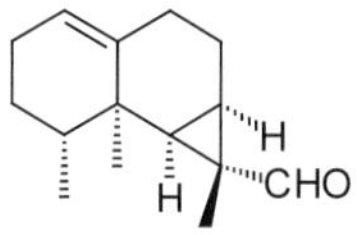

Ishwaranes represent 7-11/10-12-bicycloeremophilanes. These rare tetracyclic sesquiterpenes are found in Aristolochiaceae, for example ishwarane itself and ishwaranol in the roots of *Aristolochia indica* and 3-ishwaranone from *Aristolochia debilis*.

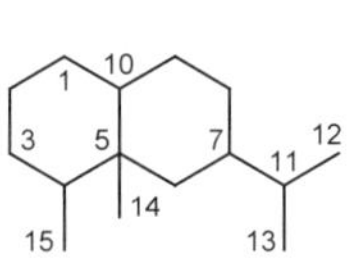

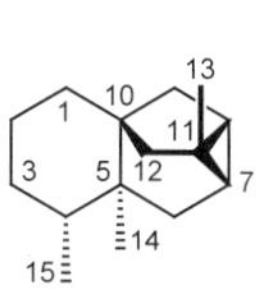

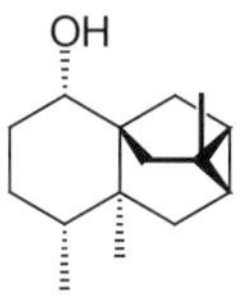

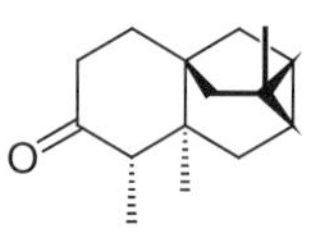

3.3.4 Cadinanes

More than 200 naturally abundant cadinanes (Table 3)[2] formally arise from ring closure of C-1 and C-6 as well as C-5 and C-10 of farnesane. The generally accepted numbering system, however, is not derived from farnesane, but from germacrane. Depending on the relative configuration at C-1, C-6 and C-7, the *trans*-decalines cadinane and bulgarane are distinguished from muurolane and amorphane, each with the *iso*-propyl group in β- or α- position at C-7. Calamenenes contain one benzenoid ring; cadalene incorporates the naphthalene bicycle.

(−)-4,9-Cadinadiene (α-cadinene) from the oil of hops (*Humulus lupulus*, Cannabaceae) as well as (−)-3,9-cadinadiene, known as β-cadinene, widely spread in plants, spicy smelling and isolated from the oil of cade obtained by distillation of the wood of Mediterranean juniper *Juniperus oxycedru* (Cupressaceae), exemplify the cadinanes. Berries of juniper species *Juniperus communis* and *J. oxycedrus* contain bulgaranes such as (−)-4,9-bulgaradiene ($β_1$-bulgarene) and (−)-4(15),10(14)-bulgaradiene (ε-bulgaren, Table 3).

Muurolanes include (+)-4(15),10(14)-muuroladiene (ε-muurolene) from Swedish turpentine and ylang-ylang oil obtained by steam distillation of freshly picked flowers of the cananga tree *Cananga odorata* (Annonaceae) growing in Madagascar and the Phillipine islands. They are pleasantly smelling and used in delicate perfumes. (−)-4,10(14)-Muuroladiene (γ-muurolene) also occurs in the expectorant oil of pine needles (*Pinus silvestris*, Pinaceae). Amorphanes are represented by (−)-4,11-amorphadiene from *Viguiera oblongifolia*; its 12-carboxylic acid, also referred to as artemisic or qinghao acid, is isolated from *Artemisia annua* (Asteraceae) and exhibits antibacterial activity (Table 3).

Benzenoid (−)-(7*S*,10*S*)-calamenene is isolated from *Ulmus thomasii* (Ulmaceae), and (+)-3,8-calamenenediol from *Heterotheca subaxillaris*. The naphthalene sesquiterpene cadalene occurs in conifers, for example in the resin of fir *Abies sibirica* (Pinaceae). The wood of several trees contains 3-cadalenol. Cadalen-2,3-quinone, also known as mansonone C, is a constituent of *Mansonia altissima* and *Ulmus lactinata* (Table 3).

Hibiscones and various reddish-brown hibiscoquinones from *Hibiscus elatus* (Malvaceae) represent furanoid derivatives of cadinane.

(+)-hibiscone A   (+)-hibiscone B   (−)-hibiscone C   hibiscoquinone D

3.3 Polycyclic Farnesane Sesquiterpenes

Table 3. Cadinanes.

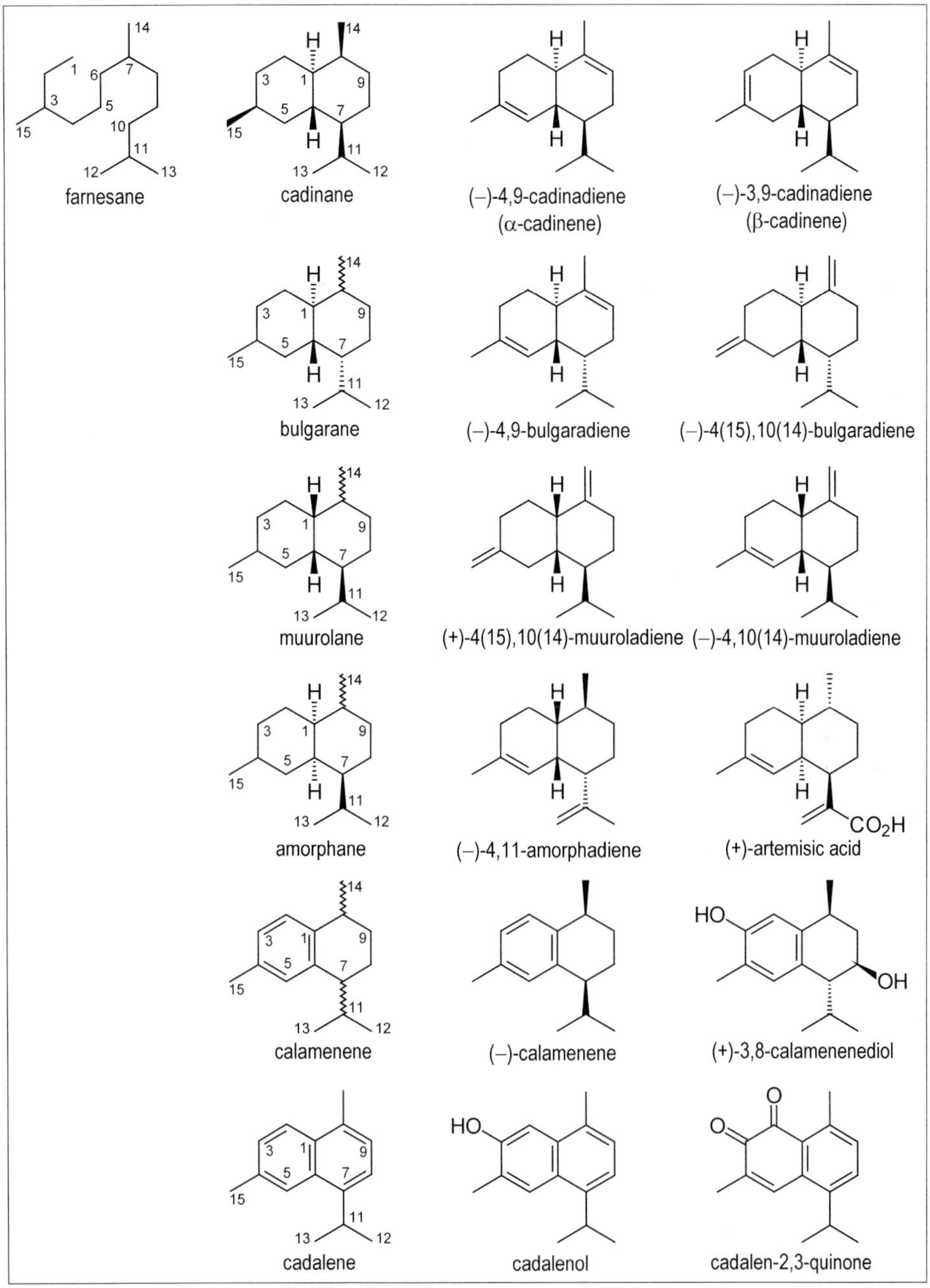

farnesane	cadinane	(−)-4,9-cadinadiene (α-cadinene)	(−)-3,9-cadinadiene (β-cadinene)
	bulgarane	(−)-4,9-bulgaradiene	(−)-4(15),10(14)-bulgaradiene
	muurolane	(+)-4(15),10(14)-muuroladiene	(−)-4,10(14)-muuroladiene
	amorphane	(−)-4,11-amorphadiene	(+)-artemisic acid
	calamenene	(−)-calamenene	(+)-3,8-calamenenediol
	cadalene	cadalenol	cadalen-2,3-quinone

Antimalarials derived from 4,5-*seco*-cadinane are found as constituents of the traditional Chinese medicinal herb *Artemisia annua* (Asteraceae), well-known as qinghao. Artemisinine, also referred to as qinghaosu, is a 3,6-peroxide of the acylal formed by 4,5-*seco*-cadinane-5-aldehyde-12-oic acid. Dihydroqinghaosu and the 11(13)-dehydro derivative artemisitene are the active substances which, nowadays, are applied as semisynthetic esters and ethers (e.g. artemether) to cure malaria [19,20]. These peroxides probably eliminate singlet oxygen, which damages the membrane of the pathogens and disturbs their nucleic acid metabolism.

3.3.5 Drimanes

Bond formation between C-2 and C-7 as well as C-6 and C-11 of farnesane formally leads to the drimane basic skeleton of sesquiterpenes [2]. The accepted numbering system is derived from decalin and not from farnesane. The parent hydrocarbon 5α,8α,9β,10β-drimane with *trans*-decalin as core structure occurs in paraffin oil.

The name of this class of sesquiterpenes stems from *Drimys winteri* (Magnoliaceae, Winteraceae); (–)-7-drimen-11-ol (drimenol), which is active as a plant growth regulator, and the lactone (–)-7-drimen-11,12-olide (drimenine) have been isolated from the bark of this tree. The constituents of tobacco (*Nicotiana tabacum*, Solanaceae) include (+)-8-drimen-7-one, and 6,14,15-trihydroxy-8-drimen-12,11-olide (astellolide A) is a metabolite of *Aspergillus variecolor* and some other mold species. 11,15-Nordrimanes such as α- and δ-ambrinol are found among the constitu-

3.3 Polycyclic Farnesane Sesquiterpenes

ents shaping the pleasant mossy sandalwood odor of waxy gray ambergris [18], found on tropical seashores, produced by the sperm whale *Physeter macrocephalus* in order to seal the wounds caused by food in the intestinal tract. It is used in perfumery for fixing delicate odors.

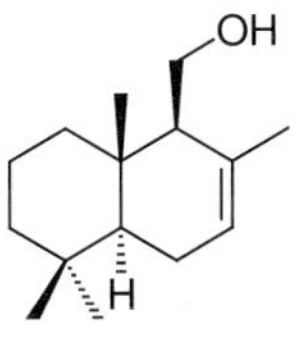

(−)-drimenol 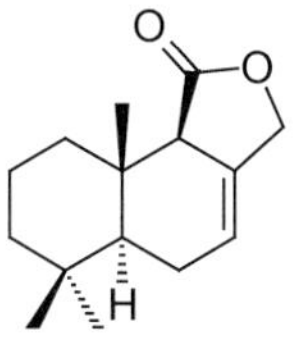   (−)-drimenine 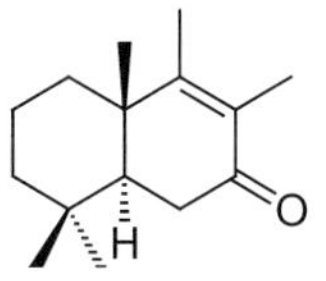   (+)-8-drimen-7-one 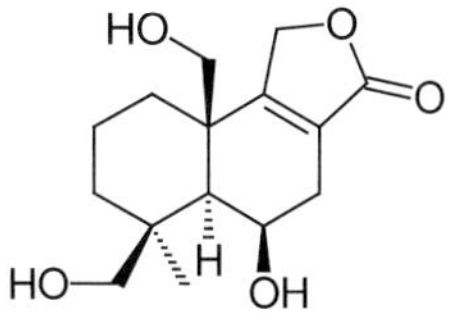   (−)-astellolide A

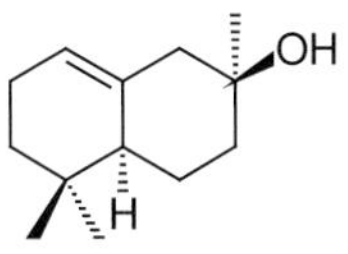

α-ambrinol 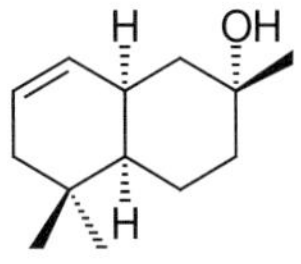   δ-ambrinol

3.3.6 Guaianes and Cycloguaianes

Bond formation from C-1 to C-10 and C-2 to C-6 of farnesane formally produces the bicyclic skeleton of more than 500 guaianes isolated so far from higher plants [2] with the numbering system adopted from that of decalin. Guaianes are also referred to as proazulenes because their naturally occurring derivatives frequently undergo dehydration to terpenoid azulenes (guaia-1,3,5,7,9-pentaenes) upon heating or steam distillation. Deep blue-violet oily guaiazulene (guaia-1,3,5,7,9-pentaene) obtained as an artifact upon work-up of the oils of camomile and guaiac wood from *Guajacum* species (Zygophyllaceae) is a well-known example. The milky juice of the delicious fungus *Lactarius deliciosus* turns from orange to greenish upon damaging the fungal body when the genuine yellow 15-stearoyloxyguaia-1,3,5,7,9,11-hexaene is decomposed enzymatically to violet lactaroviolin (guaia-1,3,5,7,9,11-hexaen-4-aldehyde) [2].

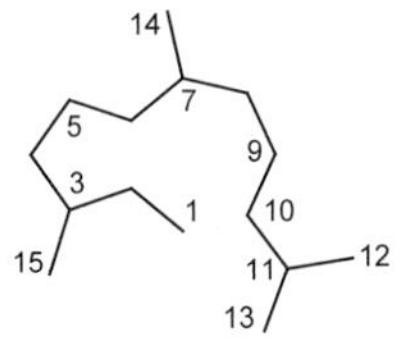

farnesane 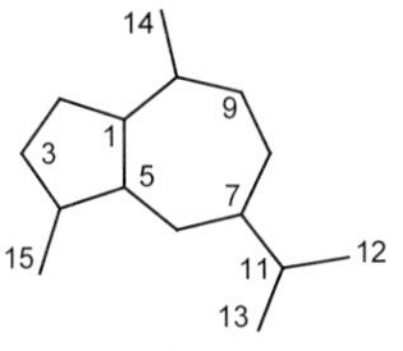   guajane 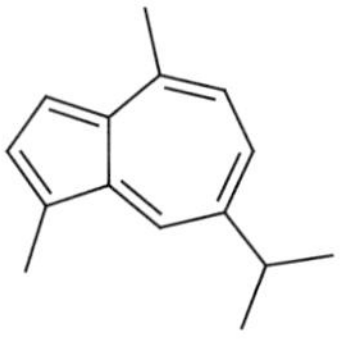   guajazulene 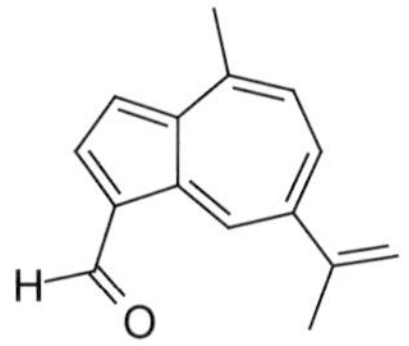   lactaroviolin

(+)-1(5),6-Guaiadiene and its 4β-stereoisomer occur in balsamum tolutanum, a balm obtained from the tree *Myroxylon balsamum* (Leguminosae) which grows in the northern areas of South America. (−)-1(5),11-, (−)-1(10),11- and (−)-1(10),7(11)-guaiadiene [2] are found among the constituents of guaiac wood oil from the tree *Guajacum officinale* (Zygophyllaceae) native in central America, and of pleasantly smelling patchouli oil obtained by steam distillation of fermented leaves of the patchouli shrub *Pogostemon patchouli* (Labiatae) cultivated for perfumery in tropical countries.

(+)-1(5),6-guaiadiene

(−)-1(5),11-guaia-
diene (α-guaiene)

(−)-1(10),11-guaia-
diene (α-bulnesene)

(−)-1(10),7(11)-guaia-
diene (β-bulnesene)

The air-dried milky exudation of the roots of *Ferula galbaniflua* (Umbelliferae), collected in Iran and known as galbanum resin or gum galbanum, as well as guaiac wood oil, contain the tertiary alcohols (−)-1(5)-, (+)-1(10)- and (+)-9-guaien-11-ol with a spicy odor of leaves and wood. (−)-Kessoglycol, a guaiane tricycle, is found in the rhizome of valerian *Valeriana officinalis* (Valerianaceae).

(−)-1(5)-guaien-
11-ol (guaiol)

(+)-1(10)-guaien-
11-ol (bulnesol)

(+)-9-guaien-11-ol

(−)-kessoglycol

12,6-Guaianolides as a class of tricyclic sesquiterpene lactones [2,21] are found in a large variety of the plant family Asteraceae (formerly denoted as Compositae). Well-known examples include (−)-artabsin containing a cyclopentadiene partial structure and its DIELS-ALDER dimer absinthin as the chief bitter principle (bitterness threshold 1 : 70 000) of wormwood (absinthium) *Artemisia absinthum*. This is used predominantly in perfumery but, because ingestion of the extracts may cause stupor, convulsions and even death, much less often it is applied to flavor alcoholic beverages (e.g., vermouth). (+)-Arglabin from *Artemisia glabella*, tanaparthin-α-peroxide from the daisy flower *Tanacetum parthenium*, achillicin from yarrow (*Achillea millefolium*), matricin and its 8-*O*-acetyl derivative from camomile *Matricaria chamomilla* (Asteraceae) [2] are additional representatives.

(+)-1(10)-epoxy-3,11(13)-guaiadien-12,6-olide (arglabin)

(−)-1,4-epidioxy-9,10-dihydroxy-2,11(13)-guaiadien-12,6-olide (tanaparthin-α-peroxide)

R=OH: 8,10-dihydroxy-1,4-guaiadien-12,6-olide (achillicin)
R=H, 10β-OH: (−)-artabsin

(−)-4,8-dihydroxy-1(10),2-guaiadien-12,6-olide (matricin)

Pseudoguaianes formally arise from guaianes by methyl shift from C-4 to C-5, represented by ambrosic acid and the antineoplastic pseudoguaianolide ambrosin, both isolated from the herb *Ambrosia maritima* (Asteraceae) and other *Ambrosia* species. Helenalin from the flowers of *Helenium autumnale* and *Arnica montana* (Asteraceae) [19,21] acts as an abortive, antiinflammatory, antineoplastic, antirheumatic, and antipyretic; external contact may cause skin irritations and sneezing, however, while ingestion may initiate vomiting, diarrhoea, vertigo, and heart pounding up to circulatory collapse and death.

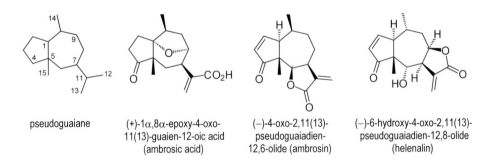

pseudoguaiane

(+)-1α,8α-epoxy-4-oxo-11(13)-guaien-12-oic acid (ambrosic acid)

(−)-4-oxo-2,11(13)-pseudoguaiadien-12,6-olide (ambrosin)

(−)-6-hydroxy-4-oxo-2,11(13)-pseudoguaiadien-12,8-olide (helenalin)

6,11-Cycloguaianes are referred to as **aromadendranes**. Various aromadendrenes and aromadendradienes belong to the constituents of balsamum tolutanum from *Myroxylon balsamum* (Leguminosae). The oil of sage from *Salvia sclarea* (Labiatae), cultivated for perfumery in the south of Europe, contains (+)-1(10)-aromadendren-7-ol [18], also known as isospathulenol.

aromadendrane   (+)-1-aromadendrene   1(5),3-aroma-dendradiene   (+)-1(10)-aromadendren-7-ol (isospathulenol)

Cubebanes and **ivaxillaranes** represent 1,6- and 8,10-cycloguaianes, respectively. The fruits of Java pepper *Piper cubeba* (Piperaceae) contain (−)-4-cubebanol. The name ivaxillarane stems from *Iva axillaris* with (−)-ivaxillarin as a constituent.

cubebane   (−)-4α-cubebanol   ivaxillarane   (−)-ivaxillarin

Patchoulanes represent 1,11-cycloguaianes and rearranged derivatives of those, occurring in the the oils of cypress, guaiac wood and patchouli. Examples include α-patchoulene and (−)-patchoulenone from the oil of cypress (*Cupressus sempervirens*, Pinaceae) as well as the rearranged derivatives β-patchoulene and (−)-patchoulialcohol (patchoulol) from patchouli oil (*Pogostemon patchouli*, Labiatae, p. 38) [18].

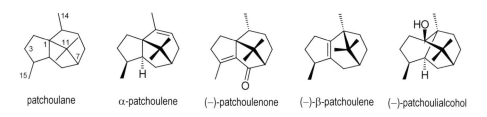

patchoulane   α-patchoulene   (−)-patchoulenone   (−)-β-patchoulene   (−)-patchoulialcohol

Valerenanes is the common name of 8(7-6)-abeoguaianes, in which the seven-membered ring of guaiane has contracted to cyclohexane, involving a migration of C-8 from C-7 to C-6. Few valerenanes are known to date, including valerenol,

3.3 Polycyclic Farnesane Sesquiterpenes

valerenal and (–)-valerenoic acid, all of which belong to the constituents of the rhizome and roots of valerian *Valeriana officinalis* (Valerianaceae).

3.3.7 Himachalanes, Longipinanes, Longifolanes

Bonds between C-1 and C-6 as well as C-1 and C-11 formally convert farnesane into the bicyclic skeleton of himachalane with the numbering system adopted from that of farnesane. Several himachalanes such as α-himachalene and himachalol are constituents of the oil of cedar wood from *Cedrus deodara* (Pinaceae).

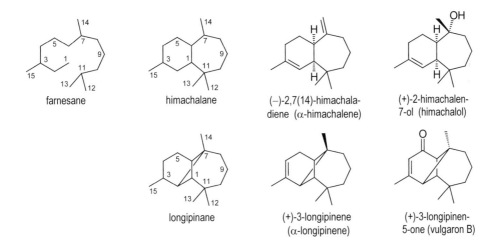

2,7-Cyclohimachalanes are known as longipinanes. They occur in various oils of pine wood and some Asteraceae, exemplified by 3-longipinene from *Pinus* species (Pinaceae) and 3-longipinen-5-one from *Chrysanthemum vulgare* (Asteraceae).

Longipinanes are differentiated from longifolanes which formally and biogenetically also emerge from farnesane by cleaving the C-3–C-4 bond and closing the bonds C-1–C-6, C-2–C-4, C-3–C-7, and C-1–C-11 to the tricycle. Examples are the isomers longicyclene and longifolene, widely spread in ethereal oils, the latter present to an extent of up to 20% in Indian turpentine oil which is produced commercially from the Himalayan pine *Pinus longifolia* (Pinaceae) for the synthesis of a widely used chiral hydroboration agent.

farnesane longifolane longicyclene (+)-longifolene

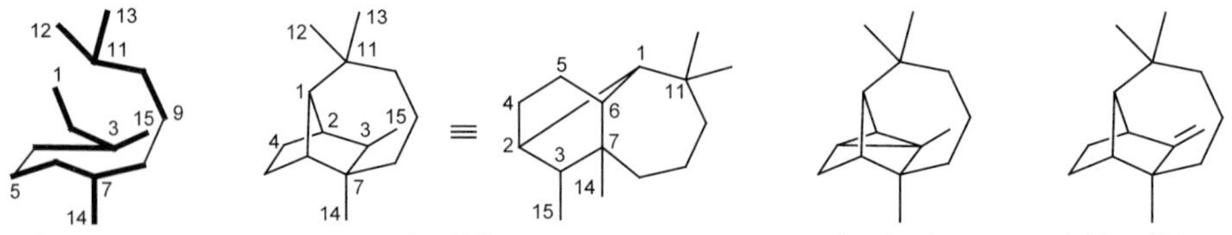

3.3.8 Picrotoxanes

Farnesane is formally converted into picrotoxane by making the bonds C-3–C-7 and C-2–C-10 (numbering system of farnesane) also involving methyl migration from C-6 to C-13 (numbering system of picrotoxane). About 15 toxic sesquiterpene alkaloids with picrotoxane skeleton such as (–)-dendrobin are among the constituents of the orchid *Dendrobium nobile* (Orchidaceae), the stems of which are used as an antipyretic and tonic in China and Japan. (–)-Picrotoxinin is one of the bitter and ichthyotoxic (fish-killing) constituents of picrotoxin produced from the fruits and seed *Anamirta cocculus* (syn. *Menispermum cocculus*, Menispermaceae); picrotoxin is used as a CNS and respiratory stimulant as well as an antidote to barbiturates.

farnesane picrotoxane (–)-dendrobin (–)-picrotoxinin

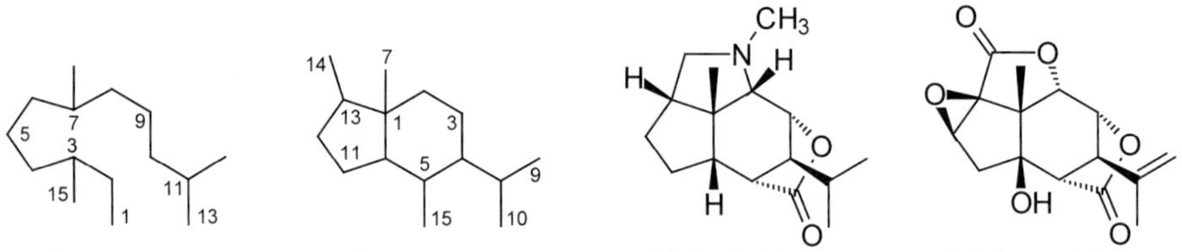

3.3.9 Isodaucanes and Daucanes

Bonds from C-1 to C-7 and from C-1 to C-10 in farnesane formally build up the sesquiterpene skeleton of isodaucanes which are converted to daucanes by migration of one methyl group (C-14) from C-7 to C-8.

3.3 Polycyclic Farnesane Sesquiterpenes

Isodaucanes such as (+)-6,10-epoxy-7(14)-isodaucene and 7(14)-isodaucen-10-one are constituents of the oil of sage from *Salvia sclarea* (Labiatae). The name daucane stems from the carrot *Daucus carota* (Umbelliferae), from which (+)-4,8-daucadiene, (+)-8-daucen-5-ol and (−)-5,8-epoxy-9-daucanol have been isolated.

3.3.10 Protoilludanes, Illudanes, Illudalanes

The tricyclic skeleton of protoilludane arises formally from farnesane by making bonds from C-1 to C-11, C-2 to C-9 and C-3 to C-6. Cleavage of the C-3–C-4 bond results in the formation of *seco*-illudane, also referred to as illudalane; protoilludane formally converts into spirocyclic illudane by migration of C-4 from C-3 to C-6.

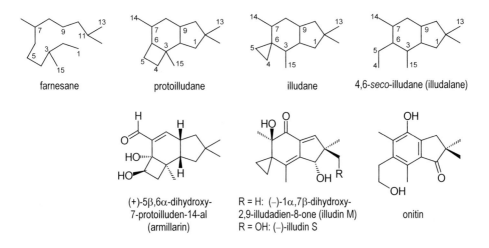

Phytopathogenic fungi *Armillariella mellea* (Basidomycetae), for example, produce the antifungal (+)-armillarin with protoilludan skeleton as an ester of *o*-orsellinic acid (2,4-dihydroxy-6-methylbenzoic acid). Two anti-tumor antibiotics (–)-illudin M and S isolated from the poisonous and luminous fungus *Clitocybe illudens* (Basidomycetae), represent the illudanes. The six-membered ring of natural illudalanes is benzenoid for the most part; variously substituted derivatives occur in the fern *Pteridium aquilinum* (Polypodiaceae). Onitin from *Onychium auratum* and *O. siliculosum* acts as a mild muscle relaxant.

3.3.11 Marasmanes, Isolactaranes, Lactaranes, Sterpuranes

Bond formation from C-1 to C-11, C-2 to C-9, C-3 to C-6 and disconnection of the C-4–C-5 bond in farnesane formally leads to marasmane. Isolactarane arises from the latter by cleavage of the C-3–C-4 and connection of the C-5–C-7 bond which, on its part, formally expands to lactarane by migration of C-3 from C-6 to C-4 involving disconnection of the C-5–C-7 bond. The names are derived from those of the fungal genera *Marasmius* and *Lactarius*.

(+)-Isovelleral is a strong antibiotic with a marasmane skeleton isolated from the fungus *Lactarius vellereus* (Basidomycetae) and closely related species; due to the sharp taste, the fungus uses isovelleral as an antifeedant against animals. Marasmic acid, an antibacterial and mutagenic constituent from *Marasmius conigenus* and other Basidomycetae, represents an acylal of a dialdehyde acid. Merulidial, a metabolite of the fungus *Merulius tremellosus* (Basidomycetae) with the isolactarane skeleton, acts as an antibacterial and antimycotic. Various lactarane derivatives referred to as blennins, such as the lactone (+)-blennin D, have been isolated from *Lactarius blennius*.

3.3 Polycyclic Farnesane Sesquiterpenes

Culture of the phytopathogenic fungus *Stereum purpureum* (Basidomycetae), a parasite of some trees, produces unsaturated and hydroxylated sesquiterpenes with the tricyclic skeleton of sterpurane, formally arising from isolactarane by opening the C-5–C-6- and closing the C-4–C-5 bond, or directly from farnesane by linking the bonds C-1–C-11, C-2–C-9 and C-4–C-7.

R1	R2	R3	
H	H	H	(+)-2-sterpurene
OH	H	H	(+)-2-sterpuren-6-ol
OH	OH	OH	(+)-2-sterpuren-6,12,15-triol

3.3.12 Acoranes

Connecting the bonds C-1–C-6 and C-6–C-10 in farnesane formally produces the spiro[4,5]decane basic skeleton of acorane. The name of this class of sesquiterpenes stems from the *Acorus* species. (−)-4-Acoren-3-one, for example, has been isolated from *Acorus calamus* (Calamus, Araceae) and from the carrot *Daucus carota* (Umbelliferae). The oil of calamus (oil of sweet flag) from the rhizome of *Acorus calamus* with its warm and spicy odor and pleasant bitter taste is predominantly used in perfumery and as a minor (possibly carcinogenic) ingredient of vermouth, some flavored wines and liqueurs. (+)-3,7(11)-Acoradiene is a constituent of juniper *Juniperus rigida*; its enantiomer occurs in *Chamaecyparis nootkatensis* (Cupressaceae).

3.3.13 Chamigranes

A spiro[5,5]undecane skeleton characterizes the chamigranes which formally originate from linking the bonds C-1–C-6 and C-6–C-11 of farnesane. More than 50 naturally occurring representatives with halogens as substituents are predominantly isolated from algae [2]. Examples include (−)-10-bromo-1,3-,7(14)-chamigratrien-9-ol (obtusadiene) and 3-bromo-2α-chloro-7-chamigren-9-one (laurencenone A) from the red alga *Laurencia obtusa*. (−)-Chamigra-3,7(14)-diene found in *Chamaecyparis taiwanensis* (Cupressaceae) is one of the rare chamigrenes of plant origin.

3.3.14 Cedranes and Isocedranes

Cedranes are formally derived from farnesane by connection of bonds between C-1 and C-6, C-2 and C-11 as well as C-6 and C-10. Cedrane is formally converted into isocedrane by migration of the C-15 methyl group from C-3 to C-5.

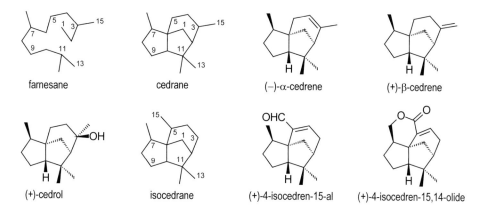

Cedrane derivatives such as (−)-3-cedrene (α-cedrene) and (+)-3(15)-cedrene (β-cedrene) are wide-spread among *Juniperus* species (Cupressaceae). (−)-α-Cedrene (content up to 25%) and (+)-cedrol (content 20-40%) are the chief constituents of the oil of cedar wood used in perfumery and as an insect repellant, obtained from *Juniperus virginiana* growing in the south-east of USA. (+)-Cedrol shapes the

3.3 Polycyclic Farnesane Sesquiterpenes

pleasant woody balsamic odor of this volatile oil which intensifies upon acetylation [18] of this sesquiterpenol.

Isocedrane derivatives occur predominantly in *Jungia* species, exemplified by (+)-4-isocedren-15-al in *Jungia malvaefolia* and (+)-4-isocedren-15,14-olide and other isocedrenes in *Jungia stuebelii*.

3.3.15 Zizaanes and Prezizaanes

Bonds between C-2 and C-11, C-6 and C-10 as well as C-6 and C-15 of farnesane formally link up the tricyclic sesquiterpene skeleton prezizaane, which converts into zizaane by methyl shift.

The oil of vetiver with its aromatic to harsh, woody odor, steam-distilled for perfumery from roots of vetiver grass *Vetiveria zizanoides* (Poaceae) which is grown chiefly in Haiti, India, and Java, contains some prezizaanes and zizaanes such as (+)-prezizaene, (−)-7-prezizaanol, (+)-6(13)-zizaene and (+)-6(13)-zizaen-12-ol, also known as khusimol.

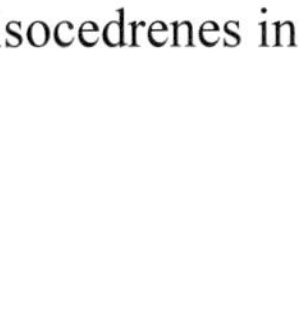

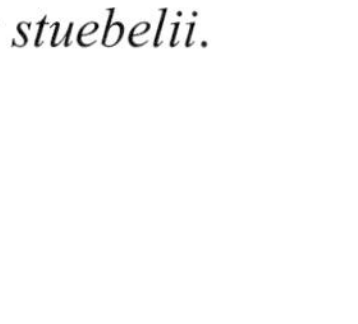

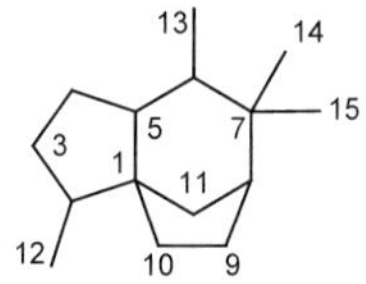

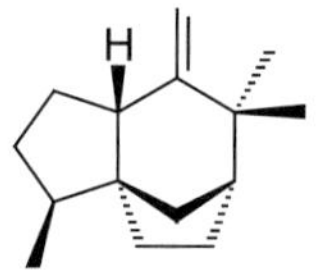

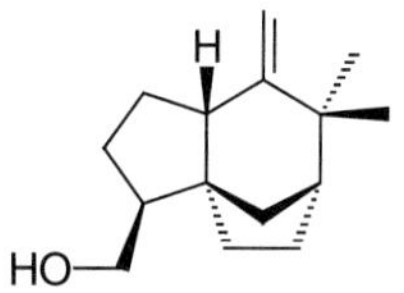

3.3.16 Campherenanes and Santalanes

Bonds from C-1 to C-6 and C-3 to C-7 of farnesane formally build up the bicyclo[2.2.1]heptane core structure of campherenane. (−)-Campherenol and (−)-

campherenone are constituents found in the camphor tree *Cinnamomum camphora* (Lauraceae) and in the ethereal oil of freshly pressed lemon juice.

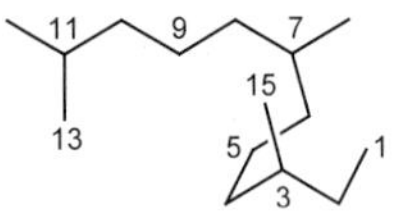

farnesane 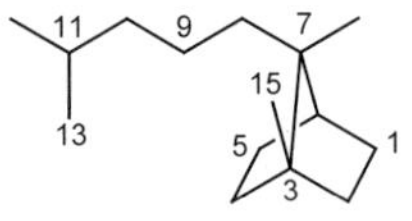   campherenane 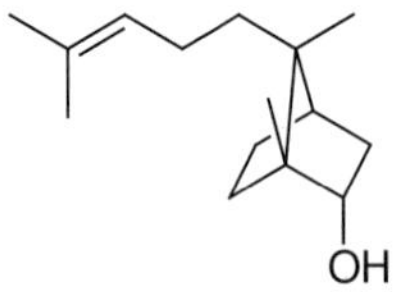   (−)-campherenol 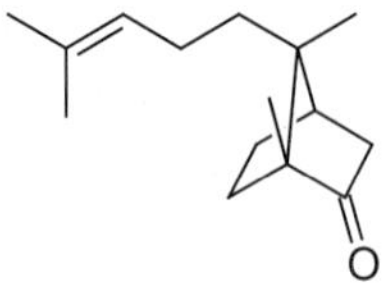   (−)-campherenone

An additional bond from C-2 to C-4 of campherenane formally leads to tricyclic α-santalane, the basic skeleton of some constituents found in sandalwood oil [18] (oil of santal) with a woody, sweet, balsamic odor, used in perfumery and as a urinary antiinfective [19], obtained from dried heartwood of the tree *Santalum album* (Santalaceae) grown in east India. Examples include (+)-(Z)-α-santalol with a slight woody odor similar to cedar wood oil and (+)-(E)-α-santalal, with a strong woody odor.

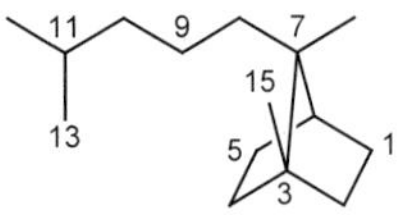

campherenane 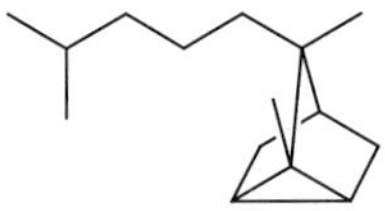   α-santalane 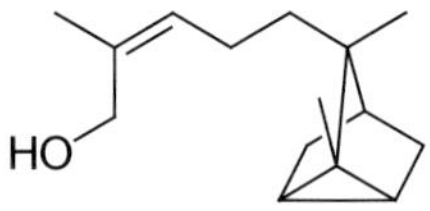   (+)-(Z)-α-santalol 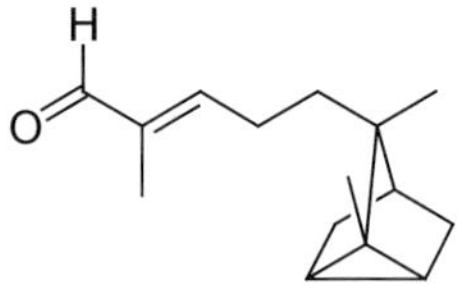   (+)-(E)-α-santalal

Cleavage of the C-3–C-4 bond of farnesane and connection of new bonds from C-2 to C-4, C-3 to C-7 and C-1 to C-4 lead to bicyclic β-santalane, which is another basic skeleton of substances found in sandalwood. Examples are (−)-β-santalene and (−)-(Z)-β-santalol with a pleasant woody, slightly urinary odor used in perfumes and detergents [18].

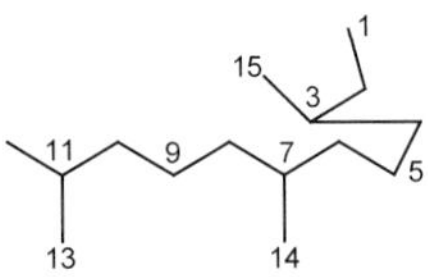

farnesane 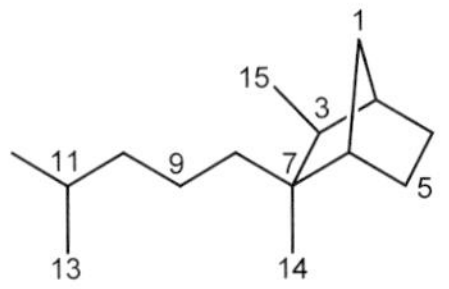   β-santalane 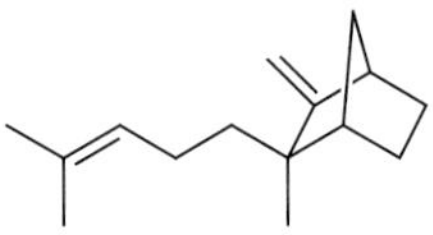   (−)-β-santalene 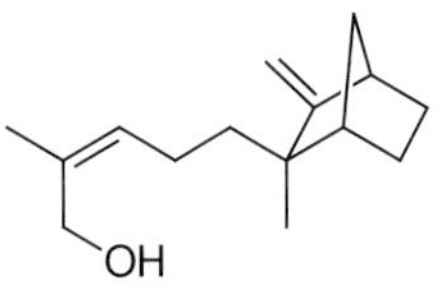   (−)-(Z)-β-santalol

3.3 Polycyclic Farnesane Sesquiterpenes

3.3.17 Thujopsanes

Farnesane formally changes to thujopsane when the C-5–C-6 bond cleaves and new bonds C-1–C-6, C-2–C-6, C-5–C-7 and C-6–C-11 connect. (−)-3-Thujopsene and (+)-15-nor-4-thujopsen-3-one (mayurone) from hiba oil obtained from the hiba live tree *Thujopsis dolabrata* (Cupressaceae) grown in Japan represent this small group of sesquiterpenes. (−)-3-Thujopsene does not shape the odor, but is, in addition to (−)-α-cedrene and (+)-cedrol, one of the chief constituents (up to 25 %) of the Texan oil of cedar wood from *Juniperus virginiana* (Cupressaceae)[18].

3.3.18 Hirsutanes

Connecting the bonds C-3–C-7, C-2–C-9 and C-1–C-11 in farnesane and subsequent shifts of the methyl groups C-14 and C-15 formally lead to the triquinane skeleton of hirsutanes occurring predominantly as fungal metabolites. 4(15)-Hirsutene is a hydrocarbon arising biogenetically from cyclization of humulene (section 3.2.3) and isolated from cultures of the fungus *Coriolus consors* (Basidomycetae). Epoxidized derivatives such as hirsutic acid produced by the fungus *Stereum hirsutum* (Basidomycetae) growing on the dead wood of broad-leaved trees, and coriolin A including two acylated derivatives (B, C) isolated from *Coriolus consors*, display antibiotic and antineoplastic activity.

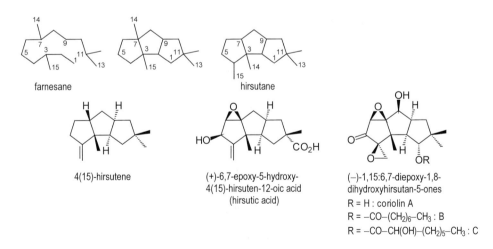

3.4 Other Polycyclic Sesquiterpenes

3.4.1 Pinguisanes

Unlike the sesquiterpenes presented so far, not all bonds between the isoprene units (boldface in the formula) within the pinguisane skeleton are attached in the usual head-to-tail manner. As a result, their structure (constitution) cannot be formally derived from cyclizations of farnesane. Moreover, one methyl group of the isopropyl residue has migrated from the side chain to the six-membered ring.

pinguisane   (−)-α-pinguisene   (+)-pinguisone

Pinguisanes occur in various species of liverwort, exemplified by (−)-α-pinguisene in *Porella platyphylla*. The name of this small group of sesquiterpenes stems from *Aneura pinguis* containing (+)-pinguisone; its bitter taste protects as an antifeedant against insects.

3.4.2 Presilphiperfolianes, Silphiperfolianes, Silphinanes, Isocomanes

Presilphiperfolianes represent an additional group of sesquiterpenes with structures that do not follow the isoprene rule and consequently are not deducible from farnesane by formal cyclizations. Silphiperfoliane, as another structural variation, derives from presiphiperfoliane by migration of C-11 from C-7 to C-8; shifting C-9 from C-1 to C-8 results in the formation of silphinane. Another methyl shift from C-6 to C-7 leads from silphinane to the isocomane skeleton with individual ring numbering.

(−)-8β-Presilphiperfolanol from *Eriophyllum staechadifolium* and *Fluorensia heterolepsi*, (−)-5-silphiperfolene as well as (−)-1-silphinene from the compass plant *Silphium perfoliatum* (Asteraceae) growing in the north west of the USA and defining the name, (+)-3-silphinenone from *Dugaldia hoopesii*, and (+)-arnicenone from *Arnica parryi* (Asteraceae), exemplify this small group of tricyclic sesquiterpenes with the attractive triquinane skeleton.

3.4 Other Polycyclic Sesquiterpenes

Another triquinane hydrocarbon of natural origin referred to as (−)-isocomene merits special attention; it is a constituent of *Isocoma wrigthii* (Asteraceae) growing in the south east of the USA, better known there as rayless golden rod, and is highly toxic for cattle and sheep. It also occurs with the synonym berkheyaradulene in south African Asteraceae, such as in the roots of *Berkheya* and parts of *Senecio isatideus*.

(−)-Modhephene is an additional hydrocarbon isolated from these plants. Its exceptional [3.3.3]propellane parent skeleton arises formally from silphinane by migration of C-3 from C-4 to C-7.

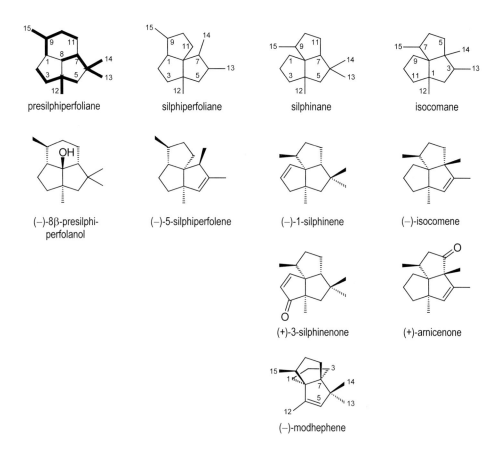

4 Diterpenes

4.1 Phytanes

Approximately 5000 naturally abundant acyclic and cyclic diterpenes derived from the parent hydrocarbon phytane are known [2]. The (3*R*,7*R*,11*R*)-enantiomer of phytane has been found in meteorites, oil slate, other sediments and, last but not least, in human liver. Oil slate additionally contains (–)-(3*R*,7*R*,11*R*)-phytanoic acid which has also been isolated from butter. 1,3(20)-Phytadiene is one among many constituents of tobacco *Nicotiana tabacum* (Solanaceae); *(E)*-1,3-phytadiene and its *(Z)*-isomer are found in zooplankton. Chlorophyll in the chloroplasts of plant cells exemplifies an ester of (+)-(2*E*,7*R*,11*R*)-2-phyten-1-ol usually referred to as phytol. 2,6,10,14-Phytatetraene-1,19-diol, better known as plaunotol, is the chief constituent of the leaves of the Thai medicinal plant *Croton sublyratus* (Euphorbiaceae) used as "plau noi" or "kelnac" as an antiulcerative.

(3*R*,7*R*,11*R*)-phytane

(–)-(3*R*,7*R*,11*R*)-phytanoic acid

1,3(20)-phytadiene

(*E*)-1,3-phytadiene

(+)-(2*E*,7*R*,11*R*)-2-phyten-1-ol (phytol)

2,6,10,14-phytatetraen-1,19-diol (plaunotol)

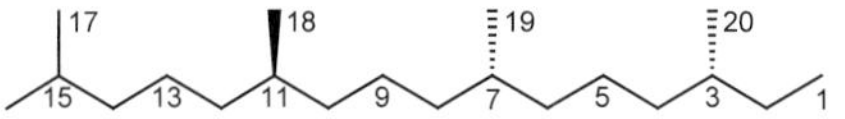

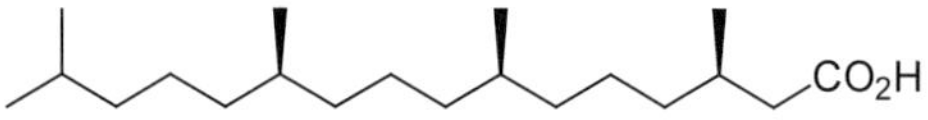

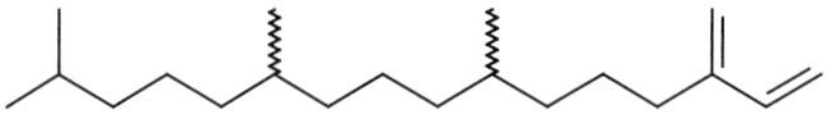

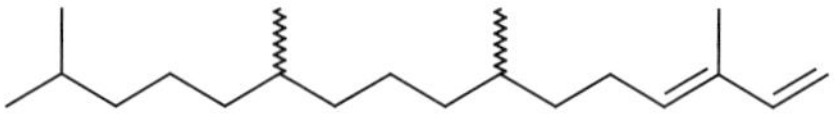

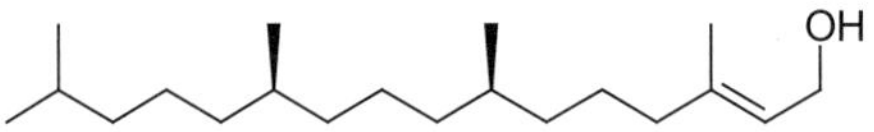

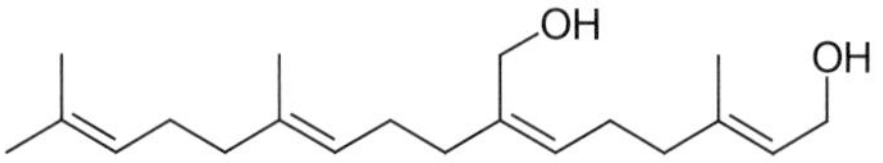

4.2 Cyclophytanes

1,6-Cyclophytanes such as 9-geranyl-α-terpineol and the aldehyde helicallenal from the straw flower *Helichrysum heterolasium* (Asteraceae) are rarities in plants.

phytane

1,6-cyclophytane
(prenylbisabolane)

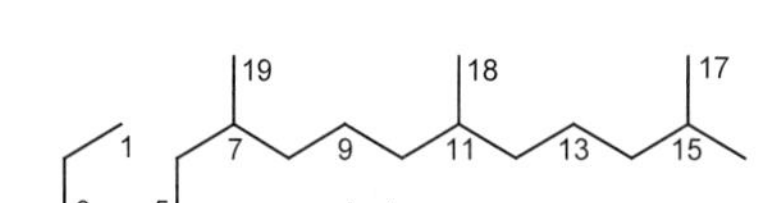

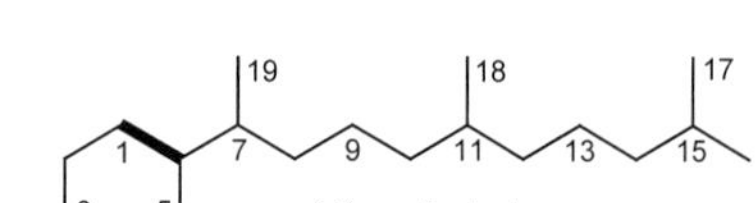

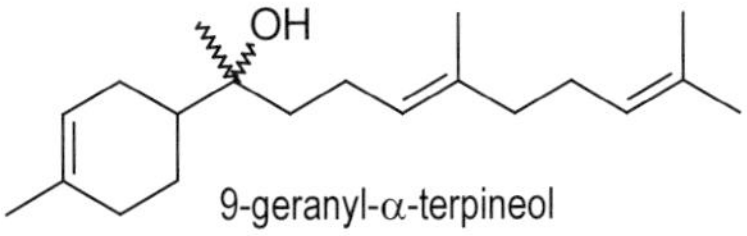

4.2 Cyclophytanes

9-geranyl-α-terpineol

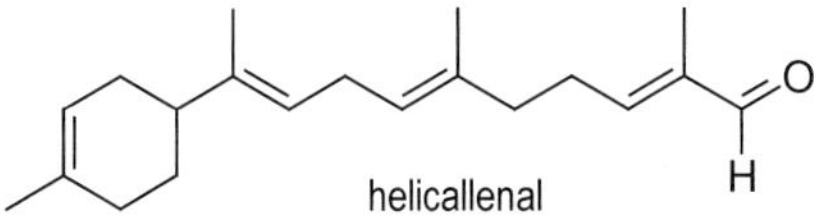

helicallenal

More frequently occurring 10,15-cyclophytanes include very important representatives of the vitamin A series such as axerophthene, retinol, retinal and tretinoin. From those, the *all-trans*-isomers as drawn are the most stable among all 16 possible *cis-trans* configurations. 11-*cis*-Retinal (vitamin A aldehyde) attaches as an imine to an L-lysine moiety of the apoprotein opsin within the photoreceptor protein rhodopsin (visual purple) found in the rods of the retina. The photoisomerization of 11-*cis*-retinal in rhodopsin induces a conformational change of the protein, resulting in a nerve pulse during the visual process in the eyes.

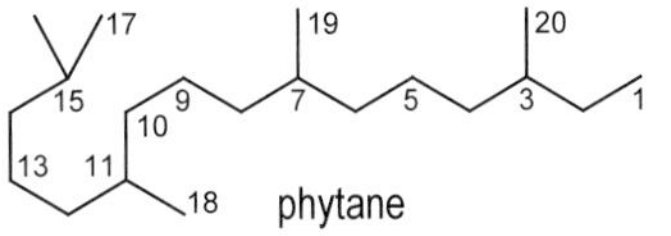

phytane

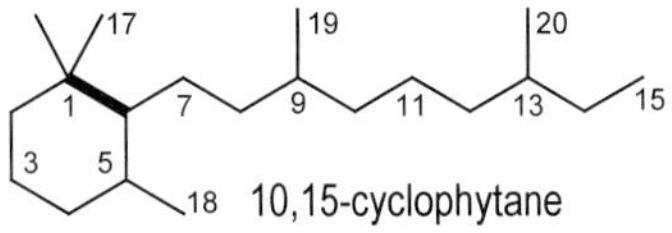

10,15-cyclophytane

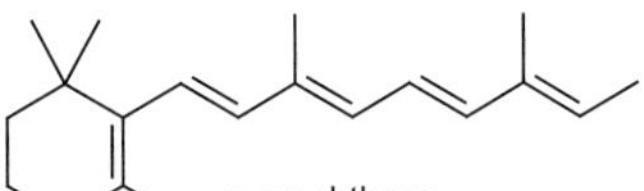

axerophthene

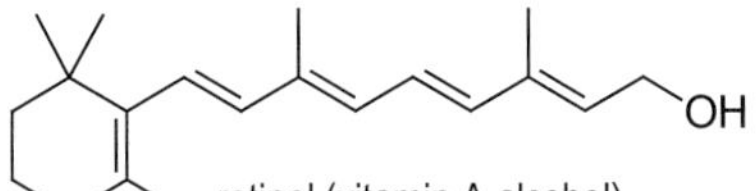

retinol (vitamin A alcohol)

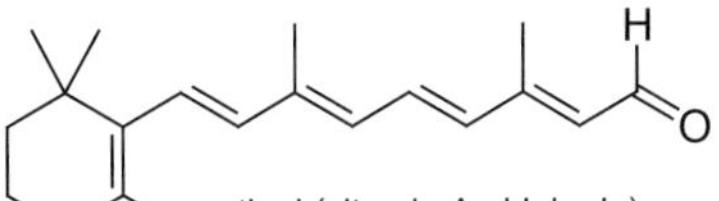

retinal (vitamin A aldehyde)

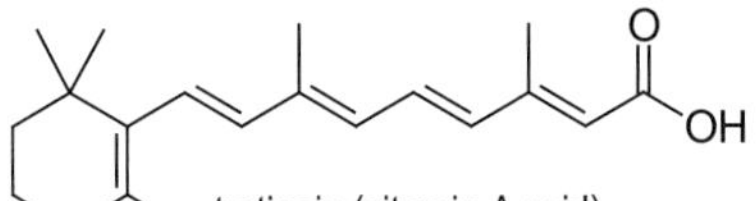

tretinoin (vitamin A acid)

Agelasines and agelasidines have been isolated from the Okinawa sponge *Agelas nakamurai*. They represent partially hydrogenated and rearranged axerophthene derivatives substituted by adenyl- and β-guanidylethylsulfonyl moieties at the end of the chains with antibacterial and anticonvulsant activities.

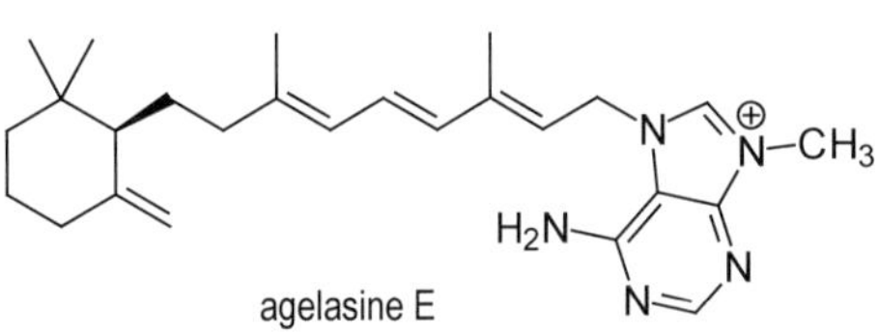

agelasine E

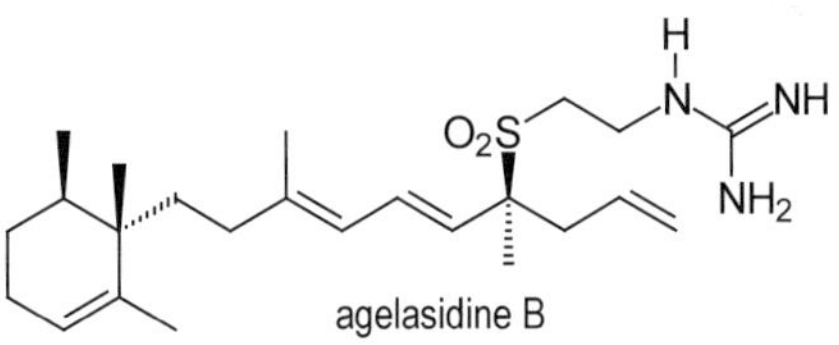

agelasidine B

4.3 Bicyclophytanes

4.3.1 Labdanes

More than 500 labdanes predominantly isolated from higher plants are known to date [2]. They represent 8,11-10,15-cyclophytanes that contain the decalin bicycle as a core structure, which also defines the usually accepted ring numbering.

The name labdane stems from *Cistus labdaniferus*:(Cistaceae) growing in Mediterranean countries (southern parts of France and Italy, Spain). This shrub and other *Cistus* species excrete the dark brown labdanum resin; this has a pleasant smell like ambergris and contains not only α-pinene but also labdan-8α,15-diol and 8β-hydroxylabdan-15-oic acid.

Constituents of Pinaceae and Cupressaceae include 8,5,18-labdanetriol, (–)-labdanolic acid and (+)-6-oxocativic acid. Numerous derivatives have been found in conifers such as pine (*Pinus*), fir (*Abies*), larch (*Larix*) and juniper (*Juniperus*). Selected examples are (+)-12,15-epoxy-8(17),12,14-labdatriene (pumiloxide) from *Pinus pumila*, (+)-12,14-labdadien-8-ol (abienol) from *Pinus strobus* and various *Abies* species, (+)-11,13-labdadien-8-ol (neoabienol) from *Abies sibirica*, (–)-13(16),14-labdadien-8-ol (isoabienol), (+)-8(17)-labdene-15,18-dioic acid from the needles of *Pinus silvestris* (Pinaceae) and 14,15-dihydroxy-8(17),13(16)-labdadien-19-oic acid from *Juniperus communis* (Cupressaceae).

Grindelic acid represents a spirotricyclic oxygen-bridged labdenoic acid and occurs in the expectorant resin of *Grindelia robusta* (Asteraceae). The tobacco plant *Nicotiana tabacum* (Solanaceae) contains (+)-abienol, (–)-13-labdene-8,15-diol and 11,14-labdadiene-8,13-diol. (–)-Sclareol obtained from sage species such as *Salvia sclarea* and *S. schimperi* (Labiatae) serves as a starting material for the partial synthesis of some ambergris fragrances [18]. (–)-Forskolin, a tricyclic labdane derivative with a tetrahydropyranone partial structure, isolated from the Indian medicinal plant *Coleus forskolii* (Labiatae), has been the target of several total syntheses because of its positive inotropic and vasodilatory activities [19].

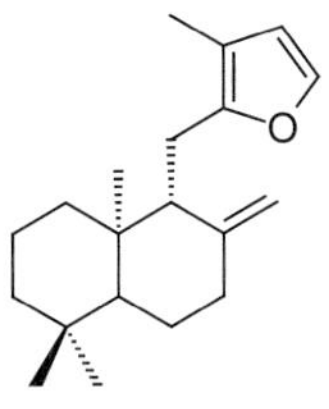

4.3 Bicyclophytanes

(+)-12,15-epoxy-8(17),12,14-labdatriene (pumiloxide)

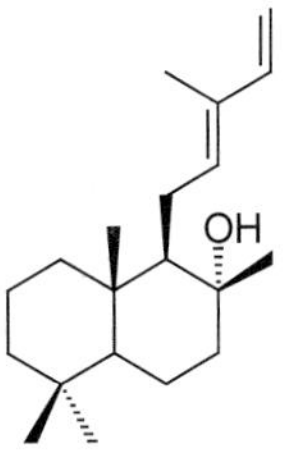

(+)-12,14-labdadien-8-ol (abienol)

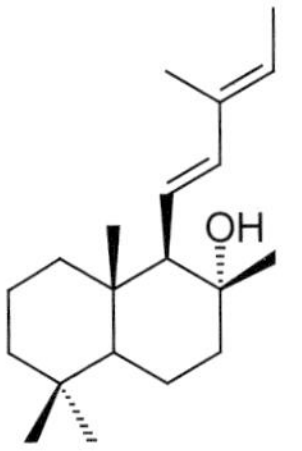

(+)-11,13-labdadien-8-ol (neoabienol)

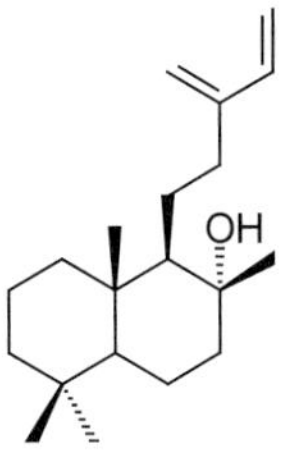

(−)-13(16),14-labdadien-8-ol (isoabienol)

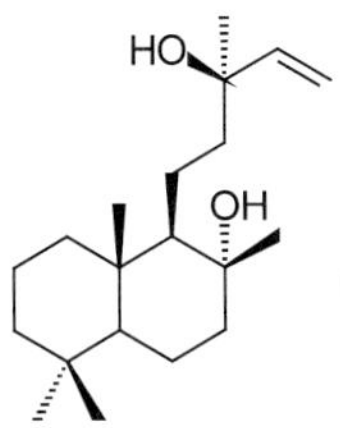

(−)-14-labdene-8,13-diol (sclareol)

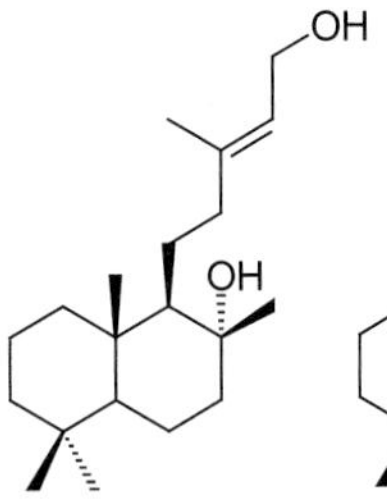

(−)-13-labdene-8,15-diol

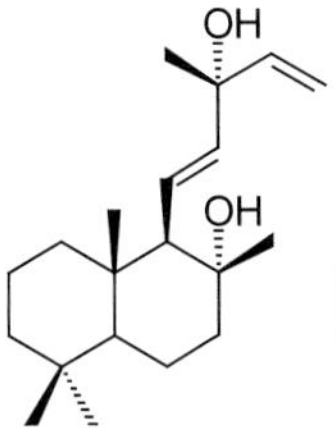

11,14-labdadiene-8,13-diol

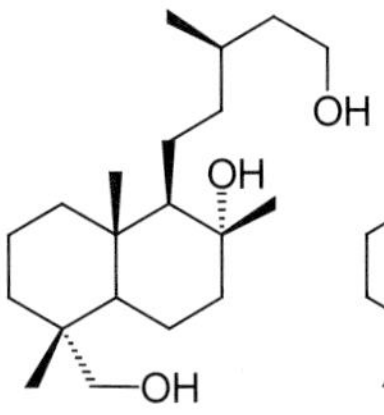

5,8,18-labdanetriol

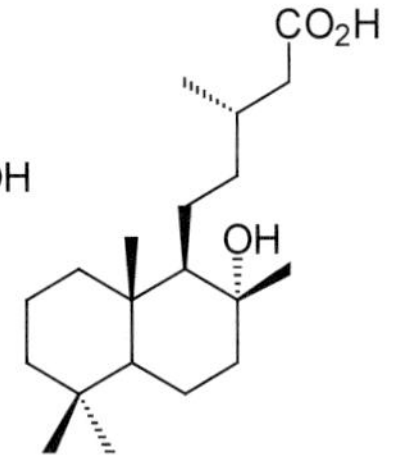

(−)-labdanolic acid

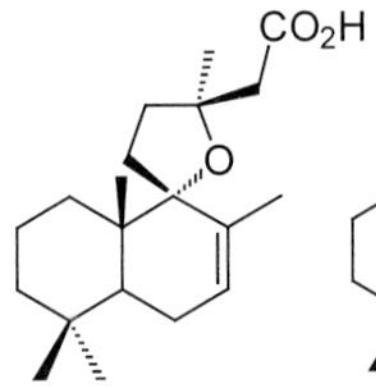

(−)-9,13-epoxylabd-7-en-15-oic acid (grindelic acid)

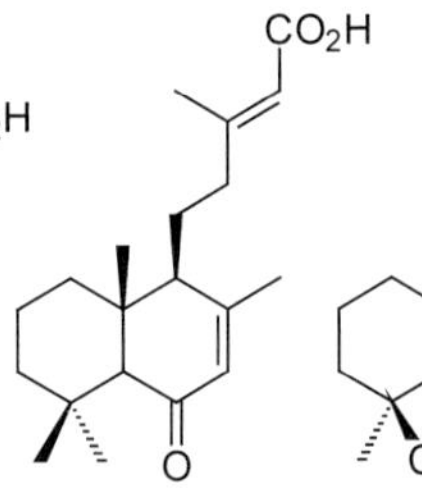

(+)-oxocativic acid

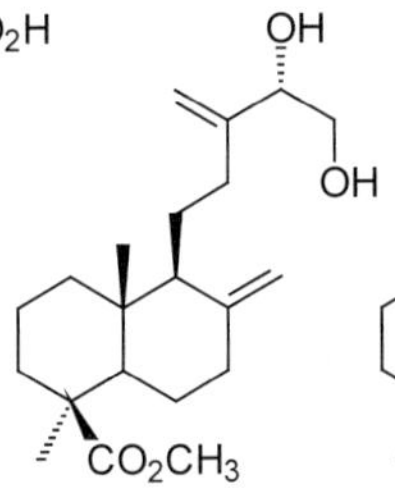

(+)-14,15-dihydroxy-8(17),13(16)-labdadien-19-oic acid methylester

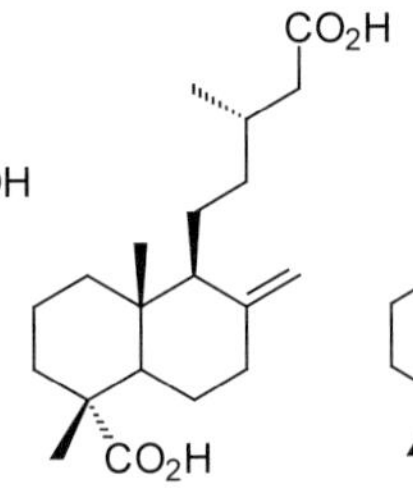

(+)-8(17)-labdene-15,18-dioic acid (pinifolic acid)

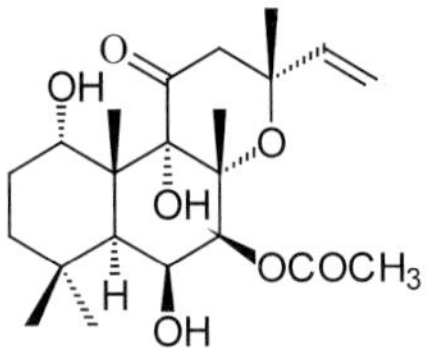

(−)-forskolin

4.3.2 Rearranged Labdanes

Labdane isomerizes to halimane by migration of the methyl group (C-20) from C-10 to C-9. An additional shift of the methyl group (C-19) from C-4 to C-5 leads from halimane to clerodane.

The name halimane stems from *Halimium viscosum* and *Halimium umbellatum* (Cistaceae). (+)-14,15-Dihydroxy-1(10),13(16)-halimadien-18-oic acid, its methyl ester as well as (+)-15-oxo-1(10),13-halimadien-18-oic acid have been isolated from these shrubs. An antibacterial recovered from the Okinawa sponge *Agelas nakamurai* named agelasin C represents another halimane derivative.

Clerodanes are found among the constituents of various *Solidago* species (Asteraceae) exemplified by (–)-junceic acid from the golden rod *Solidago juncea*. Other representatives have been isolated from the leaves of some Labiatae; (–)-teugin from *Teucrium fragile*, (–)-2,7-dihydroxy-3,13-clerodadien-16,15:18,19-olide from the sage species *Salvia melissodora* as well as (–)-ajugareptansone A from *Ajuga reptans* (Labiatae) are selected examples. All clerodane derivatives are antifeedants against insects; some of them exhibit antibacterial activity.

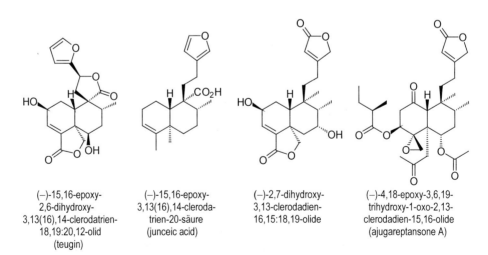

4.4 Tricyclophytanes

4.4.1 Pimaranes and Isopimaranes

Pimaranes and isopimaranes [2] are 13,17-cyclolabdanes with the perhydrophenanthrene basic skeleton, differing only in their configuration at C-13. Rearranged and cyclized pimaranes include the rosanes (shift of the methyl group C-20 from C-10 to C-9), the parguaranes (3,18-cyclopimaranes), the erythroxylanes (shift of the methyl group C-19 of rosane from C-4 to C-5), and the devadaranes (4,19-cycloerythroxylanes). Podocarpanes are formally derived from pimaranes by omitting the carbon atoms 15-17 (15,16,17-trinorpimaranes).

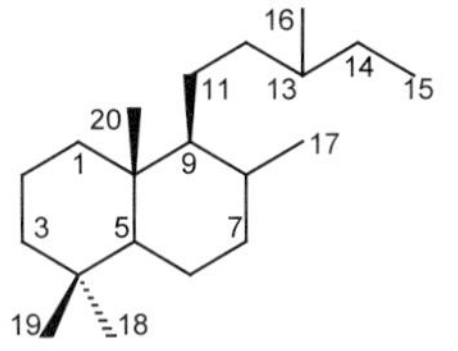

labdane 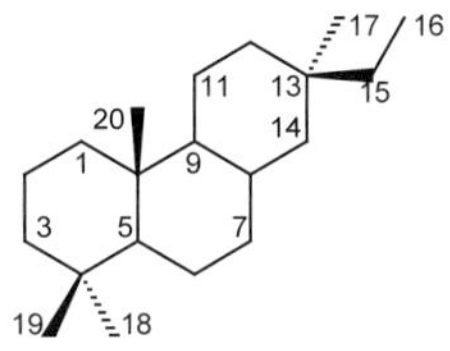   pimarane 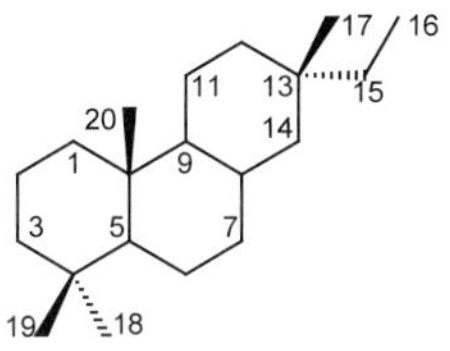   isopimarane 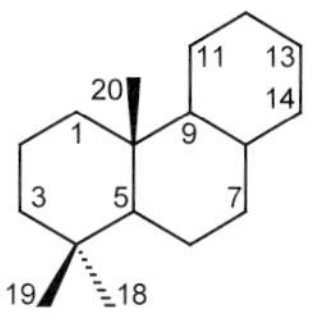   podocarpane

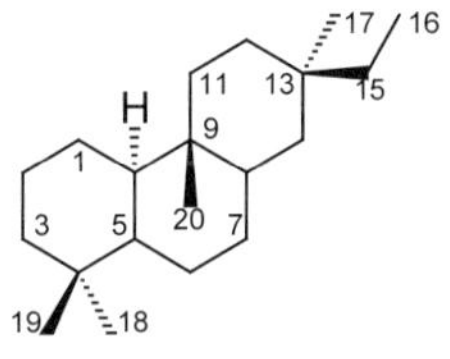

rosane 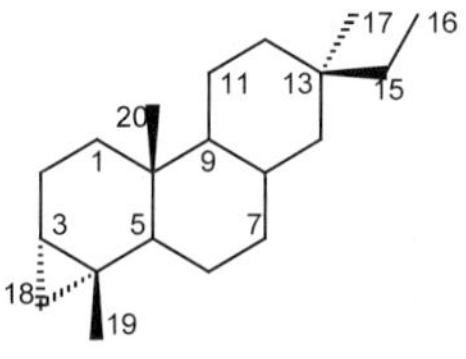   parguarane 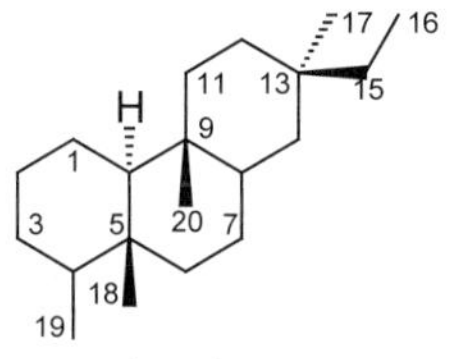   erythroxylane 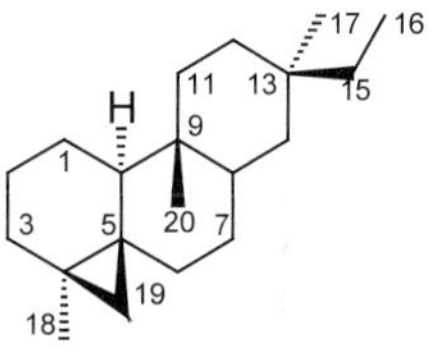   devadarane

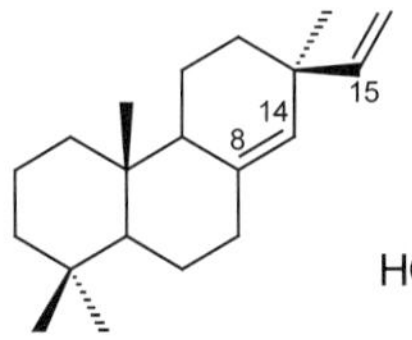

(+)-8(14),15-pimaradiene 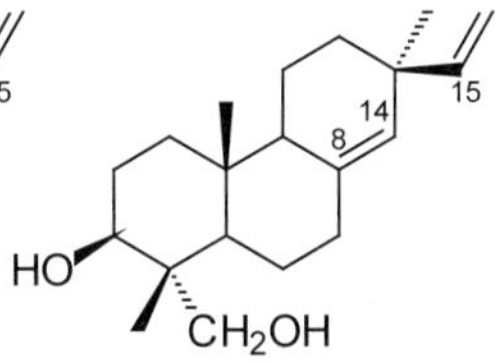   (+)-8(14),15-pimaradiene-3,18-diol 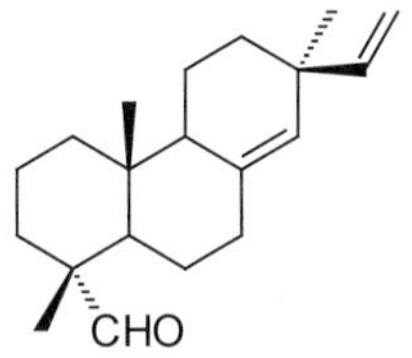   (+)-8(14),15-pimaradien-18-al 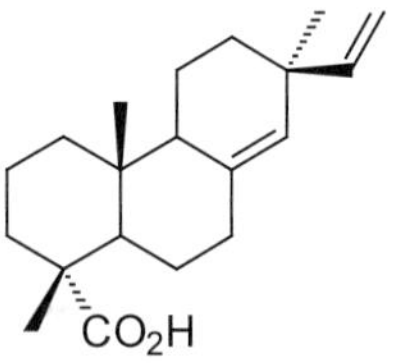   (+)-8(14),15-pimaradien-18-oic acid

Pine trees such as *Pinus silvestris* (Pinaceae), which are wide-spread in Europe, contain pimarane derivatives, e.g. (+)-8(14),15-pimaradiene-3,18-diol, 8(14),15-pimaradien-18-al also referred to as cryptopinone, and 8(14),15-pimaradien-18-oic acid denoted as pimaric acid, isolated from American rosin and belonging to the resin acids of turpentine. The parent (+)-8(14),15-pimaradiene is found among the constituents of *Erythroxylon monogynum* and *Aralia racemosa* (Araliaceae).

Isopimaranes occur in some pine (Pinaceae) and juniper species (Cupressaceae). Examples include 7,15-isopimaradiene and 8,15-isopimaradien-18-oic acid (Δ^8-isopimaric acid) from *Pinus silvestris* as well as 8(14),15-isopimaradiene-3,18-diol and 8,15-isopimaradiene-3,7,19-triol from *Juniperus thurifera*.

(+)-7,15-iso-pimaradiene

(+)-8,15-isopimara-dien-18-oic acid (Δ^8-isopimaric acid)

(+)-8(14),15-iso-pimaradiene-3,18-diol

(+)-8,15-isopimara-diene-3,7,19-triol

Representatives of podocarpanes include podocarpinol in *Podocarpus totara*, and podocarpic acid, the dominant acidic constituent of podocarpus resin obtained from Javanese *P. cupressina* (Cupressaceae), as well as the bitter-tasting phenolic compounds nimbiol and nimbione. These are found in the bark and isolated from neem oil expressed from the seed-kernels of the Indian neem tree *Azadirachta indica* (Meliaceae). Extracts of this bark are added to some mouthwashes and skin creams; neem oil is used as an agricultural insect repellant and antifeedant.

8,11,13-podocarpa-triene-12,19-diol (podocarpinol)

(+)-podocarpic acid

(+)-12-hydroxy-13-methyl-8,11,13-podocarpatrien-7-one (nimbiol)

nimbione

Rosanes are toxic metabolites arising biogenetically from the cyclization of labdane precursors produced by the fungus *Trichothecium roseum*. This fungus may intoxicate food; the toxic constituents are (−)-rosein III and its cytotoxic 11-deoxy derivative (rosenonolactone). Some parguaranes are found in algae such as (−)-15-bromo-

4.4 Tricyclophytanes

9(11)-parguerene-2α,7α,16-triol in the alga *Laurencia obtusa*. The wood of *Erythroxylon monogynum* (Araliaceae) contains derivatives of erythroxylane and devadarane such as (+)-4(18)-erythroxylene-15,16-diol and (−)-devadarane-15,16-diol.

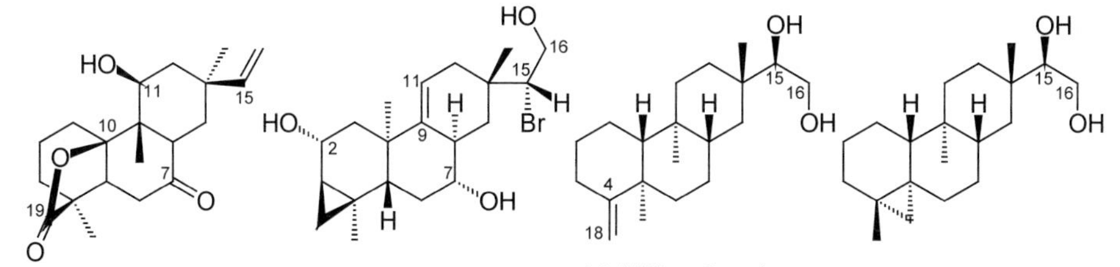

(−)-11β-hydroxy-7-oxo-15-rosen-19,10-olide (rosein III)

(−)-15-bromo-9(11)-parguerene-2α,7α-16-triol

(+)-4(18)-erythroxylene-15,16-diol

(−)-devadarane-15,16-triol

4.4.2 Cassanes, Cleistanthanes, Isocopalanes

Cassanes probably originate from pimaranes by migration of a methyl group (C-17) from C-13 to C-14. The corresponding migration of the ethyl group, on the other hand, results in the formation of cleistanthanes, which on their part may undergo additional methyl shifts to isocopalanes.

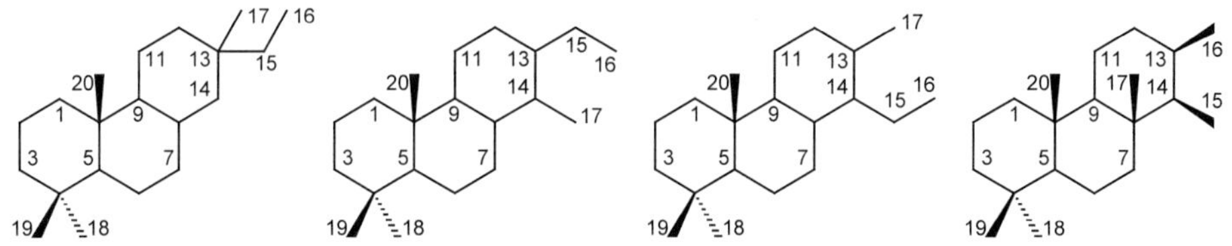

pimarane / isopimarane   cassane   ceistanthane   isocopalane

Cassane is the basic skeleton of cassaic acid, the parent diterpenoid of alkaloids from the bark of *Erythrophleum guinese* and other *Erythrophleum* species (Fabaceae). These act as local anaesthetics, cardiotonics, antihypertonics, and also induce cardiac arrest. *Erythrophleum* alkaloids such as cassaidine, cassain and cassamine are 2-(*N,N*-dimethylamino)ethyl esters of cassaic acid derivatives.

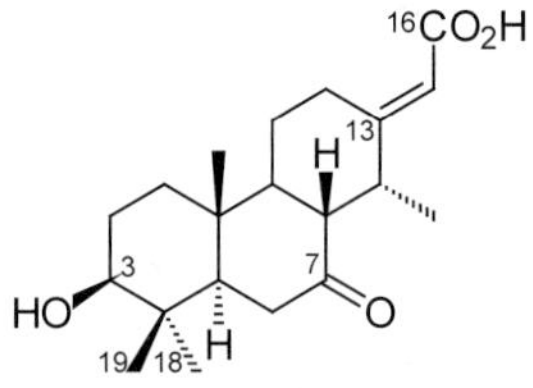

(−)-cassaic acid

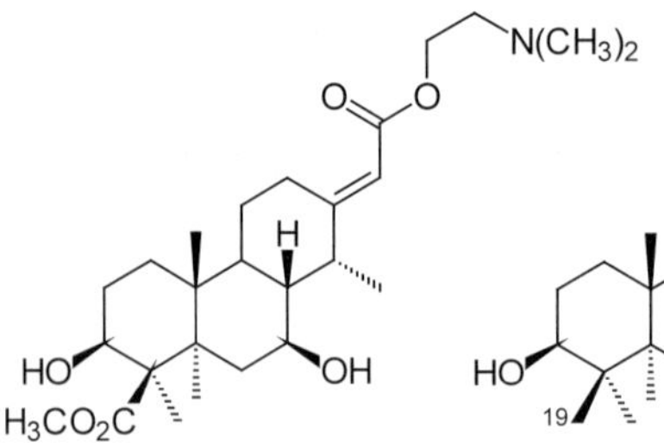

(−)-cassaidine 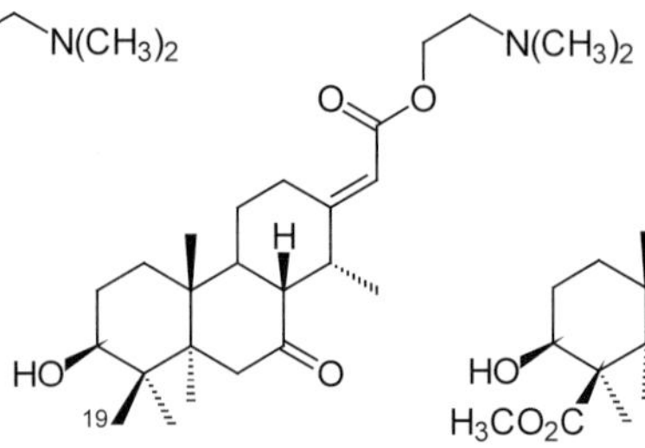   (−)-cassaine 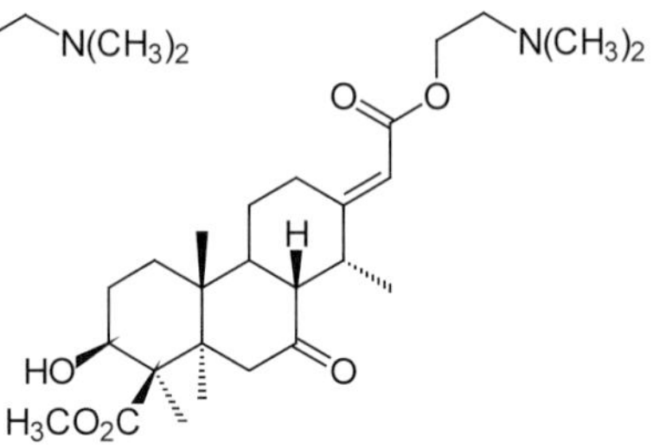   (−)-cassamine

Representatives of naturally occurring cleistanthanes include (+)-13(17),15-cleistanthadiene from *Amphibolis antarctica*, (+)-auricularic acid from *Pogostemon auricularis* (Labiatae), the antineoplastic (−)-spruceanol from *Cunurea spruceana* and (−)-cleistanthol from *Cleistanthus schlechteri* (Euphorbiaceae).

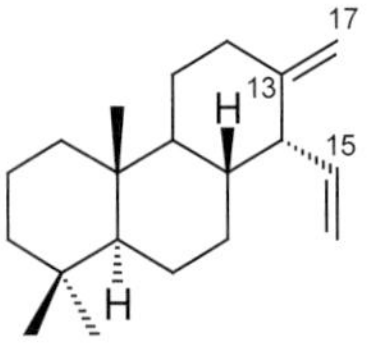

(+)-13(17),15-cleistanthadiene

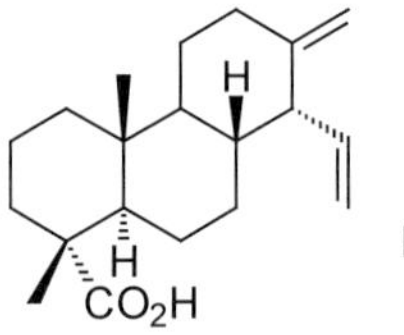

(+)-13(17),15-cleistanthadien-18-oic acid (auricularic acid)

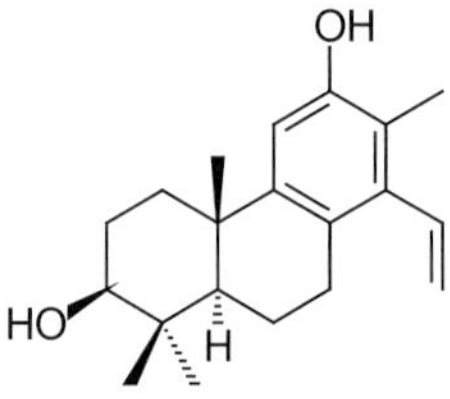

(−)-8,11,13,15-cleistanthatetraene-3,12-diol (spruceanol)

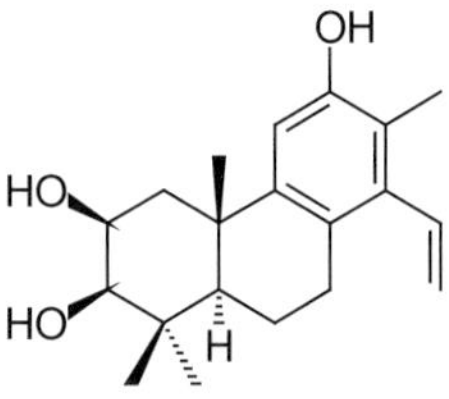

(−)-8,11,13,15-cleistanthatetraene-2,3,12-triol (cleistanthol)

Isocopalane and its tetracyclic furan derivatives, referred to as spongianes, are found in various sponges such as *Spongia officinalis*, and in the naked snails eating these sponges. Such compounds include (+)-isocopalene-15,16-dial, the cyclohemiacetal (−)-spongiane-15,16-diol, the lactone (+)-11β-hydroxy-12-spongiene-16-

4.4 Tricyclophytanes

one as well as (−)-13(16)14-spongiadiene and its 2α,19-dihydroxy-3-oxo-derivative which is reported to act as an antileukemic and antiviral.

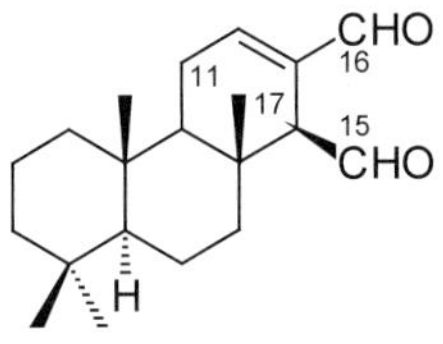

(+)-12-isocopalene-15,16-dial

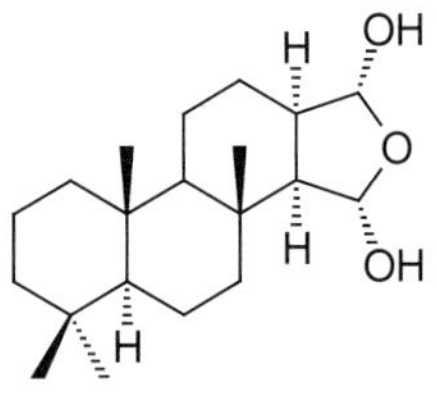

(−)-spongiane-15,16-diol

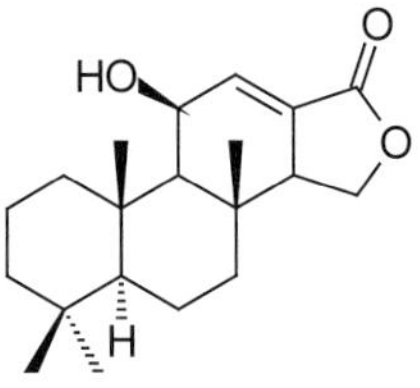

(+)-11β-hydroxy-12-spongien-16-one

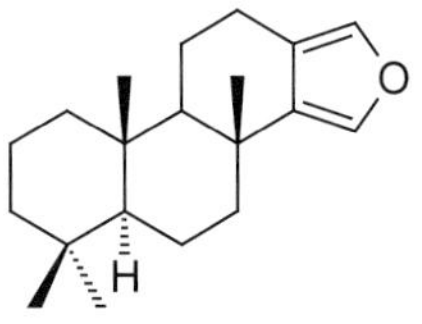

(−)-13(16),14-spongiadiene

4.4.3 Abietanes and Totaranes

Abietanes may formally be derived from pimaranes by migration of the methyl group C-17 from C-13 to C-15. In plants, however, they emerge from cyclization of geranylgeranyl diphosphate. Related parent diterpene hydrocarbons include 13,16-cycloabietanes, 17(15-16)-*abeo*-abietanes in which the methyl group C-17 has shifted from C-15 to C-16, and totaranes. The latter formally arise from abietane when the isopropyl group migrates from C-13 to C-14.

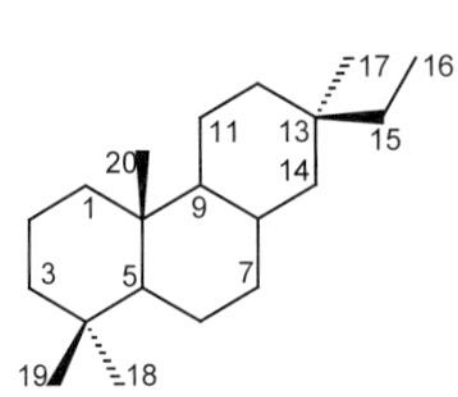

pimarane

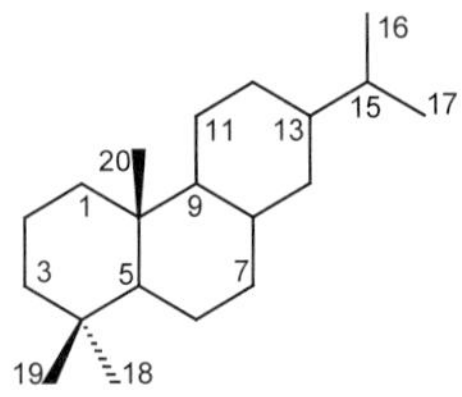

abietane

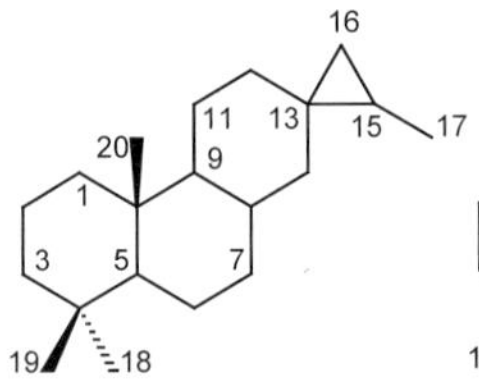

13,16-cycloabietane

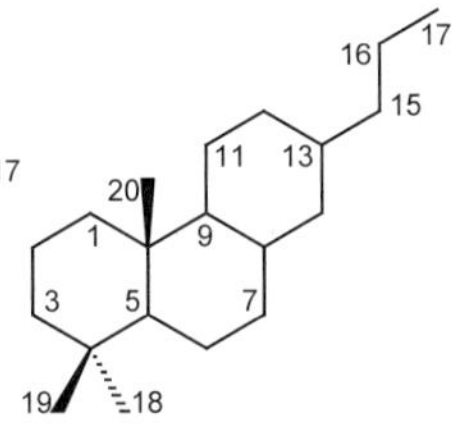

17(15-16)-*abeo*-abietane

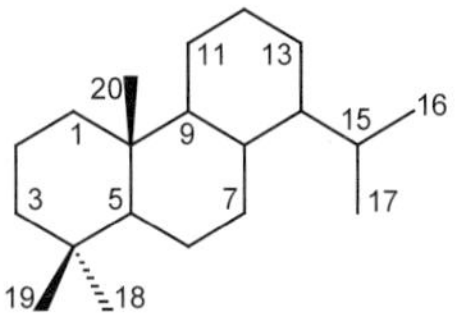

totarane

More than 200 diterpenes with an abietane skeleton are reported to exist naturally [2]. Numerous representatives occur in conifers. Selected examples include palustradiene, also referred to as (−)-8,13-abietadiene, from the pine tree *Pinus palustris*, from the so-called berries of the sade tree *Juniperus sabina* (Cupressaceae) and other species of juniper trees, (−)-abietenol from the pine *Pinus silvestris* and the fir *Abies sibirica*, (−)-abietic acid belonging to the resin acids of turpentine [18] and wide-spread in conifers such as various pines (*Pinus*), larch trees (*Larix*) and firs (*Abies*), as well as (+)-palustric acid from the balm and the roots of *Pinus palustris*, isolated from gum rosin.

(−)-8,13-abietadiene (palustradiene)

(+)-8,13-abietadien-18-oic acid (palustric acid)

(−)-7,13-abietadien-18-ol (abietenol)

(−)-7,13-abietadiene-18-oic acid (abietic acid)

Other abietane derivatives with benzenoid ring C are among the active substances in some well-known medicinal herbs. The parent hydrocarbon (−)-8,11,13-abietatriene occurs in the pine tree *Pinus pallasiana* (Pinaceae). (+)-Carnosolic acid and the 20,7β-lactone (-olide) of its hydroxy-derivative referred to as carnosol belong to the bitter substances of the oil of sage from *Salvia carnosa* (Labiatae) and related species. (−)-Rosmanol, a 20,6β-lactone, is an antioxidant isolated from the leaves of rosemary *Rosmarinus officinalis* (Labiatae), which also contain carnosol.

(+)-8,11,13-abietatriene

(+)-11,12-dihydroxy-8,11,13-abtietatrien-20-oic acid (carnosolic acid)

(−)-11,12-dihydroxy-8,11,13-abitatrien-20,7β-olide (carnosol)

(+)-7α,11,12-trihydroxy-8,11,13-abietatrien-20,6β-olide (rosmanol)

4.5 Tetracyclophytanes

More than 20 quinoid 13,16-cycloabietanes, named as coleones A–Z, occur in the yellow glands in the leaves of African *Coleus* species (Labiatae). Lanugone A isolated from *Plectranthus lanuginosis* (Labiatae) represents a quinoid 17(15)-*abeo*-abietane. Totaranes such as (+)-totarol and the derived biphenyl-type dimer (+)-podototarine are found in the wood of *Podocarpus totara* (Cupressaceae).

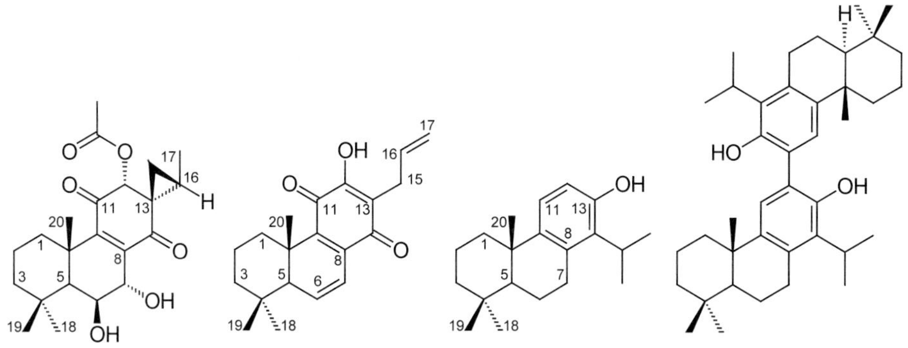

13,16-cycloabietane: coleone P

17(15-16)-*abeo*-abietane: lanugone A

(+)-8,11,13-totara-trien-13-ol (totarol)

(+)-podototarine

4.5 Tetracyclophytanes

4.5.1 Survey

Tetracyclophytanes with individual ring numbering are, for the most part, derived from the skeleton of pimarane. Beyerane as an example is an 8,16-cyclopimarane, from which kaurane formally arises by migration of the methyl group C-17 from position C-13 to C-16. Villanovane and atisane formally originate from pimarane when the ethyl group (C-15–C-16) shifts from C-13 to C-12 and rings close subsequently by forming bonds between C-9 and C-16 for villanovane and between C-8 and C-16 for atisane, respectively. Gibberellane with the tetracyclic gibbane parent skeleton arises from kaurane by cleavage of the C-7–C-8-bond and closing the new one, C-6–C-8. Similarly, leucothol is formed by cleaving the C-1–C-10- and connecting the C-1–C-6-bond. Disconnection of the C-1–C-6-bond of leucothol and attachment of C-5 to C-6 formally produces the basic skeleton of grayanotoxane.

pimarane beyerane kaurane villanovane

grayanotoxane leucothol giberellane atisane

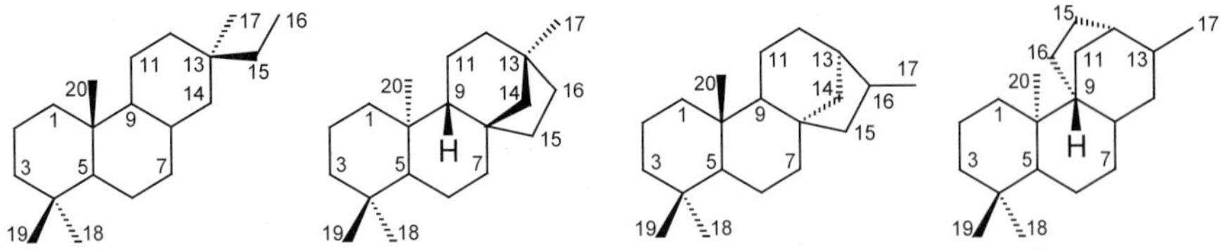

4.5.2 Beyeranes

The name beyerane stems from the Australian plant *Beyeria leschenaultii*, which contains (+)-17-*O*-cinnamoyl-15-beyerene-3,17,19-triol, also referred to as the cinnamic acid ester of beyerol. (+)-15-Beyerene as the parent hydrocarbon is one of the constituents of *Erythroxylon monogynum* (Araliaceae); its (–)-enantiomer is found in the ethereal oils of some conifers such as *Thujopsis dolabrata* and *Cupressus macrocarpa* (Cupressaceae). (+)-15-Beyeren-3-one occurs in the Tambooti wood from *Spirostachys africana*, and (+)-7-hydroxy-15-beyeren-19-oic acid was isolated from *Stevia aristata* (Asteraceae).

(+)-15-beyerene (+)-15-beyeren-3-one (+)-15-beyerene-3α,17,19-triol (beyerol) (+)-7β-hydroxy-15-beyeren-19-oic acid

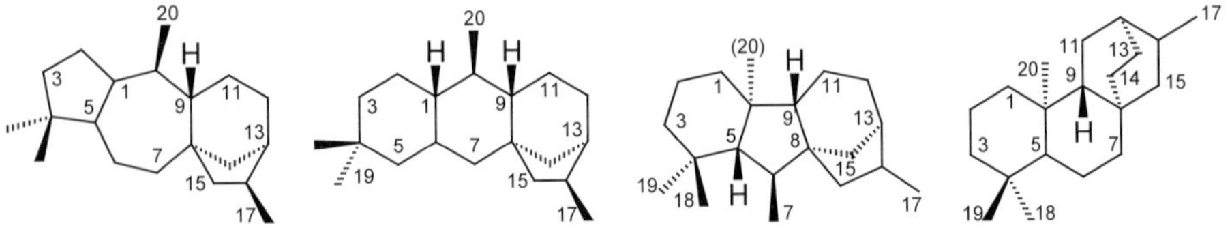

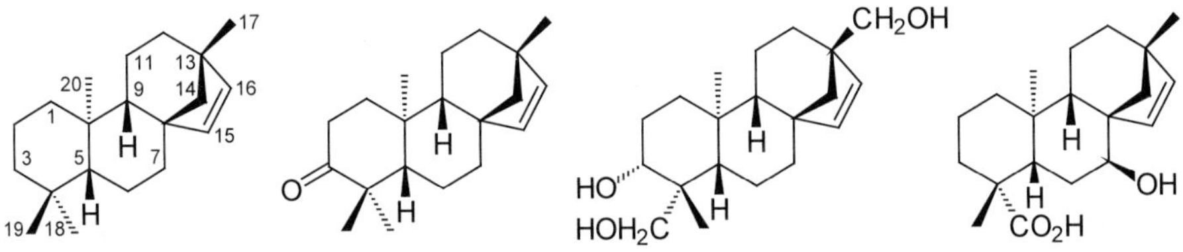

4.5.3 Kauranes and Villanovanes

More than 200 kauranes are reported to exist in plants [2]. The parent hydrocarbon (–)-kaurane occurs in *Aristolochia triangularis*, and its 16α-stereoisomer in various sediments. (+)-16,17-Dihydroxy-9(11)-kauren-19-oic acid is one of the constituents

4.5 Tetracyclophytanes

of roasted coffee. (−)-1,7,14-Trihydroxy-16-kauren-15-one, which has antibacterial and antineoplastic activities *in vitro*, was isolated with other kaurane derivatives from various *Rabdosia* species. The fungus *Gibberella fujikuroi* produces not only the gibberellines but also (−)-16-kaurene, (−)-17,18-dihydroxy-16-kauren-19,6β-olide and other kauranolides. Some rearranged furanokauranes such as (−)-cafestol and (−)-kahweol are the antiinflammatory constituents of coffee also present in the green coffee oil obtained from *Coffea arabica* (Rubiaceae).

Villanovanes are rare diterpenes. Some representatives were first isolated from *Villanova titicaensis*, exemplified by 3,13,17- and 13,17,19-villanovantriol, both as esters of isobutyric acid. (+)-Villanovan-13α,19-diol is found in *Stemodia maritima*.

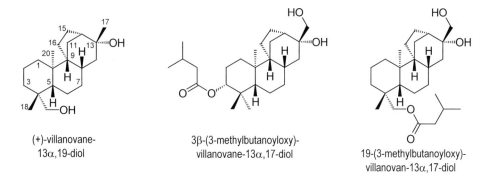

4.5.4 Atisanes

Atisane is the basic skeleton of various diterpene alkaloids (aconitum-alkaloids) found in the plant families of Ranunculaceae and Garryaceae. (−)-Atisine, as a typical representative, was isolated from the Atis plant *Aconitum heterophyllum* (Ranunculaceae), which also contains (−)-15,20-dihydroxy-16-atisen-19-oic acid as the lactone (19,20-olide). *Erythroxylon monogynum* and related species are reported to contain (−)-16-atisene. Euphorbiaceae such as *Euphorbia acaulis* and *E. fidjiana* produce (−)-16α,17-dihydroxyatisan-3-one. (+)-13-Atisene-16β,17-diol is known as serradiol due to its natural occurrence in *Sideritis serrata* (Labiatae).

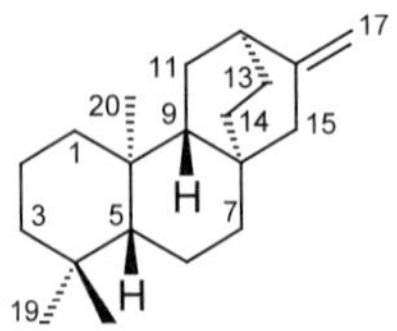

(−)-16-atisene

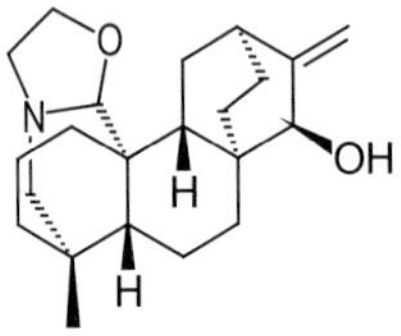

(−)-atisine

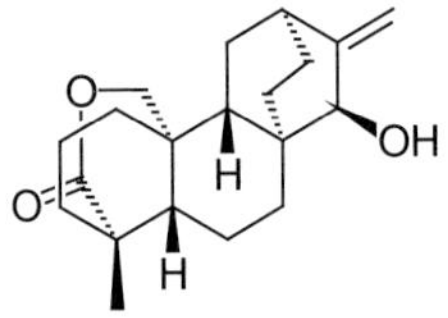

(−)-15,20-dihydroxy-16-atisene-19,20-olide

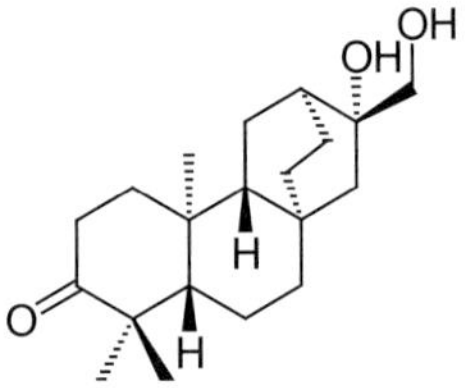

(−)-16α,17-dihydroxy-atisan-3-one

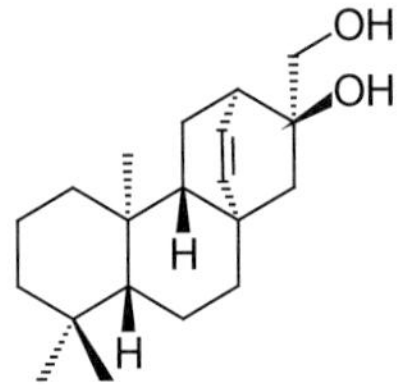

(+)-13-atisene-16β,17-diol (serradiol)

4.5.5 Gibberellanes

More than 60 gibberellanes isolated from higher plants and fungi to date [2] are, for the most part, C-20-norditerpenes. They play an essential role as plant growth hormones, and also regulate the degradation of chlorophyll as well as the formation of fruits, and thus are used in agriculture. Large amounts of (+)-gibberellin A_3, known as gibberellic acid, are isolated from the culture filtrate of the Japanese fungus *Gibberella fujikuroii*.

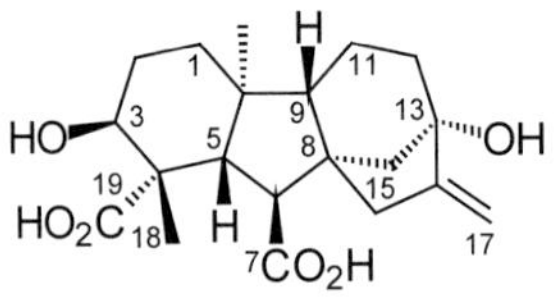

(−)-gibberelline A_{18}

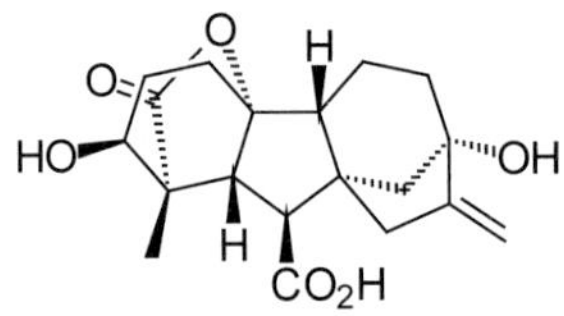

(+)-gibberelline A_1

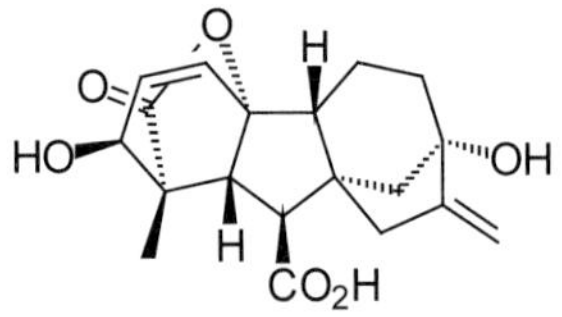

(+)-gibberelline A_3 (giberellic acid)

4.5 Tetracyclophytanes

This fungus excessively stimulates the growth of rice seedlings (Bakanae disease) due to various gibberellines it produces from geranylgeranyldiphosphate *via* (−)-16(17)-kaurene (section 4.5.3). These include dihydrogibberellic acid, better known as (+)-gibberelline A_1, that also occurs in many higher plants. Unripe seeds of the rabbit clover *Lupinus luteus* (Leguminosae) contain (−)-gibberelline A_{18} with a methyl group (C-20) present.

4.5.6 Grayanotoxanes

Leucothoe grayana (Ericaceae) defined the name of two other classes of tetracyclic diterpenes, the rare leucothols and the more abundant grayanatoxins. Examples include (−)-leucothol C, 10(20),15-grayanatoxadiene-3β,5β,6β-triol and 10(20)-grayanotoxene-2β,5β,6β,14β,16α-pentol as neurotoxic constituents of the leaves of *Leucothoe grayana* (Ericaceae). Other neurotoxic grayanotoxanes occur in the leaves and in the honey from flowers of some *Rhododendron* species; these are exemplified by 2,3-epoxygrayanotoxane-5β,6β,10α,14β,16β-pentol from *Rhododendron japonicum* (Ericaceae).

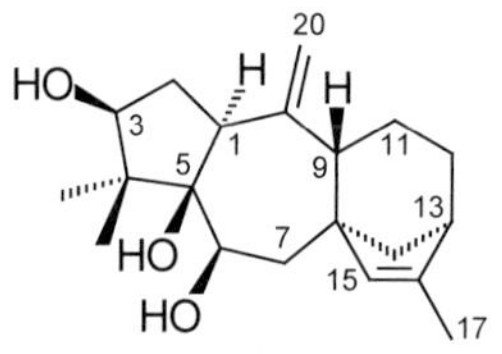

(−)-10(20),15-grayano-
toxadiene-3β,5β,6β-triol

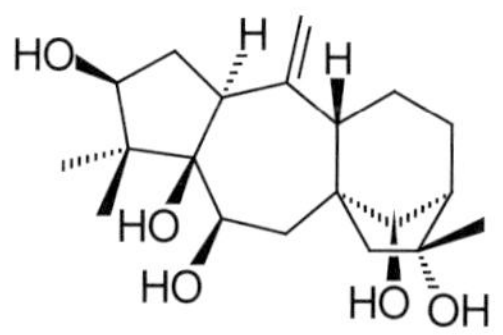

(−)-10(20)-grayanotoxene-
3β,5β,6β,14β,16α-pentol
(grayanotoxin II)

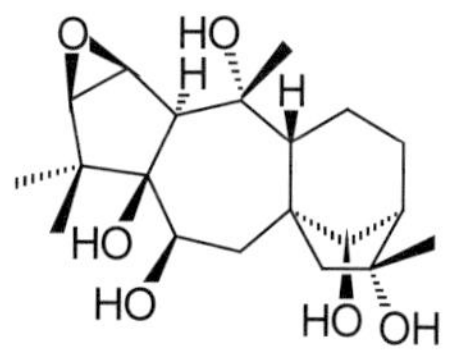

2β,3β-epoxygrayano-
toxane-5β,6β,10α,14β,16β-
pentol (rhodojaponin III)

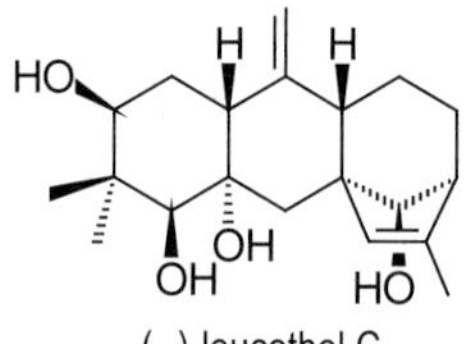

(−)-leucothol C

4.6 Cembranes and Cyclocembranes

4.6.1 Survey

Various bi- and tricyclic diterpenes are derived from the monocyclic cembrane (Table 4). Casbane, for example, is simply 2,15-cyclocembrane; another bond between C-6 and C-10 leads to lathyrane, from which the jatrophanes arise by opening the C-1–C-2 bond.

Table 4. Polycyclic diterpene basic skeletons derived from cembrane (Part 1).

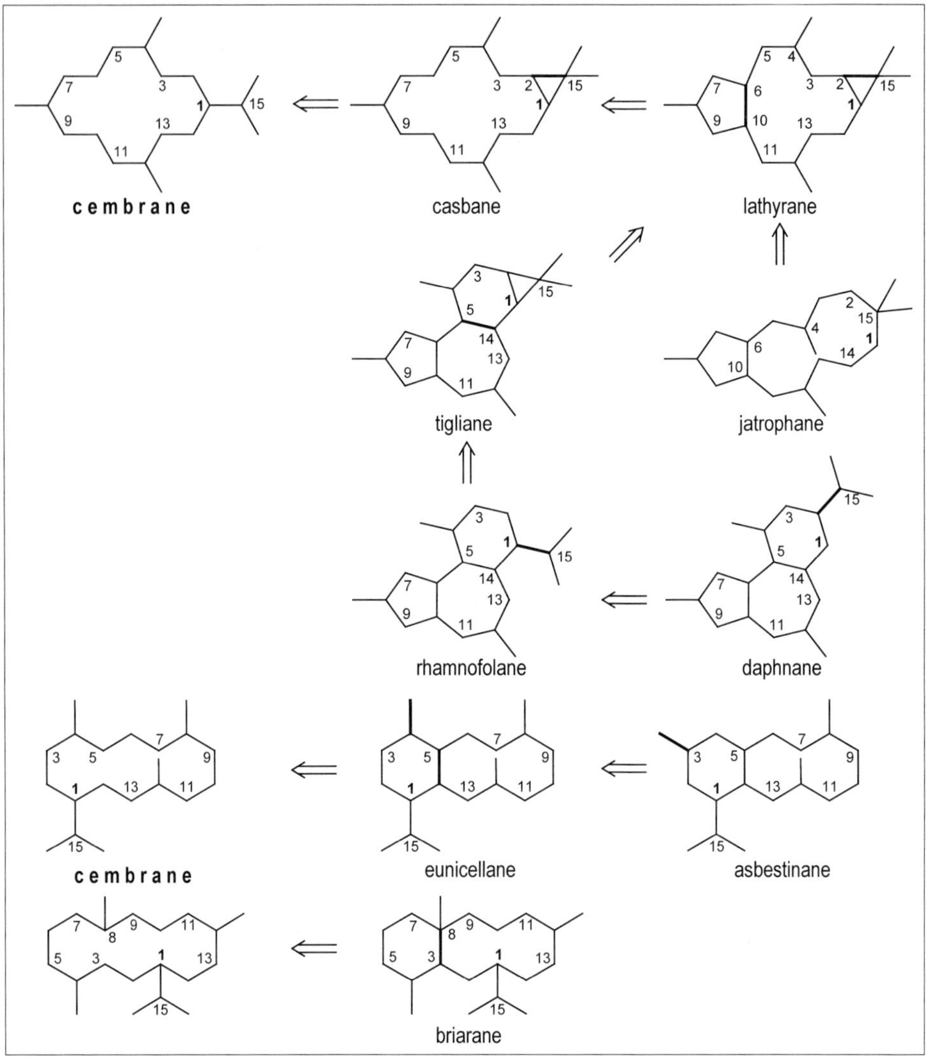

4.6 Cembranes and Cyclocembranes

Table 4. Polycyclic diterpene basic skeletons derived from cembrane (Part 2).

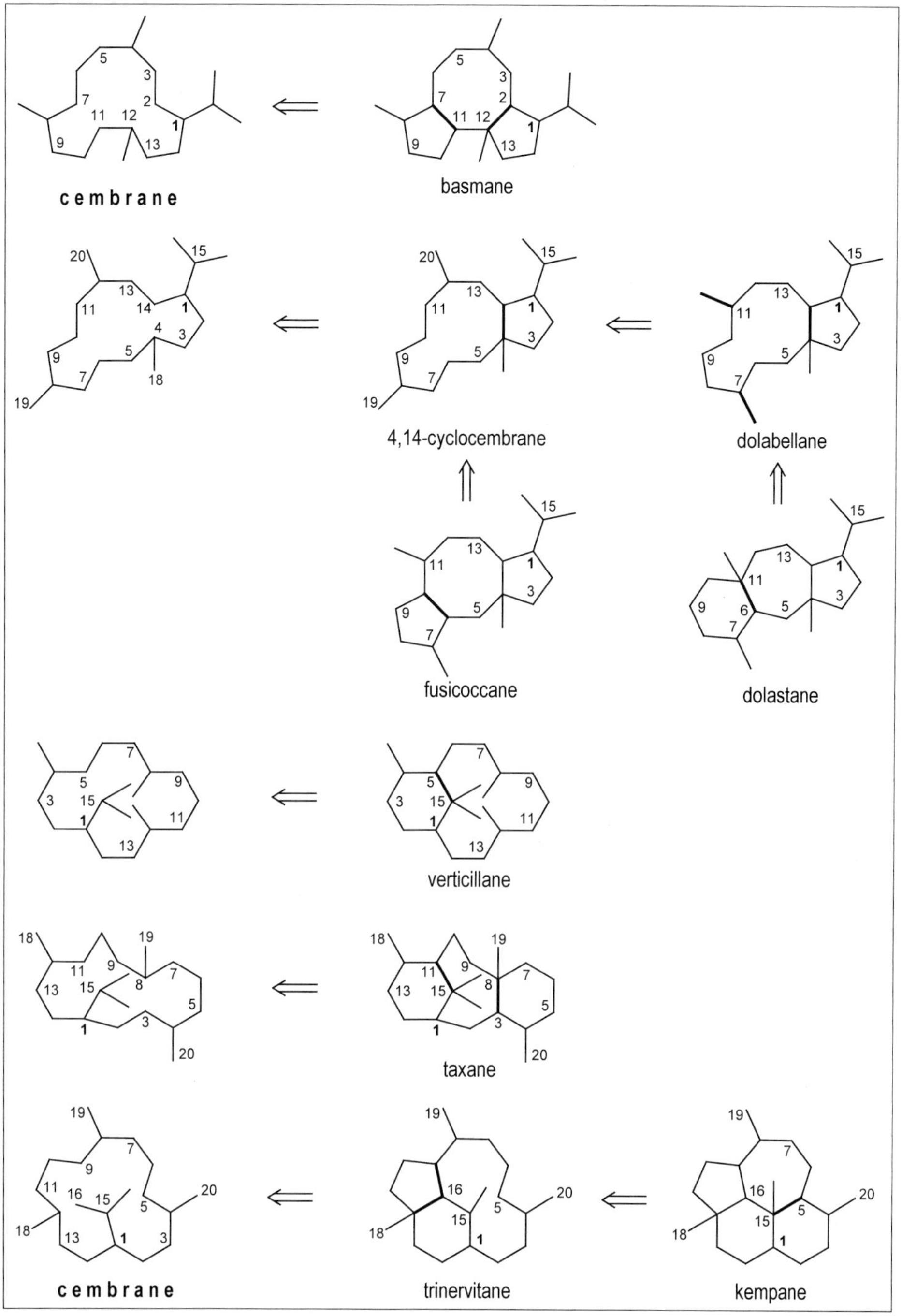

cembrane ⇐ basmane

4,14-cyclocembrane ⇐ dolabellane

⇕ ⇕

fusicoccane — dolastane

verticillane ⇐ (cembrane derivative)

taxane ⇐ (cembrane derivative)

cembrane ⇐ trinervitane ⇐ kempane

Lathyrane formally cyclizes to tigliane by closing the C-5–C-14-bond which, on its part, converts into rhamnofolane and daphnane by opening the 2,15- and the 1,15-bond, respectively (Table 4). Eunicellanes represent 2,5-cyclocembranes, which formally undergo C-18-methyl shift from C-4- to C-3 resulting in the formation of asbestinanes. Briaranes are 3,8-cyclocembranes (Table 4).

Bonds between the carbon atoms C-2 and C-12 as well as C-7 and C-11 of cembrane close two cyclopentane rings in basmane. Dolabellanes formally originate from 4,14-cyclocembrane involving two methyl shifts (C-20 from C-12 to C-11; C-19 from C-8 to C-7). Fusicoccanes represent 6,10-cyclodolabellanes; dolastanes arise from 6,11-cyclization of dolabellanes. 5,15-Cyclocembranes are referred to as verticillanes. Taxane (Table 4), the tricylic diterpene skeleton of the *Taxus* alkaloids, arises from cembrane by connecting bonds from C-3 to C-8 and from C-11 to C-15. Trinervitanes are formally built up from cembrane by closing two additional rings due to bonds from C-9 to C-16 and from C-12 to C-16. The traditionally accepted ring numbering of cyclocembranes reviewed in Table 4 does not always follow that defined for cembrane.

4.6.2 Cembranes

More than 100 cembrane derivatives have been reported to occur in plants [2]. Among those, (–)-3,7,11,15-cembratetraene (referred to as cembrene A) is widely spread among higher plants. Moreover, it serves as a pheromone of various termites and belongs, together with (–)-3,7,11-cembratrien-1β-ol (better known as serratol), to the odorless constituents of incense (olibanum) from *Boswellia serrata* (Burseraceae). Some species of tobacco also contain cembranes; β-cembrenediol, more precisely (+)-2,7,11-cembratriene-4β,6α-diol, from *Nicotiana tabacum* (Solanaceae) is an example.

(–)-3,7,11,15-cembratetraene
(cembrene A)

(–)-3,7,11-cembratriene-1β-ol
(serratol)

(+)-2,7,11-cembratriene-4β,6α-diol
(β-cembrenediol)

(+)-7,8-epoxy-4-basmen-6-one

4.6 Cembranes and Cyclocembranes

(+)-7,8-Epoxy-4-basmen-6-one from Greek tobacco is the only naturally occurring basmane (2,12:7,11-bicyclocembrane) derivative reported to date.

4.6.3 Casbanes

Casbanes are also rare in higher plants. Casbene, for example, occurs as an antifungal in the seed and sprout of *Ricinus communis* (Euphorbiaceae). (+)-Crotonitenone from *Croton nitens* (Euphorbiaceae) is another representative.

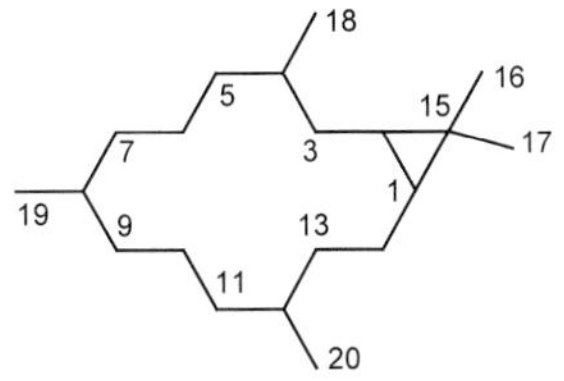

casbane

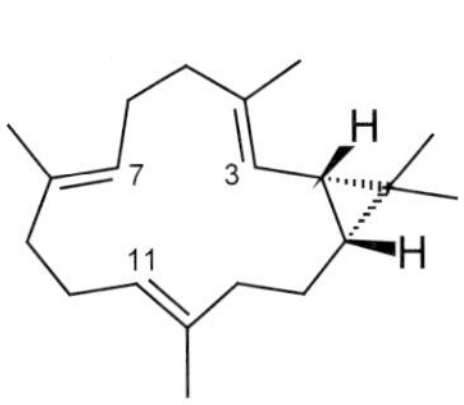

all-*trans*-casba-3,7,11-triene (casbene)

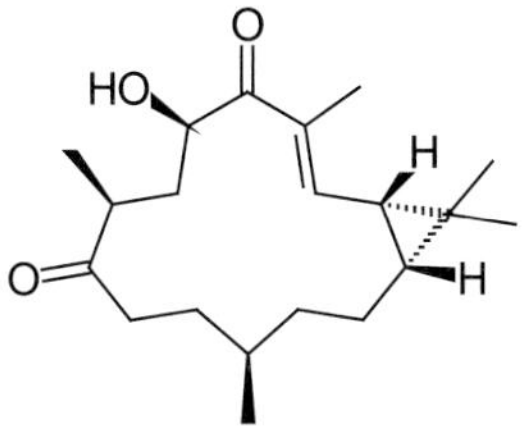

(+)-6β-hydroxy-*trans*-3-casbene-5,9-dione (crotonitenone)

4.6.4 Lathyranes

Lathyranes are, for the most part, isolated from Euphorbiaceae. Representatives include 7β-hydroxylathyrol from *Euphorbia lathyris*, the cinnamic acid esters denoted as jolkinols from *E. jolkini*, and ingol, a skin-irritating and antineoplastic hydrolyzate obtained from *E. ingens* and *E. kamerunica*.

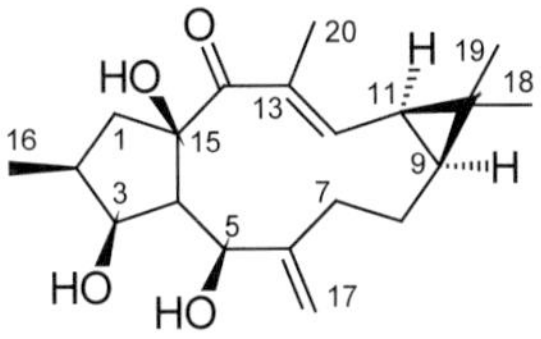

7β-hydroxylathyrol

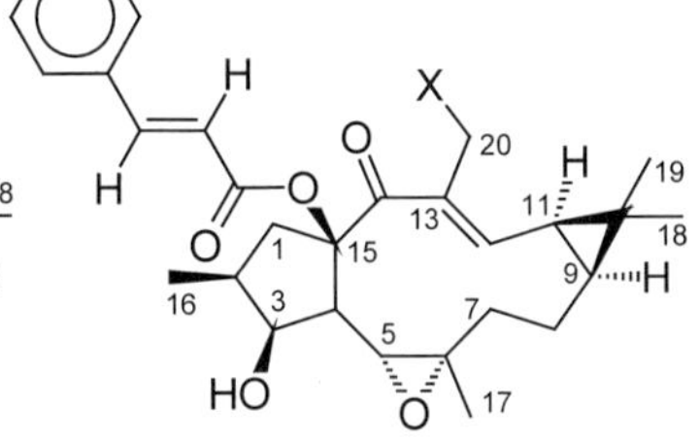

X = OH : (−)-jolkinol A
X = H : (−)-jolkinol B

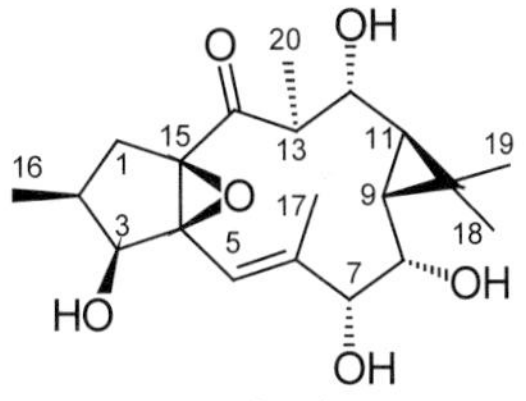

ingol

4.6.5 Jatrophanes

The name jatrophane stems from *Jatropha gossypiifolia* (Euphorbiaceae) which was found to contain the antineoplastic and antileukemic (+)-jatrophone. Various differently substituted jatrophanes are isolated from Euphorbiaceae, such as the esulones from *Euphorbia esula* and the euphornines from *E. helioscopia* and *E. maddeni*.

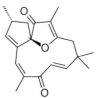

R = COCH₃ ; R' = COC₆H₅ ; R" = H :
(+)-esulon A
R = R" = COCH₃ ; R' = COC₆H₅ :
(−)-esulon B

R = COCH₃ ; R' = COC₆H₅ ; R" = H :
(−)-euphornine A
R = R" = COCH₃ ; R' = COC₆H₅ :
(−)-euphornine

(+)-jatrophone

4.6.6 Tiglianes

Polyhydroxylated tiglianes esterified with linoleic and palmitic acid are among the irritant and cocarcinogenic (tumor-promoting) constituents of various Euphorbiaceae. They occur in the purgative and counterirritant croton oil expressed from the seeds of *Croton tiglium* (Euphorbiaceae) [19]. Phorbol and isophorbol are obtained upon hydrolysis of these esters. Hydrolysis of prostratin, isolated from *Pimela prostrata*, yields 12-deoxyphorbol. Fatty acid esters of the latter occur in various Euphorbiaceae.

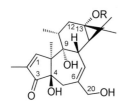

R = COCH₃ : (+)-13α-acetoxy-
4β,9α,20-trihydroxy-1,6-tigliadien-3-one
(prostratin)
R = H : 4β,9α,13α,20-trihydroxy-
1,6-tigliadien-3-one
(12-deoxyphorbol)

(+)-4β,9α,12β,13α,20-
pentahydroxy-
1,6-tigliadien-3-one
(phorbol)

(+)-4α,9α,12β,13α,20-
pentahydroxy-
1,6-tigliadien-3-one
(isophorbol)

4.6.7 Rhamnofolanes and Daphnanes

Rhamnofolanes such as (−)-20-acetoxy-9-hydroxy-1,6,14-rhamnofolatriene-3,13-dione from *Croton rhamnifolius* (Euphorbiaceae) and other constituents from various *Jatropha* species rarely occur in plants.

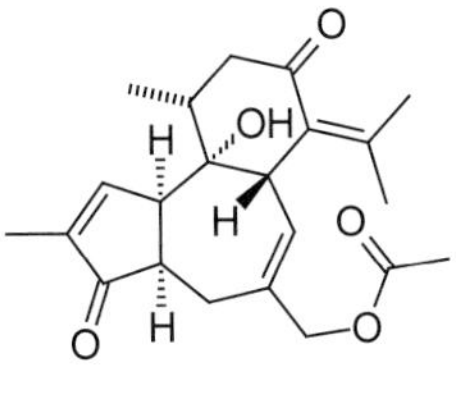

(−)-20-acetoxy-9-hydroxy-1,6,14-rhamnofolatriene-3,13-dione

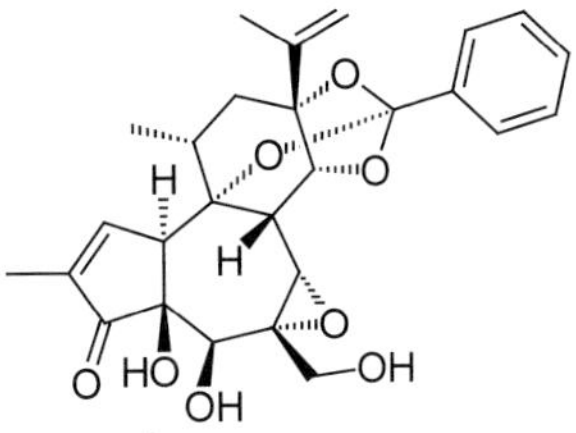

(+)-daphnetoxin

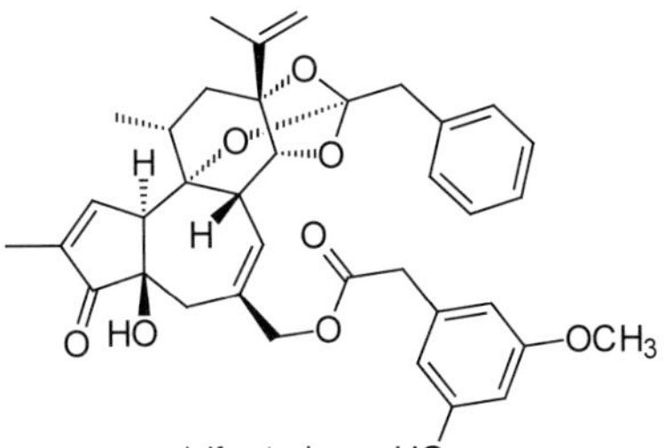

resiniferatoxin

Daphnanes are more frequently found such as daphnetoxin in flowers and bark of *Daphne* species exemplified by *Daphne mezereum* (Thymeliaceae), irritating human skin and mucous membranes, as well as (+)-resiniferatoxin from *Euphorbia resinifera* and related species. Some daphnanes are reported to have antineoplastic and antileukemic activities.

4.6.8 Eunicellanes and Asbestinanes

Eunicellanes and (infrequently) asbestinanes are diterpenes of marine origin, including eunicelline from the alga *Eunicella stricta* and cladielline found in some *Cladiella* species. Asbestinanes such as (−)-asbestinine 2 are characteristic of the gorgonia *Briareum asbestinum* (Araceae).

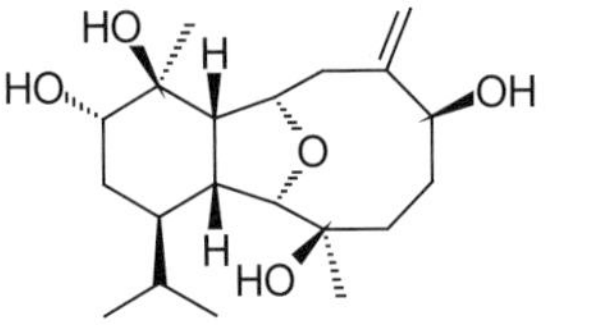

(−)-6α,13α-epoxy-8(19)-eunicellene-3α,4β,9β,12β-tetrol (eunicelline)

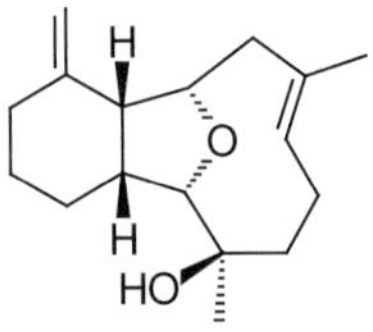

(−)-6α,13α-epoxy-4(18),8-eunicelladiene-12β-ol (cladielline)

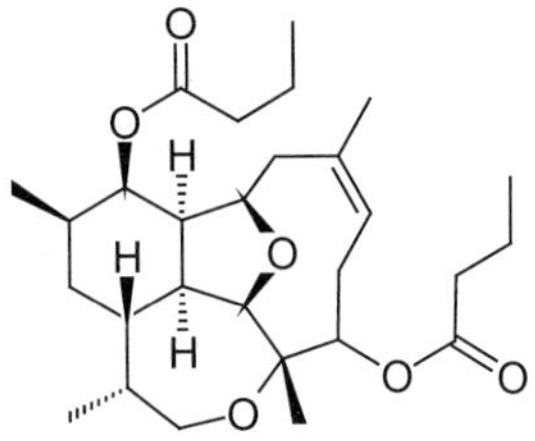

(−)-asbestinine 2

4.6.9 Briaranes

Briaranes are found in marine organisms including corals, represented by the briantheines W and X from *Briareum polyanthes*. Brianthein X is reported to be an insecticide, while the closely related (–)-solenolide A isolated from *Solenopodium* species exhibits antiviral and antiinflammatory activities. These briarane lactones are usually referred to as erythrolides.

2β,14α-diacetoxy-5,8(17),11-briaratriene-18,7-olide (briantheine W)

9β,12α-diacetoxy-6α-chloro-13α,14α-epoxy-2-hydroxy-3,5(16)-briaradiene-18,7-olide (Brianthein X)

(–)-solenolide A

4.6.10 Dolabellanes

Dolabellanes occur in various corals and algae. Cytotoxic and ichthyotoxic (poisonous for fish) 7,8:10,11-diepoxy-4(16)-dolabellene-17,18-diol known as (–)-stolondiol from *Clavularia* corals, (–)-2,6-dolabelladiene-6β,10α,18-triol from the herbivorous (plant-eating) sea rabbit *Dolabella california* and 4,8,18-dolabellatriene-3α,16-diol and other dolabellanes isolated from various *Dictyota* brown algae are representative examples.

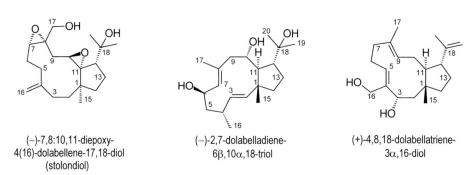

(–)-7,8:10,11-diepoxy-4(16)-dolabellene-17,18-diol (stolondiol)

(–)-2,7-dolabelladiene-6β,10α,18-triol

(+)-4,8,18-dolabellatriene-3α,16-diol

4.6.11 Dolastanes

Metabolites of various algae and corals also include dolastadienes and -trienes, exemplified by 1(15),8-dolastadiene-4β,14β-diol (amijiol), its isomer isoamijiol, and 1(15),7,9-dolastatrien-14β-ol from brown algae *Dictyota linearis* and *D. cervicornis*, as well as 1(15),17-dolastadiene-3α,4β-diol from the soft coral *Clavularia inflata*.

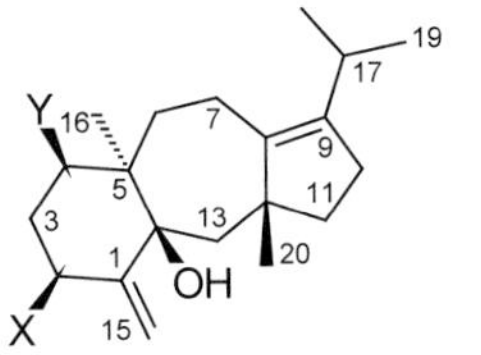

X = OH, Y = H : (−)-1(15),8-dolastadiene-2β,14β-diol (isoamijiol)
X = H, Y = OH : (−)-amijiol

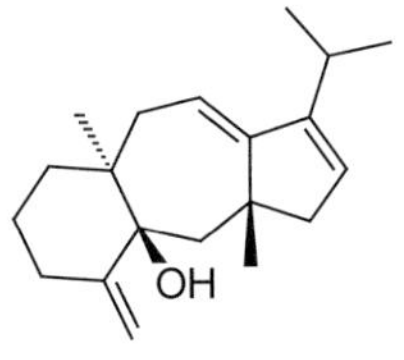

(−)-1(15),7,9-dolastatrien-14β-ol

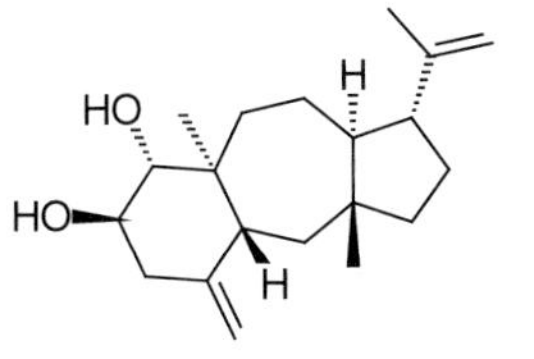

(+)-1(15),17-dolastadiene-3β,4α-diol

4.6.12 Fusicoccanes

Fusicoccanes are metabolites of some fungi, liverworts and algae. Some of these and their glycosides act as growth regulators. (+)-Epoxydictymene from the brown alga *Dictyota dichotoma*, (−)-fusicoplagin A from the liverwort *Plagiochila acanthophylla* and (+)-fusicoccin H from the fungus *Fusicoccum amygdali* are selected representatives.

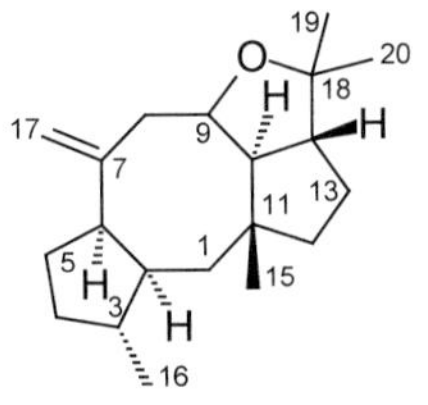

(+)-9,15-epoxy-7(17)-fusicoccene (epoxydictymene)

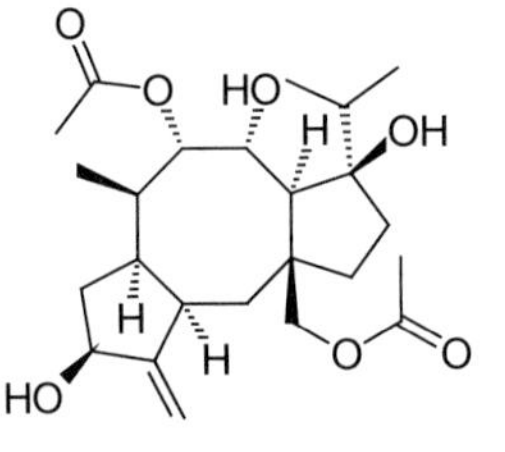

(−)-8α,18-diacetoxy-4β,9α,14β-trihydroxy-3(16)-fusicoccene [(−)-fusicoplagin A]

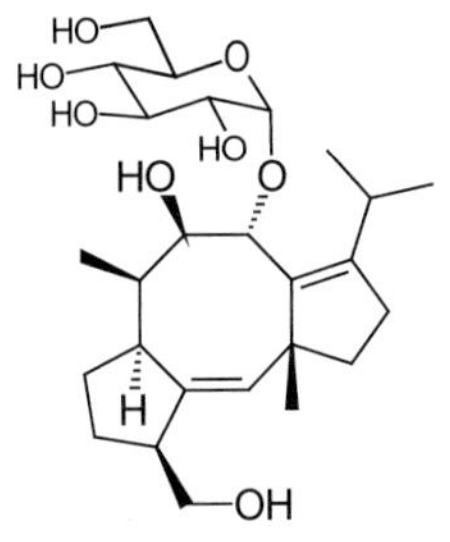

(+)-fusicoccin H

4.6.13 Verticillanes and Taxanes

Verticillanes such as (+)-verticillol from the wood of *Sciadopitys verticillata* (Taxodiaceae) rarely occur as natural products. 8,19-Cycloverticillane, however, is the core skeleton of (+)-taxine A, the crystalline chief constituent of the poisonous alkaloid mixture taxine isolated from the needles but not the red berries of the European yew tree *Taxus baccata* (Taxaceae), causing gastrointestinal irritation as well as cardiac and respiratory failure. Various taxanes are also found in the alkaloid mixtures obtained from Taxaceae, exemplified by (−)-10-deacetylbaccatin from the needles and the bark of European *Taxus baccata* and (−)-taxol [19] from the bark of Pacific *Taxus brevifolia* and *T. cuspidata*. The latter is applied for the chemotherapy of leukemia and various types of cancer.

4.6.14 Trinervitanes and Kempanes

Trinervitanes including their *seco*-, methyl- and 11,15-cyclo-derivatives (also referred to as kempanes) belong to the defense pheromones [14-17] of various termite species, represented by 1(15),8-trinervitadien-14β-ol and 1(15),8-trinervitadiene-13α,14β-diol from *Trinervitermes gratiosus* as well as 6,8-kempadiene-2α,13α-diol (known as kempene 1) from *Nasutitermes kempae*.

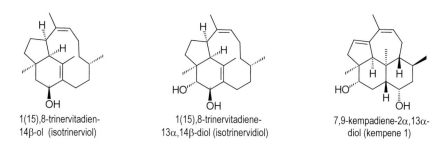

4.7 Prenylsesquiterpenes

About 200 diterpenes are reported in which an isoprenyl (= "prenyl") residue extends one of the side chains of a sesquiterpene, consequently referred to as prenylsesquiterpenes [2]. This class of diterpenes is reviewed in Table 5.

Table 5. Prenylsesquiterpenes and their parent skeletons.

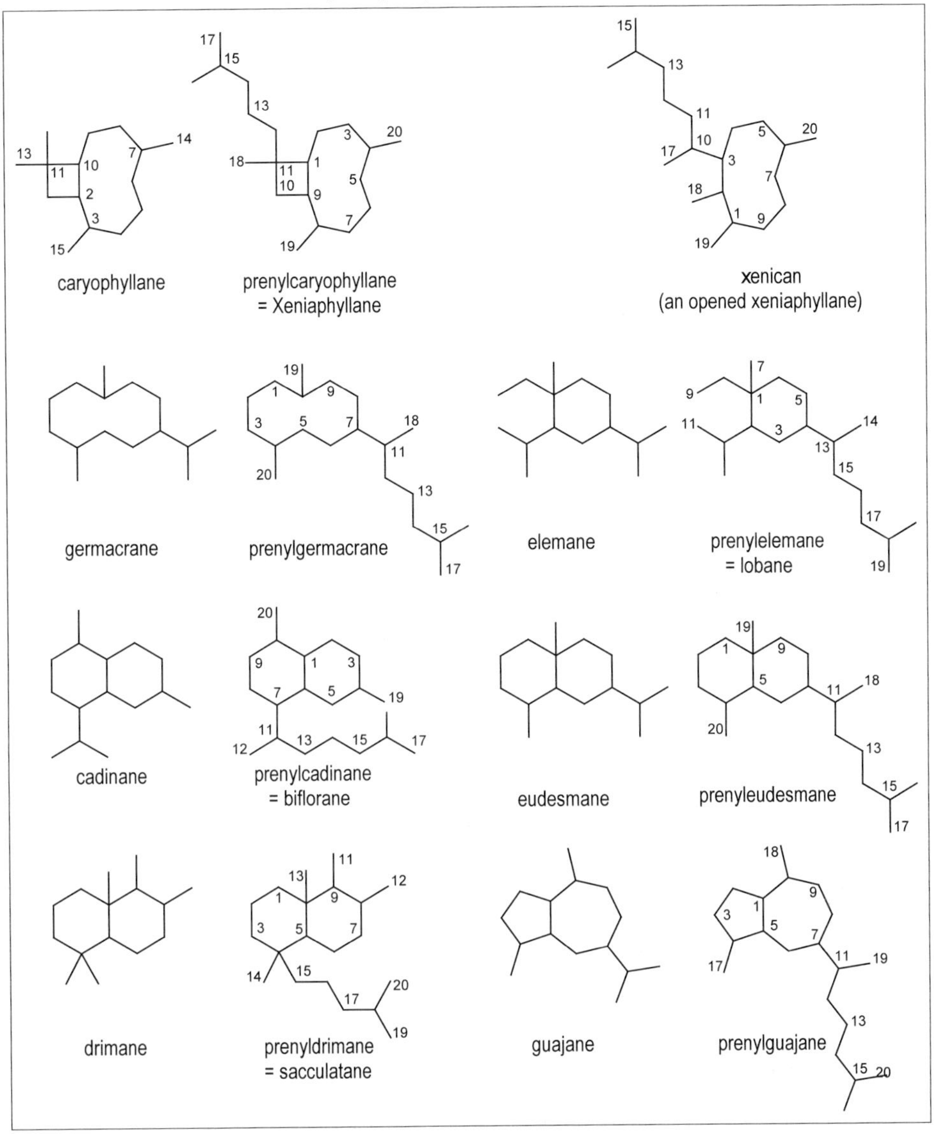

4.7.1 Xenicanes and Xeniaphyllanes

Xeniaphyllanes (prenylcaryophyllane) incorporating the cylobutane ring of the parent caryophyllane and represented by isoxeniaphyllenol in some *Xenia* corals such as *X. macrospiculata*, are rarely found in other organisms. Xenicanes with opened cyclobutane ring more frequently occur in algae and corals, for example isodictyohemiacetal in the alga *Dictyota dichotoma*, xeniolit A and xeniaacetal in the corals *Xenia macrospiculata* and *Xenia crassa*.

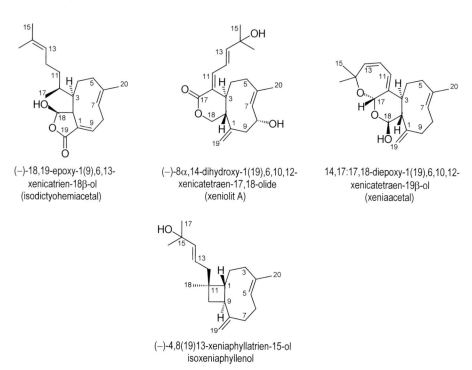

(−)-18,19-epoxy-1(9),6,13-
xenicatrien-18β-ol
(isodictyohemiacetal)

(−)-8α,14-dihydroxy-1(19),6,10,12-
xenicatetraen-17,18-olide
(xeniolit A)

14,17:17,18-diepoxy-1(19),6,10,12-
xenicatetraen-19β-ol
(xeniaacetal)

(−)-4,8(19)13-xeniaphyllatrien-15-ol
isoxeniaphyllenol

4.7.2 Prenylgermacranes and Lobanes

Prenylgermacranes and prenylelemanes, usually known as lobanes and possibly arising from COPE rearrangements of appropriate prenylgermacranes, are also metabolites of some marine organisms. Various derivatives of the prenylgermacrane dilophol, for example, occur in the algae *Dilophus ligulatus, Dictyota dichotoma* and *Pachydictyon coriaceum*. 17,18-Epoxy-8,10,13(15)-lobatriene and (+)-14,17-

4.7 Prenylsesquiterpenes

diacetoxy-8,10,13(15)-lobatrien-18-ol representing the prenylelemanes are isolated from several *Lobophytum* species.

4.7.3 Prenyleudesmanes and Bifloranes

The alga *Dictyota acutiloba* produces dictyolene, one of the rarely abundant prenyleudesmanes. Prenylcadinanes are more frequently found not only in marine organisms but also in higher plants and insects. They are usually referred to as bifloranes and, when containing a benzenoid ring, as serrulatanes. (−)-4,10(19),15-Bifloratriene, for example, is secreted by the termite *Cubitermes umbratus* as a constituent of the defense pheromone cocktail, but is also produced by the soft coral *Xenia obcuranata*. Dictyotin B is found in the brown alga *Dictyota dichotoma*. (−)-8,16-Dihydroxy-19-serrulatanoic acid, however, is isolated from *Eremophila serrulata*.

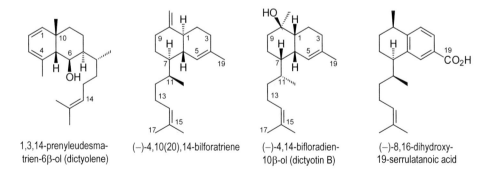

4.7.4 Sacculatanes (Prenyldrimanes)

Sacculatanes (prenyldrimanes) occur in various liverwort species. *Porella perrottetiana*, for instance, protects itself against insects by producing the strongly bitter compounds 3β-hydroxy-7,14-sacculatadien-12,11-olide and perrottetianal A; the latter also inhibits the sprouting of rice. The strongly sharp-tasting and thus antifeedant thalli of *Trichocoleopsis sacculata* may induce contact dermatitis; 18-hydroxy-7,16-sacculatadien-11,12-dial (sacculatal) and tumor-promoting 7,17-sacculatadiene-11,12-dial were found to be among the active substances.

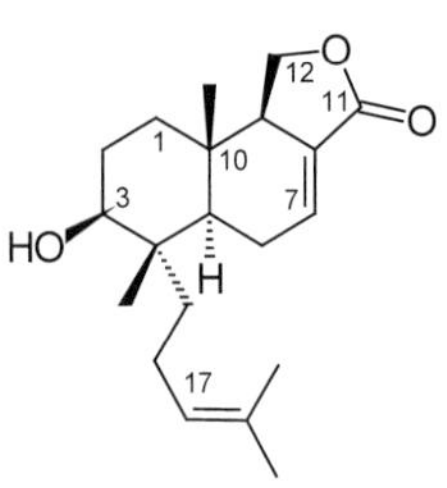

(−)-3β-hydroxy-7,17-sacculatadien-12,11-olide

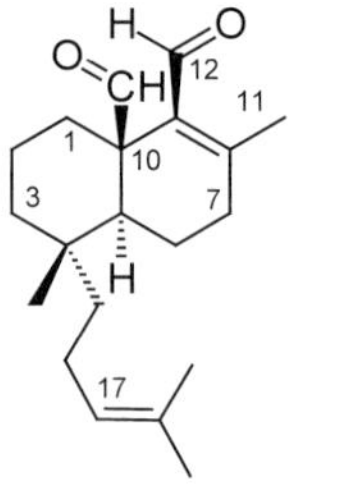

(−)-8,17-sacculatadiene-12,14-dial (perrottetianal A)

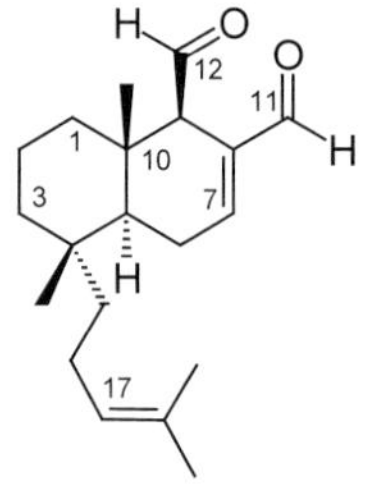

(−)-7,17-sacculatadiene-11,12-dial (sacculatal)

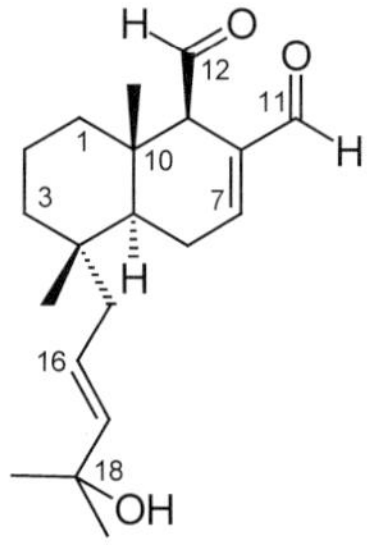

(+)-18-hydroxy-7,16-sacculatadiene-11,12-dial

4.7.5 Prenylguaianes and Prenylaromadendranes

Prenylguaianes occur predominantly in algae of the genera *Dictyota* and *Aplysia*, and less frequently in corals of the genus *Xenia*. Following their dominant origin, prenylguaianes are referred to as dictyols. (+)-Dictyol A and (+)-dictyol B from the brown alga *Dictyota dichotoma* are typical representatives. Prenylaromadendranes characterize the Cneoraceae and are consequently denoted as cneorubines. They are constituents of west Mediterranean dwarf oil tree *Cneorum tricoccon* and Canaric stone berry *Neochamaelea pulverulenta*.

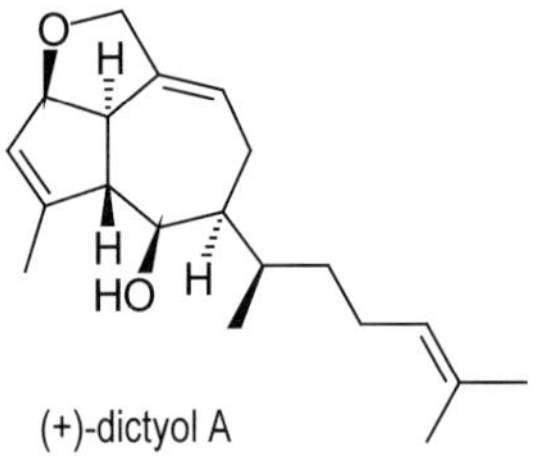

(+)-dictyol A

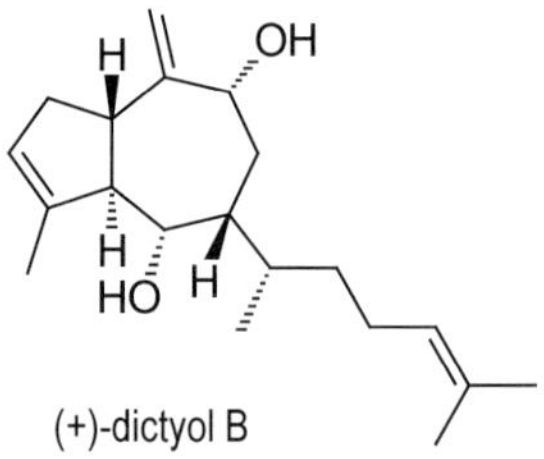

(+)-dictyol B

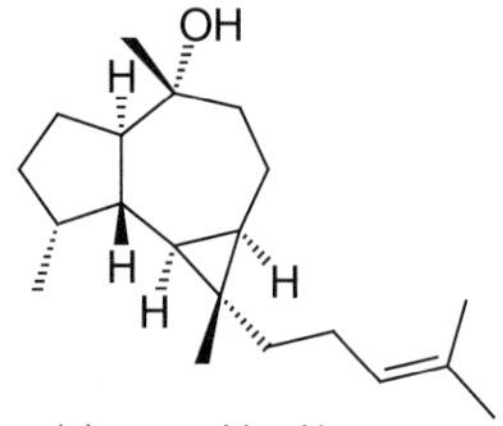

(+)-cneorubine U

4.7.6 Sphenolobanes (Prenyldaucanes)

Prenyldaucanes predominantly found in liverworts and marine organisms are commonly referred to as sphenolobanes. (+)-3α,4α-Epoxy-13(15),16-sphenolobadiene-5α,18-diol which inhibits sprouting and growth of rice, and its 5α-acetoxy derivative, both from the liverwort *Anastrophyllum minutum*, as well as (+)-3,17-sphenolobadien-13-ol (tormesol) from *Halimium viscosum*, are representatives. Reiswigins, sphenolobanediones from the deep-sea sponge *Epiupolasis reiswigi* such as reiswigin A, merit mention because of their activity against the *Herpes simplex* virus.

(+)-3α,4α-epoxy-13(15),16-sphenolobadiene-5α,18-diol

(+)-3,17-sphenolobadien-13-ol (tormesol)

(−)-3-sphenoloben-5,16-dione (reiswigin A)

4.8 Ginkgolides

The leaves of the maiden's hair tree *Ginkgo biloba*, a survivor of the Ginkgoaceae genus wide-spread during the Mesozoikum era of the Earth's history and valued as a park tree, contain flavonoids, the tetracyclic sesquiterpenelactone bilobalide A and various hexacyclic high-melting and bitter-tasting diterpene lactones known as the ginkgolides A-M, which are resistant towards acids [2]. Bilobalide and ginkgolides carry a *t*-butyl group which rarely occurs in natural compounds. Ginkgo extracts stimulate cerebral metabolism and activate functions of the cognitive area of the brain (concentration, memory); Ginkgolides A and B were found to have the cerebroprotective properties.

bilobalide A

R1	R2	R3	ginkgolide
H	OH	H	A
H	OH	OH	B
OH	OH	OH	C
OH	OH	H	J
OH	H	OH	M

5 Sesterterpenes

5.1 Acyclic Sesterterpenes

More than 150 sesterterpenes are known to date [2]. The acyclic representatives are derived from 3,7,11,15,19-pentamethylicosane $C_{25}H_{52}$. Their biogenesis primarily yields geranylfarnesyldiphosphate, involving the known pathway (section 1.3).

Sesterterpenes rarely occur in higher plants. 3,7,11,15,19-Pentamethylicosa-2,6-dien-1-ol as an example is found in the leaves of potatoes *Solanum tuberosum* (Solanaceae). About 30 sesterterpenes bridged by furan rings, however, are reported to occur in various marine sponges; these include (–)-ircinin I from *Ircinia oros* which acts as an antibacterial, and (+)-8,9-dehydroircinin I from *Cacospongia scalaris* which inhibits the division of fertilized starfish egg cells.

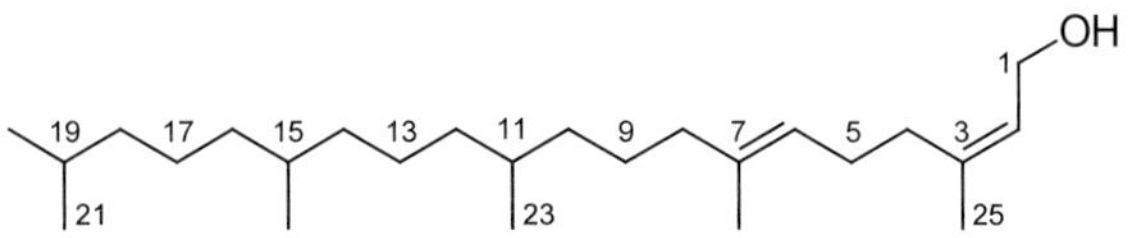

3,7,11,15,19-pentamethyl-2,6-icosadien-1ol

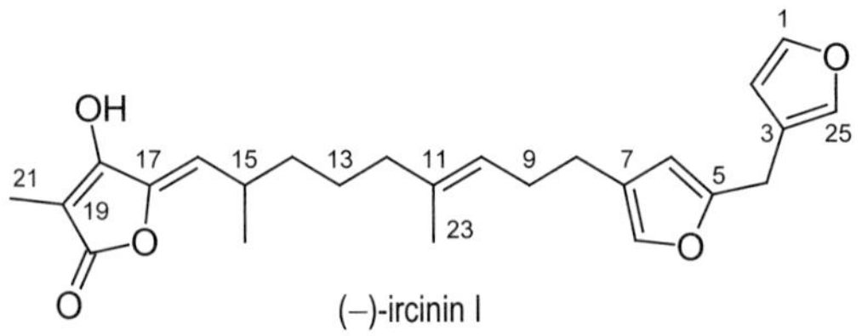

(–)-ircinin I

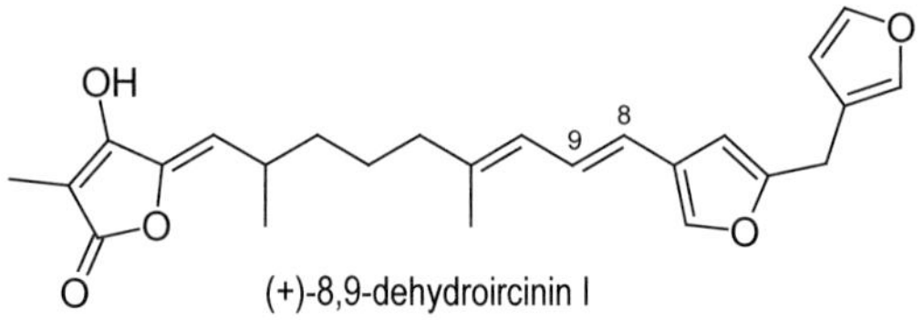

(+)-8,9-dehydroircinin I

5.2 Monocyclic Sesterterpenes

Cyclohexane sesterterpenes are formally derived from 3,7,11,15,19-pentamethylicosane by connection of the C-14–C-19 bond. A few representatives are found in various marine sponges. The lactone (–)-manoalide isolated from the sponge *Luf-*

fariella variabilis living in the Manoa valley of Oahu island (Hawai) as one of the few examples is reported to have analgesic, antiinflammatory and immunosuppressive properties.

Cyclization of 3,7,11,15,19-pentamethylicosane by connection of the C-1–C-14 bond leads formally to the cericeranes. Representatives are, for the most part, found in the waxes and secretions of insects; an example is (–)-ceriferol I from *Ceroplastes ceriferus*.

3,7,11,15,19-pentamethylicosane

cyclohexane sesterterpene

cericerane

(–)-manoalide

(–)-ceriferol 1

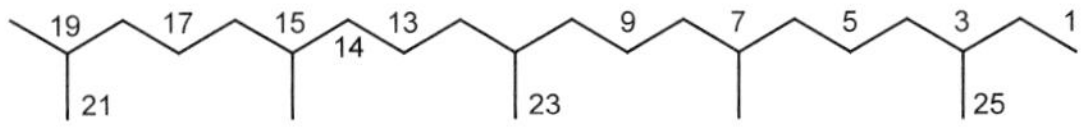

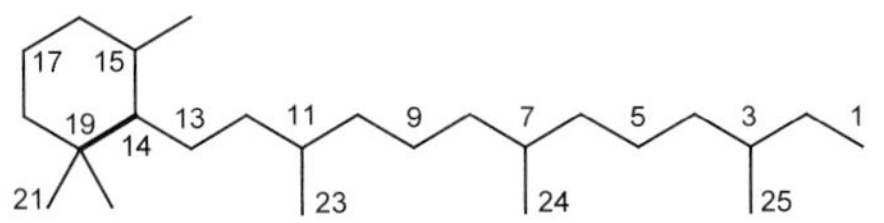

5.3 Polycyclic Sesterterpenes

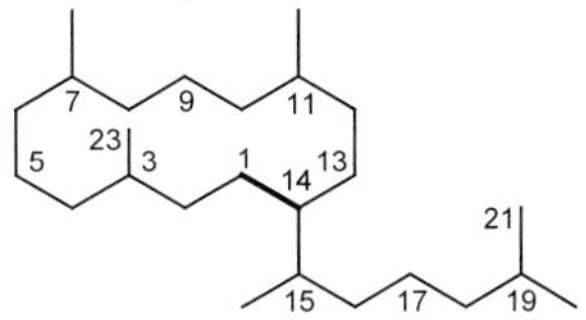

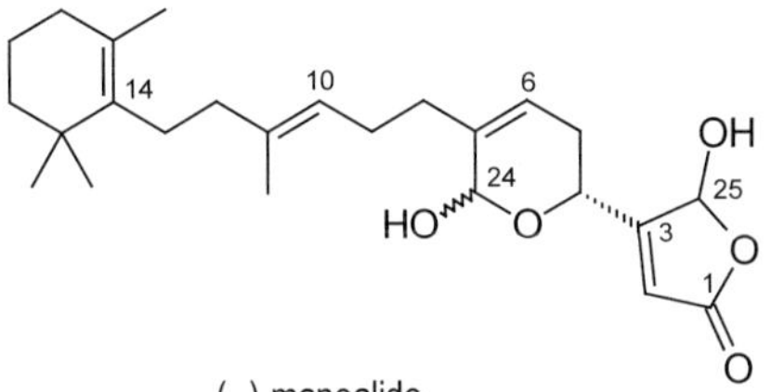

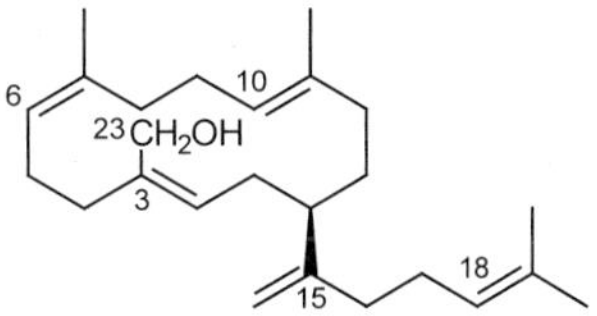

5.3.1 Bicyclic Sesterterpenes

The parent hydrocarbon skeletons of bicyclic sesterterpenes are derived predominantly from some sesquiterpenes. Diprenyldrimanes, for example, realized as (+)-dysideapalaunic acid and related lactones, partly reported to be natural inhibitors of protein phosphatase, are isolated from the Caribbean sponge *Dysidea etheria* and other *Dysidea* species. Prenyldrimanes in higher plants are represented by (–)-salvisyriacolide from *Salvia syriaca* and as (+)-salvileucolide methylester from *Salvia hypoleuca* (Labiatae).

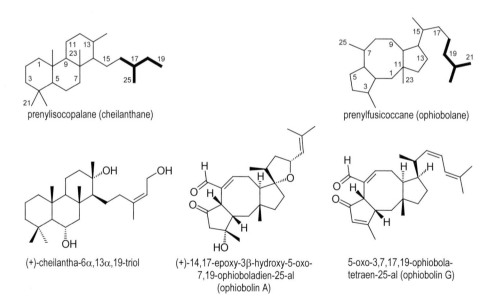

5.3.2 Tricyclic Sesterterpenes

Tricyclic sesterterpenes known to date incorporate tricyclic diterpenes as core structures which are prenylated at the side chains such as cheilanthanes and ophiobolanes, representing prenylisocopalanes and prenylfusicoccanes, respectively. (+)-Cheilanthatriol isolated from the fern *Cheilanthes farinosa* and the ophiobolins found in the phytopathogenic fungi *Ophiobolus miyabeanus*, *O. heterostrophus* and *Aspergillus ustus* are examples. Some of the ophiobolines exhibit antibacterial and phytotoxic activities.

5.3.3 Tetra- and Pentacyclic Sesterterpenes

4,4,8,17,18-Tetramethylhomoandrostane denoted as scalarane and related to the gonane nucleus of the steroids is the core structure of the majority of tetracyclic sesterterpenes known to date [2]. Various scalaranes occur in marine sponges, exemplified by (+)-scalarin in *Cacospongia scalaris*, (+)-desoxyscalarin in *Spongia officinalis*, and hyrtial in *Hyrtios erecta* with antiinflammatory activity.

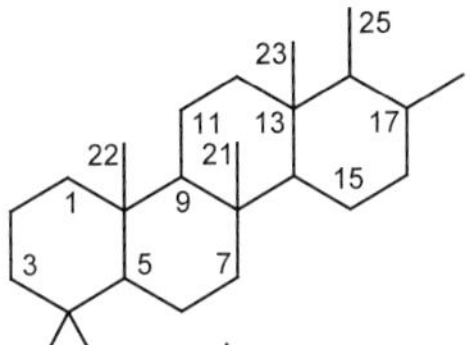

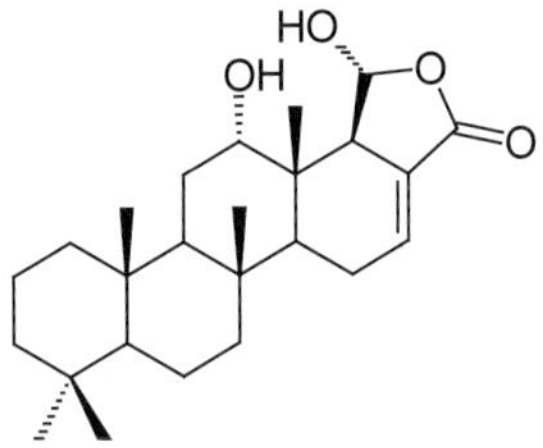

(+)-12α,25α-dihydroxy-16-scalaren-24,25-olide (scalarin)

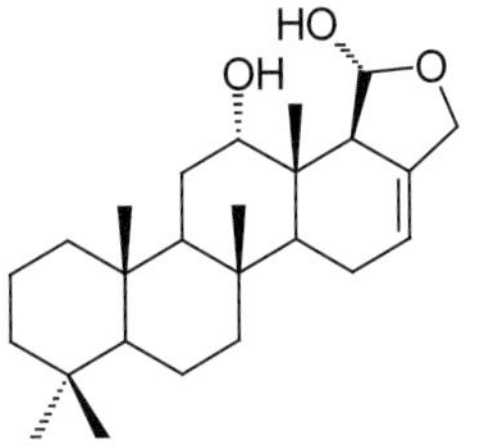

(+)-24,25-epoxy-16-scalarene-12α,25α-diol (desoxyscalarin)

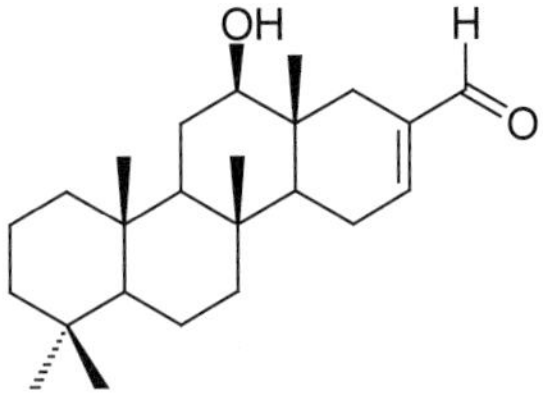

12β-hydroxy-25-nor-16-scalaren-24-al (hyrtial)

The lichen *Lobaria retigera* (Stictaceae), a parasite of some leaves, was found to contain the (−)-sesterterpenoid retigeranic acids; these should not be confused with the triterpenoid retigeric acids (section 6.4.3) isolated from the same organism. A crystal structure investigation of these unique sesterterpenes revealed the pentacyclic triquinane core structure.

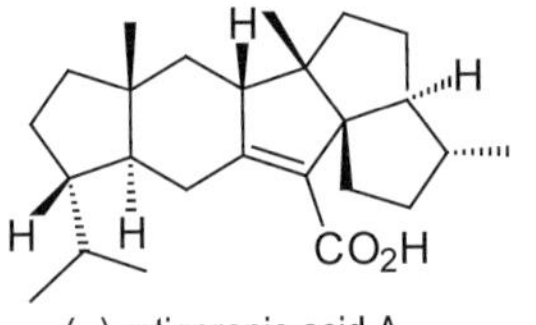

(−)-retigeranic acid A

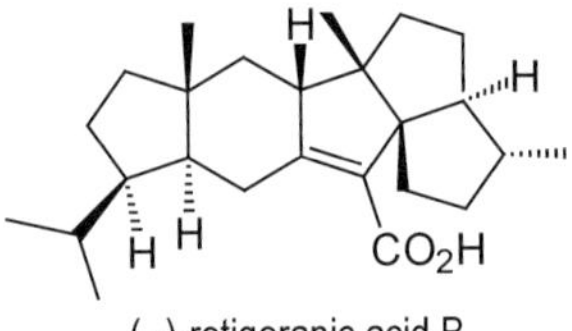

(−)-retigeranic acid B

6 Triterpenes

6.1 Linear Triterpenes

About 5000 naturally abundant triterpenes are documented in the literature [2]. Most of these are derived from squalane and squalene, with two farnesane units linked in the tail-to-tail manner.

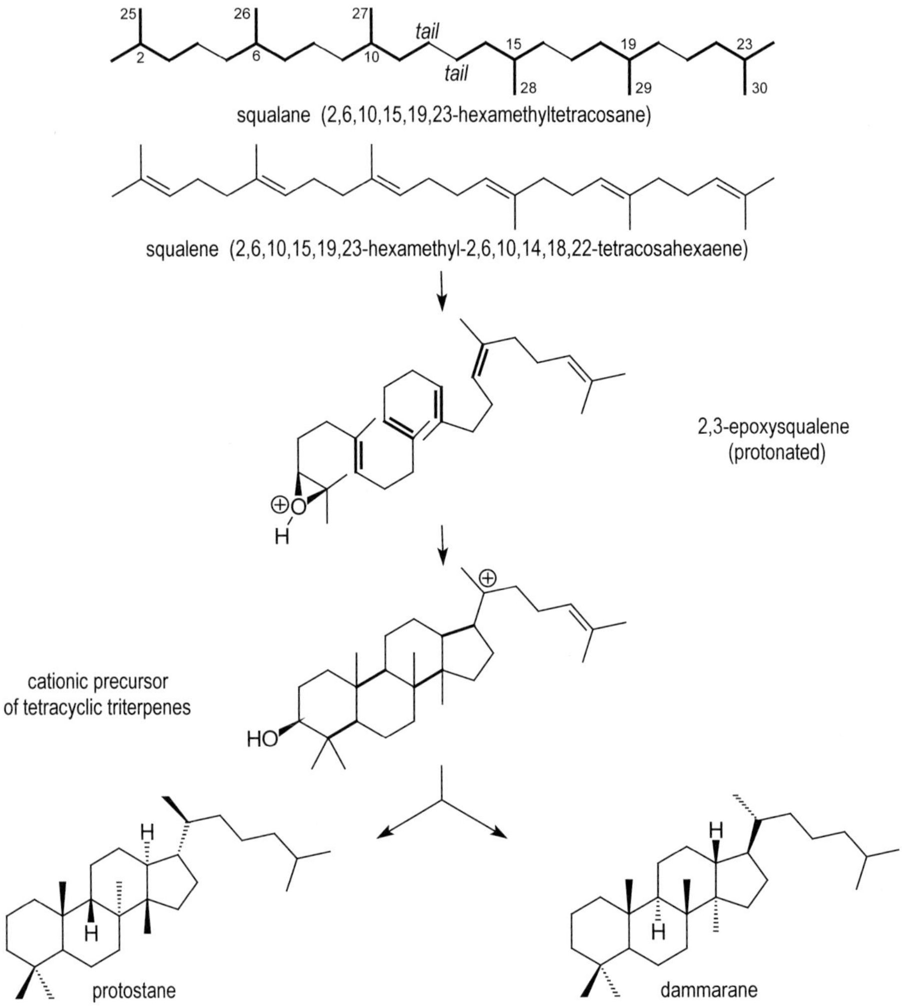

Protonated 2,3-epoxysqualene is the biosynthetic precursor of polycyclic triterpenes such as protostanes and dammaranes. The geneses of the cations of protostane and dammarane start from two different conformers of the carbenium ion, which arise intermediately from opening of the protonated 2,3-epoxysqualene.

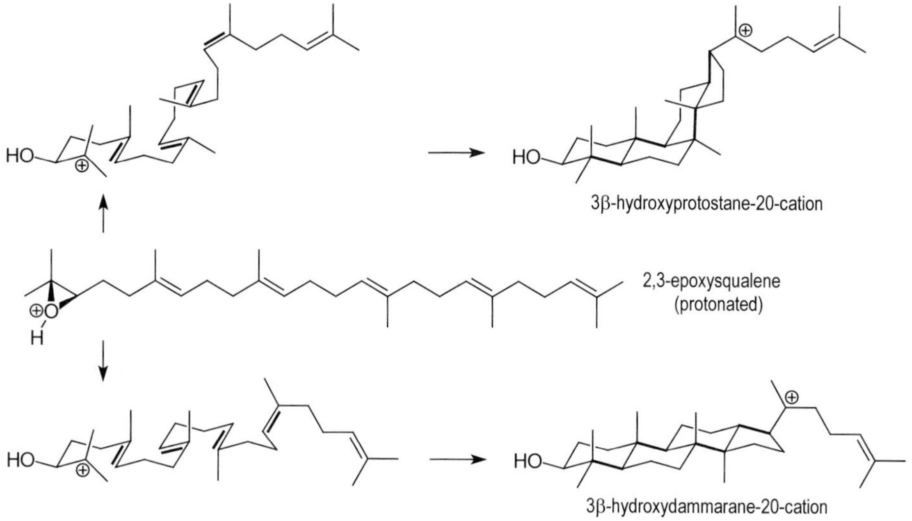

Squalene occurs in cod-liver oil and in several vegetable oils, for example those from rape and cotton seed. Red algae *Laurencia okamurai* produce (+)-1,10-epoxysqualene. Botryococcane is the parent hydrocarbon of some branched and alkylated triterpenoid polyenes found in the algae *Botryococcus braunii* and referred to as C_{30}-C_{37}-botryococcenes, depending on the number of their carbon atoms.

(+)-10,11-epoxysqualene

botryococcane
(10-Ethyl-2,6,10,13,17,21-hexamethyldocosane)

C_{30}-botrycoccene

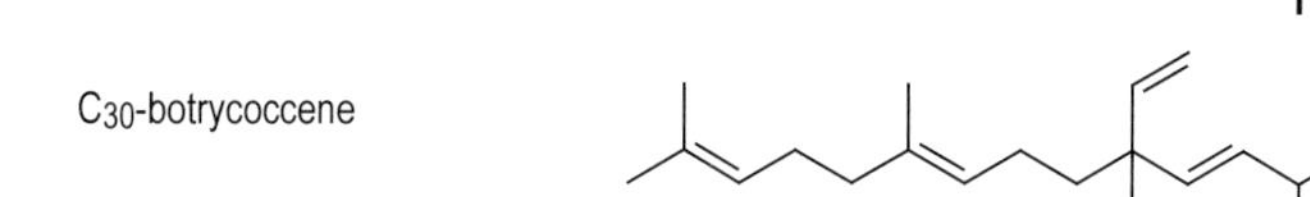

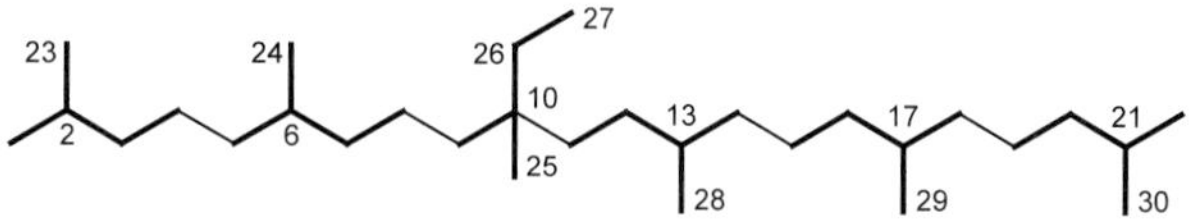

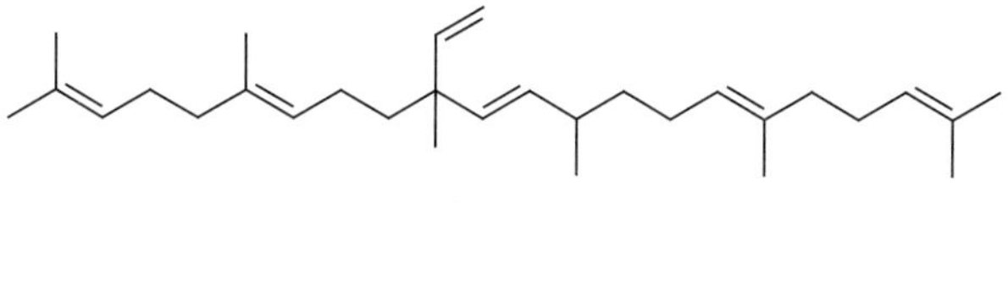

6.2 Tetracyclic Triterpenes, Gonane Type

6.2.1 Survey

The cyclization of 2,3-epoxysqualene leads to the stereoisomers protostane and dammarane with the gonane tetracycle known from steroids. 29-Norprotostanes are referred to as fusidanes. Dammarane converts into apotirucallane involving migration of the methyl group C-18 from C-14 to C-13 (Table 6).

Table 6. Tetracyclic triterpenes with the gonane skeleton.

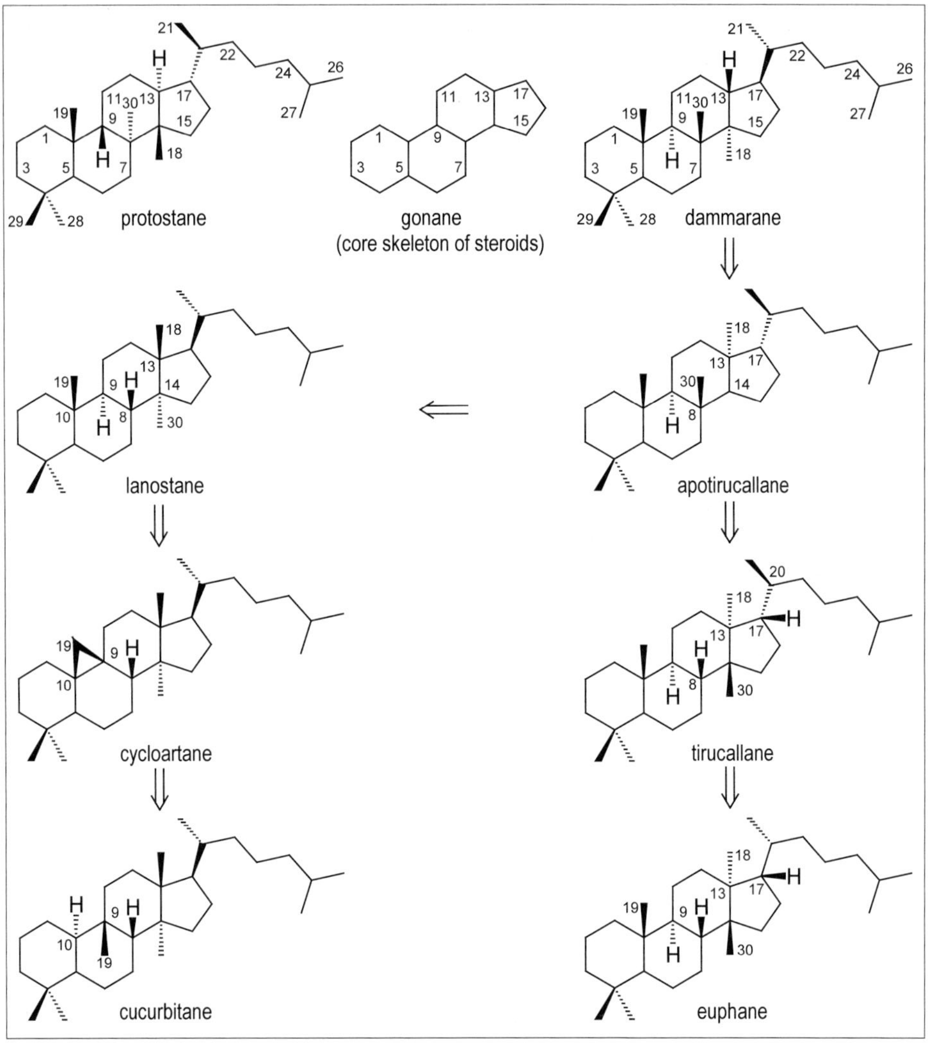

6.2 Tetracyclic Triterpenes, Gonane Type

Another methyl shift (C-30 methyl from C-8 to C-14) converts apotirucallane into tirucallane, including the C-20-epimeric euphane. Two methyl shifts involving C-18 and C-30 formally rearrange dammarane to lanostane. Cycloartane arises from lanostane by connecting C-9 and C-19 to an additional ring, and this finally opens the C-10–C-19-bond of the cyclopropane ring to cucurbitane (Table 6).

6.2.2 Protostanes and Fusidanes

Protostane and their 29-nor-derivatives denoted as fusidanes are fungal metabolites. *Cephalosporium caerulens*, for instance, produces (+)-protosta-17(20)-(Z)-24-diene-3β-ol. Fusidanes such as helvolic acid from the *Helvola* mutant of the mold *Aspergillus fumigatus* and related structures are widely used as antibacterials. Fusidic acid isolated from the fermentation broth of *Fusidium coccineum* and related tribes is an example. Its sodium salt has immunosuppressive and antibacterial properties, and is applied in ointments against multiresistant staphylococci during the therapy of wound infections [19,22].

(+)-protosta-17(20)-Z,24-dien-3β-ol

(–)-16β-acetoxy-3α,11α-dihydroxyfusida-17(20)Z,24-diene-21-oic acid (fusidic acid)

(–)-6β,16β-diacetoxy-3,7-dioxofusida-1,17(20)-Z,24-dien-21-oic acid (helvolic acid)

6.2.3 Dammaranes

Dammaranes exemplified by (+)-dammara-20,24-diene-3β,20R-diol and its (+)-20S-diastereomer belong to the constituents of the yellowish-white gum damar, the resinous exudate of the south-east Asian damar tree *Shorea wiesneri* (Dipterocarpaceae) used in plasters, varnishes, and lacquers. Multiple hydroxylated dammar-24-enes play an important role as *Ginseng* sapogenins, which are the aglyca of the *Ginseng* saponins known as ginsenosides or panaxosides from the roots of *Panax Ginseng* (Araliaceae) and related species, commonly known as power root and predominantly indigenous to Eastern Asia. Examples include (−)-dammar-24-ene-3β,12β,20S-triol and dammar-24-ene-3β,6α,12β,20R-tetrol, better known as protopanaxatriol. Infusions prepared from chewed and dried *Ginseng* roots with a slightly sweet and aromatic taste are applied as immune stimulants, for regulating cholesterol levels, and as a tonic to stimulate mental, physical, and sexual activity [19,22].

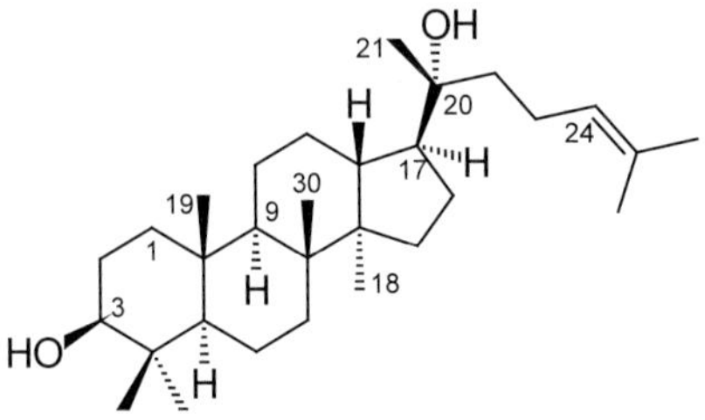

(+)-dammar-24-ene-3β-20S-diol

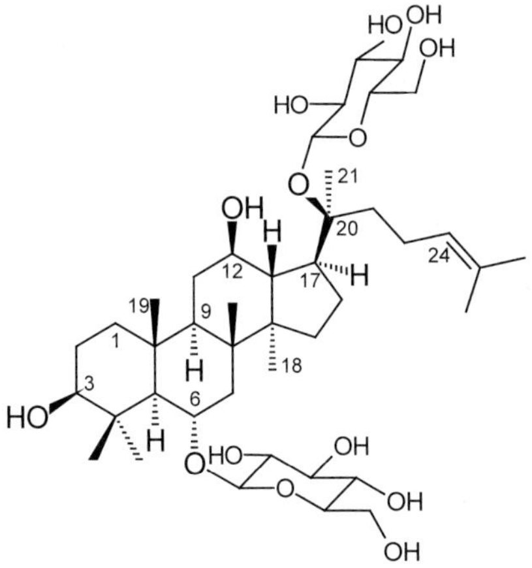

(+)-dammar-24-ene-3β,6α,12β,20R-tetrol-6,20-Di-O-β-D-glucopyranoside (panaxoside A)
one of the *Ginseng* **saponins**

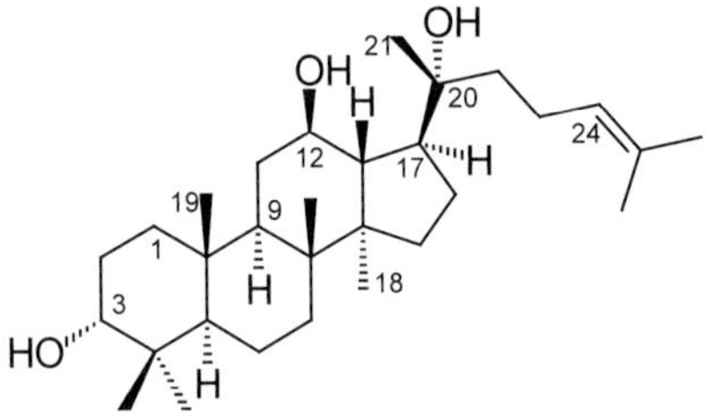

(+)-dammar-24-ene-3α,12β,20S-triol
one of the *Ginseng* **sapogenins**

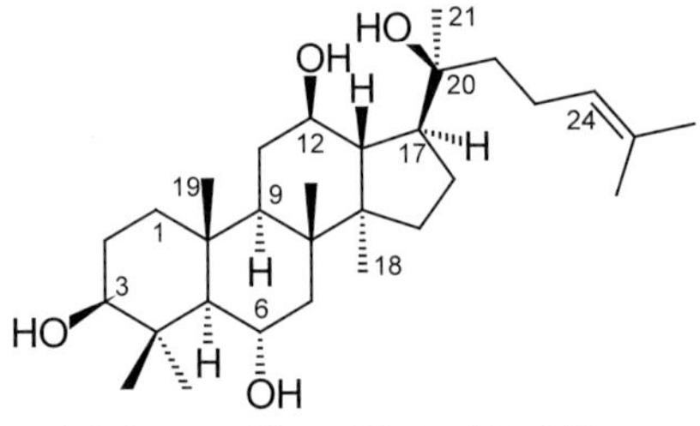

(+)-dammar-24-ene-3β,6α,12β,20R-tetrol (protopanaxatriol)
one of the *Ginseng* **sapogenins**

6.2 Tetracyclic Triterpenes, Gonane Type

6.2.4 Apotirucallanes

Apotirucallanes such as melianine A occur in Indian lilac *Azadirachta indica* (Meliaceae); infusions are reported to have insecticidal and slightly anaesthetic actions. Some tetranortriterpenes (C_{26}) and other degradation products of triterpenes referred to as quassinoids (C_{20}) originate from the apotirucallane tetracycle. Bitter-tasting quassinoids are reported to be antifeedants protecting the Simarubaceae from insects. Examples include the tetranortriterpene (+)-azadirone from the Indian neem tree *Azadirachta indica* and the insecticide bitter substance (+)-quassin (bitterness threshold 1 : 60 000) with the C_{20} skeleton referred to as picrasane from the trees *Ailanthus glandulosa*, *Picrasma excelsa* and the bitter wood *Quassia amara* (Simarubaceae) commonly known as Surinam quassia. Commercial quassin also contains some isomers and substituted derivatives of (+)-quassin; it is added to spirits as a bitter substance.

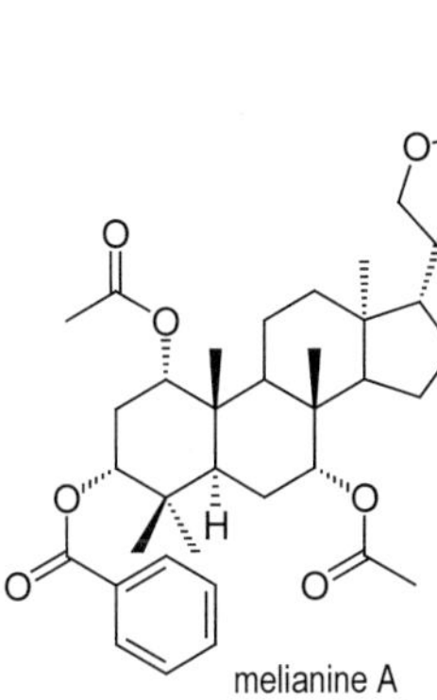

melianine A
(C_{30}-skeleton)

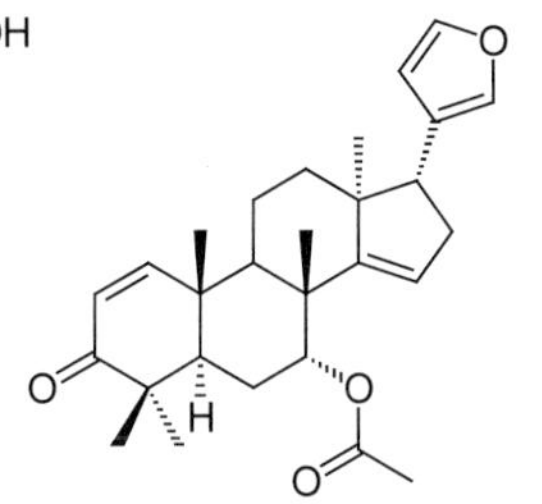

(+)-azadirone
(C_{26}-skeleton)

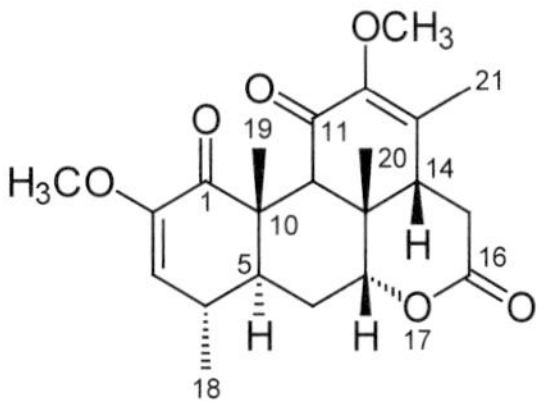

(+)-quassin
(C_{20}-skeleton, picrasane)

6.2.5 Tirucallanes and Euphanes

Tirucallanes and the 20-epimeric euphanes are frequently found in Euphorbiaceae, and are exemplified by tirucallol and its (20R)-epimer euphol from *Euphorbia tirucalli* and related species.

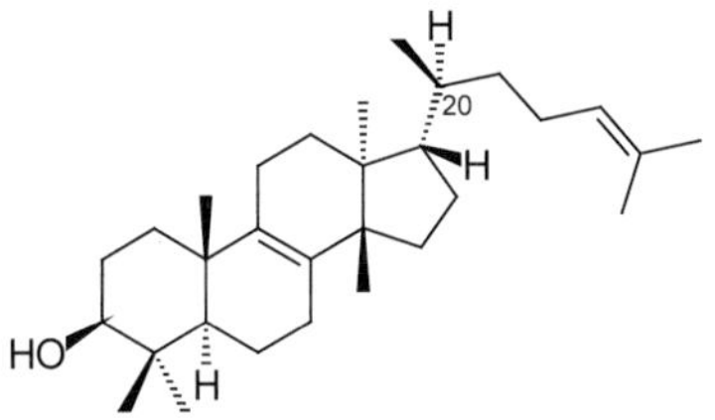

(20S)-(+)-tirucalla-8,24-dien-3β-ol (tirucallol)

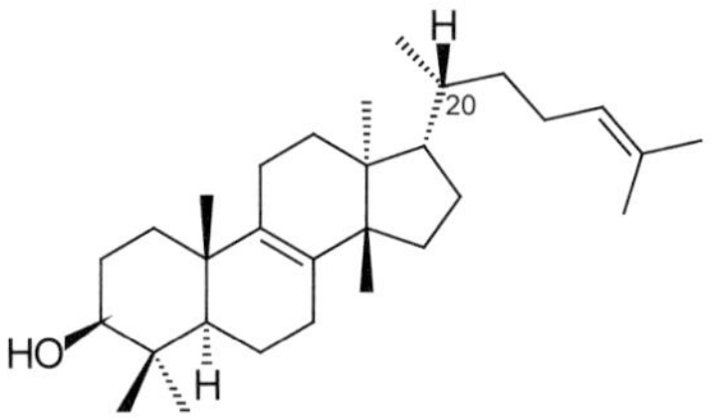

(20R)-(+)-eupha-8,24-dien-3β-ol (euphol)

(−)-Tirucalla-7,24-dien-3β-ol, a regioisomer of tirucallols, is found among the constituents of the seed of black tea (*Camellia sinensis*, Theaceae). Euphorbol isolated from various Euphorbiaceae represents a homotriterpene with a tirucallane skeleton.

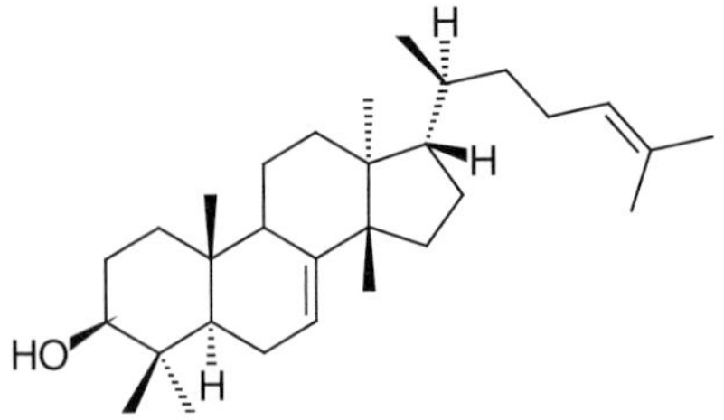

(−)-tirucalla-7,24-dien-3β-ol

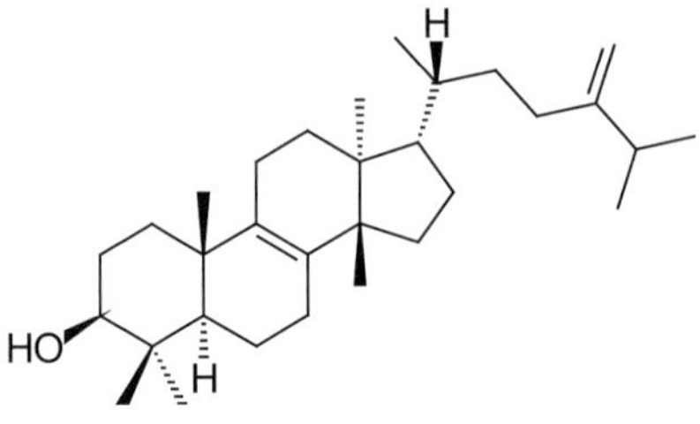

(+)-(24-methyleneupha-8-en-3β-ol (euphorbol)

6.2.6 Lanostanes

More than 200 naturally abundant lanostanes are reported [2]. They are found as constituents of some higher plants and as fungal metabolites. (+)-Lanosta-8,24-dien-3β-ol, commonly known as lanosterol, is a prominent constituent of lanolin, the wool fat of sheep used as ointment base, emulsifier, conditioner and lubricant in cosmetics. Lanosterol is also found in yeast and other fungi, in Euphorbiaceae such as *Euphorbia regis jubae*, and in various other higher plants. Moreover, it is the first isolated precursor of steroid biogenesis from 2,3-epoxysqualene in mammals. (−)-Abieslactone from the bark of fir *Abies mariesii* (Pinaceae) is another natural lanostane derivative.

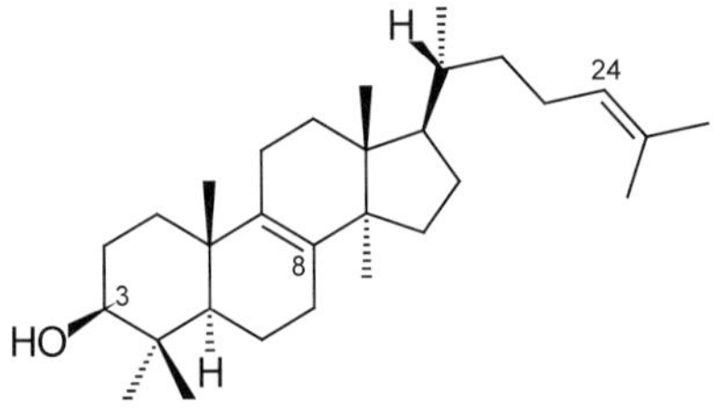

(+)-lanosta-8,24-dien-3β-ol (lanosterol)

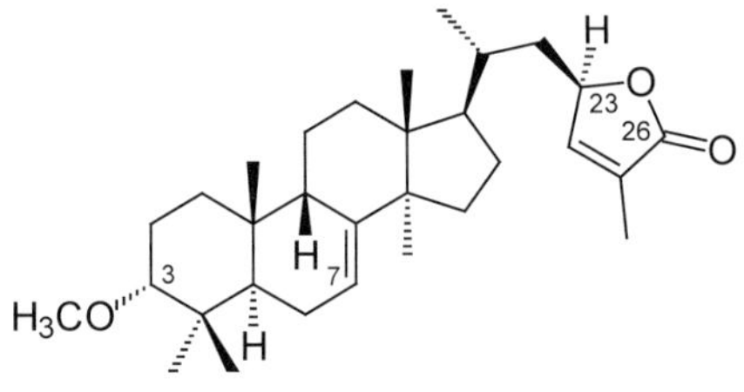

(−)-3α-methoxy-9β-lanosta-7,24-dieno-26,23R-lactone (abieslactone)

6.2 Tetracyclic Triterpenes, Gonane Type

Ganoderic acids, for example S, are found in the above- and under-ground parts of the fungus *Ganoderma lucidum* which is used in traditional Chinese and Japanese medicine for tonic preparations to increase vitality. 27-Norlanostanes, for example (−)-17,23-epoxy-28-hydroxy-27-norlanost-8-ene-3,24-dione, are isolated from the bulbs of some Mediterranean grape hyacinths such as *Muscari comosum* and *Scilla scilloides* (Liliaceae).

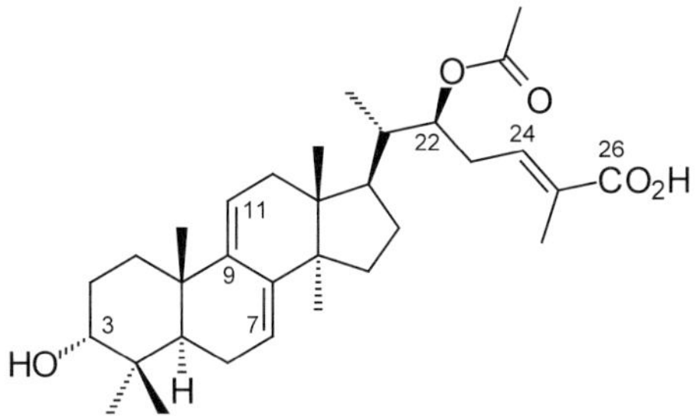

(+)-22-acetoxy-3α-hydroxylanosta-7,9(11),24-trien-26-oic acid
(ganoderic acid S)

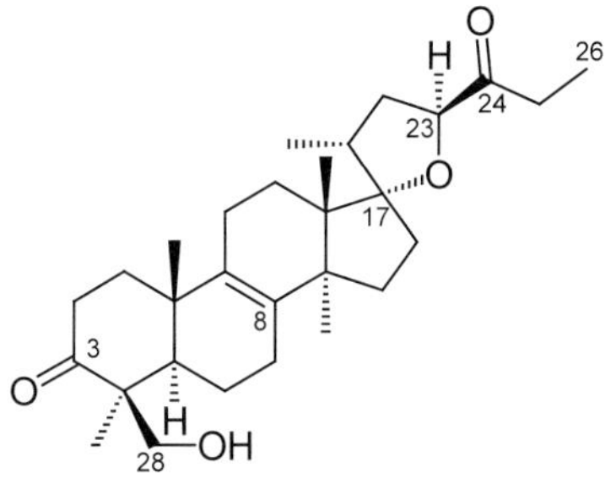

(−)-17,23-epoxy-28-hydroxy-27-norlanost-8-en-3,24-dione

6.2.7 Cycloartanes

(+)-Cycloartenol, also known as cyclobranol, from the fruits of *Strychnos nux vomica* (Loganiaceae), in the leaves of potato *Solanum tuberosum* (Solanaceae) and the seed of rice *Oryza sativa* (Poaceae), is one typical representative of more than 120 naturally abundant cycloartanes [2].

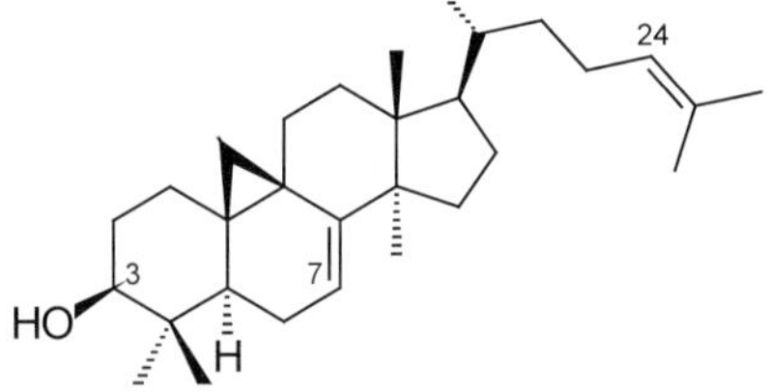

(+)-cycloarta-7,24-dien-3β-ol
(cimicifugenol)

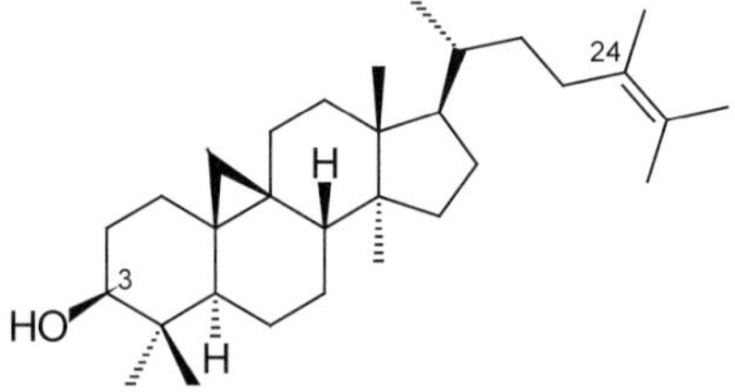

(+)-24-methylcycloarta-24-en-3β-ol
(cyclobranol, cycloartenol)

Other remarkable examples include (+)-cimicifugenol from the bugwort *Cimicifuga acerina*, *C. japonica* and *C. simplex* (Ranunculaceae) [19] applied as an estrogen substitute, (+)-ananas acid from the the wood of pineapple *Ananas comosus* (Thymeliaceae), as well as the β-glucopyranoside passiflorin from passion flowers *Passiflora incarnata* and *P. edulis* (Passifloraceae); these are used as sedatives and analgesics.

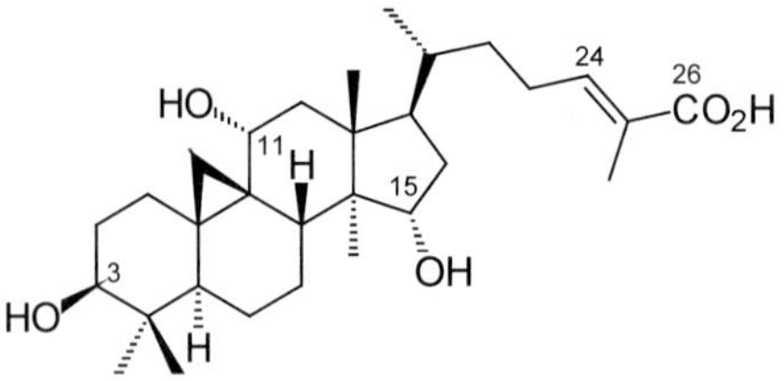

(+)-3β, 11α,15α-trihydroxycycloart-
24E-en-26-oic acid (ananas acid)

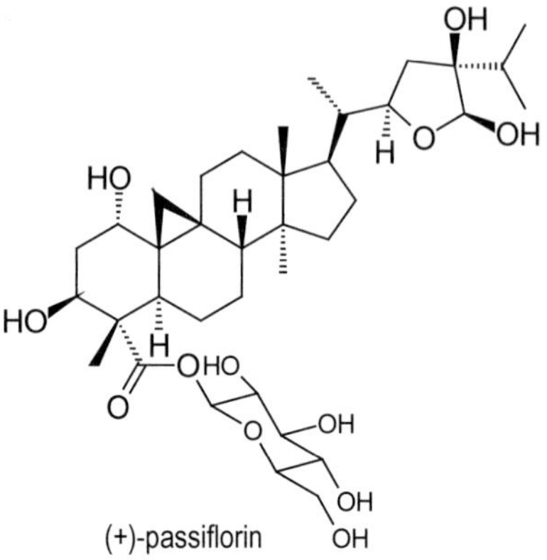

(+)-passiflorin

6.2.8 Cucurbitanes

The name of about 50 naturally abundant cucurbitanes [2] stems from Cucurbitaceae, the Latin term of cucurbitaceous plants such as cucumbers and pumpkins, known since antiquity for their beneficial and toxic properties. One of the most frequently isolated representatives is the bitter substance (+)-cucurbitacin B from *Phormium tenax* and *Ecballium elaterium* (Cucurbitaceae), also found in *Iberis* species (Cruciferae), Euphorbiaceae and Scrophulariaceae. (+)-Cucurbitacin F from *Cucumis angolensis* and *C. dinteri* is reported to inhibit the growth of human tumor cells. Toxic cucurbitacines shape the unpleasant bitter taste of salads prepared from spoiled cucumbers *Cucumis sativus*; some representatives are reported to be antihypertonic, antirheumatic, and also active against HIV.

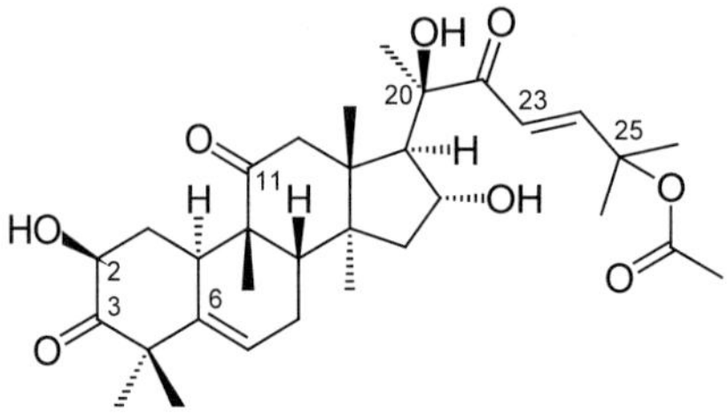

(+)-25-acetoxy-2β,16α,20-trihydroxy-
cucurbita-5,23-diene-3,11,22-trione
(cucurbitacin B)

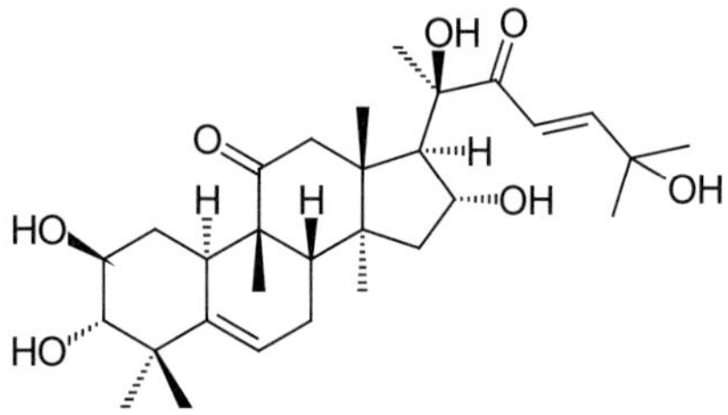

(+)-2β,3α,16α,20,25-pentahydroxy-
cucurbita-5,23-diene-11,22-dione
(cucurbitacin F)

6.3 Pentacyclic Triterpenes, Baccharane Type

6.3.1 Survey

Another cyclization of 2,3-epoxysqualene, yielding the six-membered ring D, followed by WAGNER-MEERWEIN rearrangements, builds up the intermediate cation of the tetracyclic triterpene 3β-hydroxybacchar-21-ene, which finally closes the fifth ring to the pentacyclic 3β-hydroxylupanium ion.

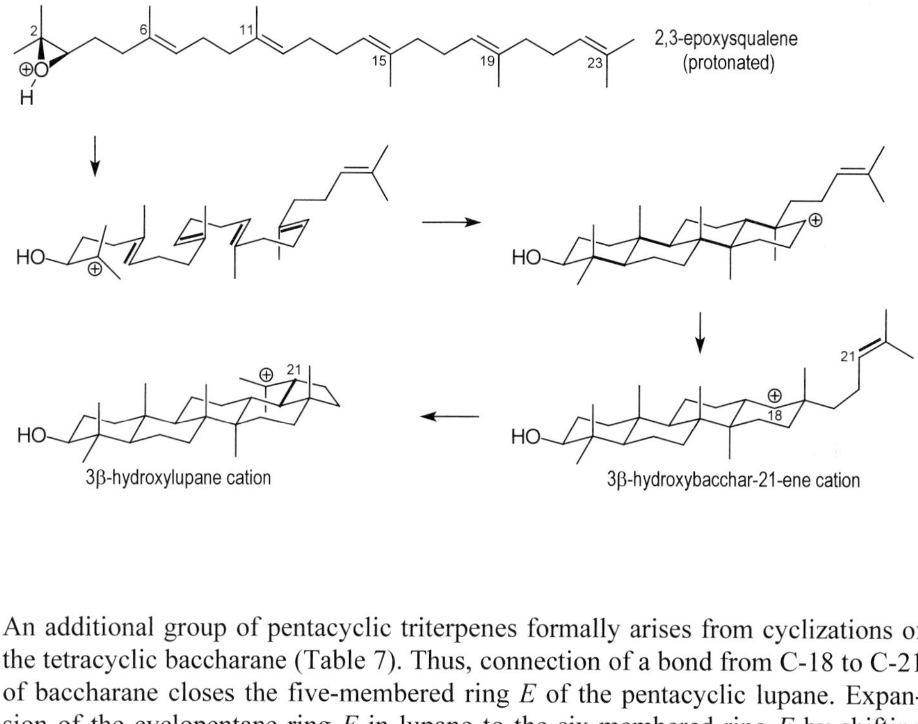

An additional group of pentacyclic triterpenes formally arises from cyclizations of the tetracyclic baccharane (Table 7). Thus, connection of a bond from C-18 to C-21 of baccharane closes the five-membered ring E of the pentacyclic lupane. Expansion of the cyclopentane ring E in lupane to the six-membered ring E by shifting carbon atom C-21 from C-19 to C-20 leads to oleanane. Oleanane formally may undergo methyl shifts to a variety of other pentacyclic triterpenes with cyclohexane ring E. This results in the formation of taraxerane (C-27 from C-14 to C-13), multiflorane (C-26 from C-8 to C-14), glutinane (C-25 from C-10 to C-9), friedelane (C-24 from C-4 to C-5), and pachysanane (C-28 from C-17 to C-16). Ursane and taraxastane, including taraxastene, arise from oleanane when methyl group C-29 migrates from C-20 to C-19; a corresponding methyl shift rearranges multiflorane to bauerane (Table 7).

Table 7. Pentacyclic triterpenes derived from baccharane.

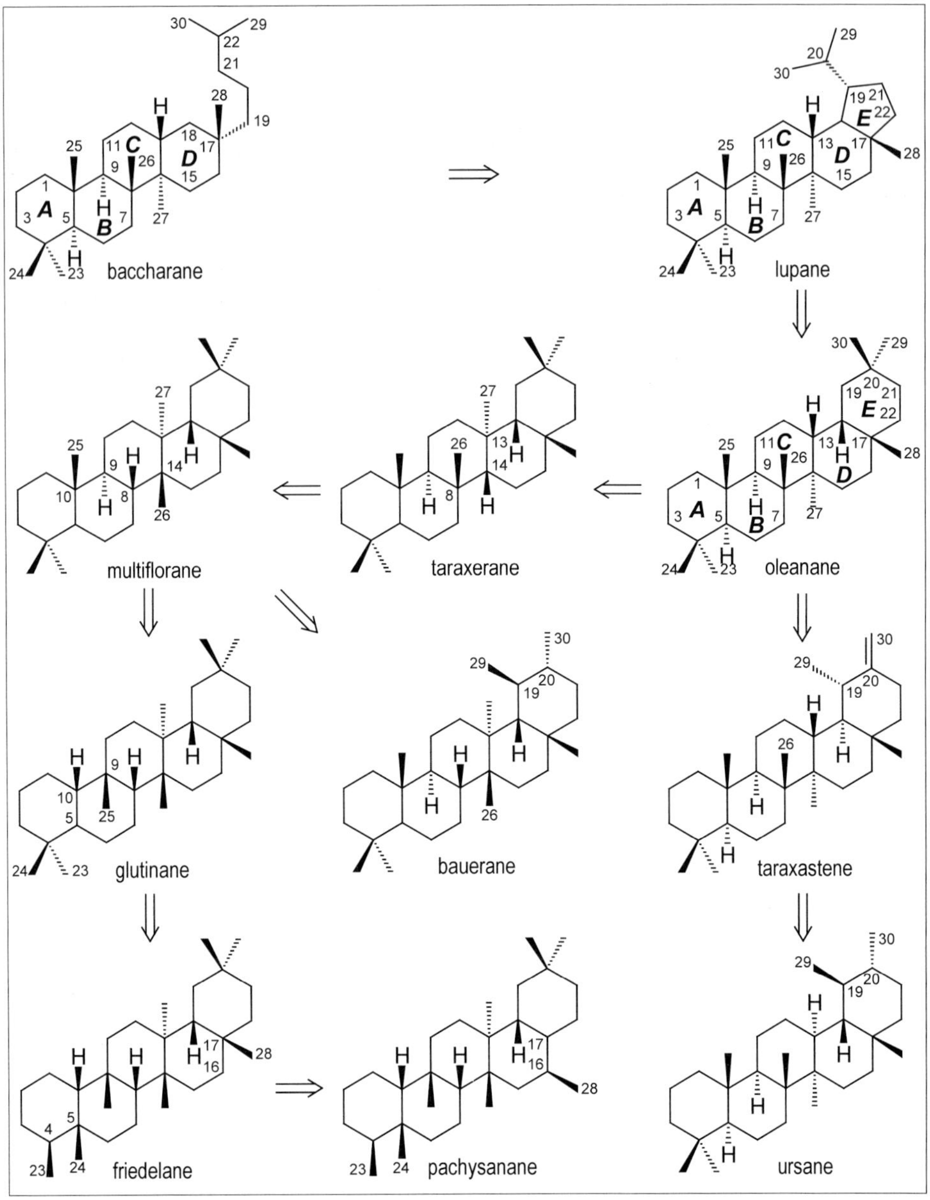

6.3.2 Baccharanes and Lupanes

Baccharane triterpenes such as (+)-12,21-baccharadiene from the fern *Lemmaphyllum microphyllum* var. *obovatum* are rare. More than 100 lupanes [2], however, are found as constituents of higher plants. The first lupane triterpene was isolated from the skin of lupin seeds *Lupinus luteus* (Leguminosae) and is therefore referred to as lupeol; this abundant plant triterpene is also found in the barks of Apocynaceae and Leguminosae, as well as in the latex of fig trees and of rubber plants. It is also detected in the cocoons of the silk worm *Bombyx mori*. Other lupane derivatives include (+)-1,11-dihydroxy-20(29)-lupen-3-one from *Salvia deserta*, (+)-20(29)-lupene-3β,11α-diol from *Nepeta hindostana* (Labiatae) and various species of sage, reported to be antibacterial and to reduce cholesterol levels, (+)-12,20(29)-lupadiene-3β,27,28-triol from oleander *Nerium oleander* (Apocynaceae), therefore referred to as oleandrol, and betulinic acid, reported to be active against HIV and melanoma [19,22], occurring in the leaves of *Szygium claviflorum* (Myrtaceae) and highly concentrated, associated with betulin, in the outer portion of the bark of white birch *Betula alba* (Betulaceae).

(+)-12,21-baccharadiene

(+)-lup-20(29)-en-3β-ol (lupeol)

(+)-12,20(29)-lupadiene-3β,27,28-triol (oleandrol)

(+)-20(29)-lupene-3β,11α-diol

(+)-20(29)-lupene-1β,11α-diol-3-one

R = CO_2H : (+)-3β-hydroxy-20(29)-lupen-28-oic acid (betulinic acid)
R = CH_2OH : (+)-lup-20(29)-ene-3β,28-diol (betulin)

6.3.3 Oleananes

More than 300 oleananes are reported to exist in plants [2]. The parent (+)-oleanane is isolated from petroleum, and its (+)-3β,11α,13β-triol derivative from *Pistazia vera* (Anacardiaceae). Oleananes frequently occur as surface-active glycosides (saponins) in plants, forming foaming aqueous solutions like soaps, and yielding the sugar-free triterpenes as aglyca (referred to as sapogenins) upon hydrolysis. Well-known representatives are (–)-priverogenin B from the cowslip *Primula veris* (Primulaceae), and (+)-soyasapogenol from the soybean, the seed of *Glycine* species (Leguminosae). *Quillaja* saponin, extracted from the bark and the wood of the soap tree *Quillaja saponaria* (Rosaceae) growing in Chile, is a powder containing a mixture of saponins which causes sneezing when dispersed in the air, and foams easily when shaken with water. It is used as a commercial foam producing raw material in the production of shampoos, tooth-paste and films, and as an emulsifier in nutrition and pharmaceutical technology. Hydrolysis liberates (+)-quillajic acid as the sapogenin. (+)-Oleanolic acid occurs in the free state in the leaves of olive trees such as *Olea europaea* (Oleaceae), sugar beet, *Ginseng* roots, and mistletoe (*Viscum album*, Viscaceae); it occurs widely as the aglycone of saponins. 3α-Hydroxy-12-oleanen-24-oic acid, isolated from incense (section 6.3.6) *Boswellia serrata* (Burseraceae) and better known as α-boswellic acid, also incorporates the oleanane skeleton.

6.3.4 Taraxeranes, Multifloranes, Baueranes

(+)-14-Taraxeren-3β-ol, one of the 30 naturally abundant taraxane derivatives and known as taraxerol from lion's tooth *Taraxacum officinale* (Asteraceae), occurs widely among higher plants such as *Skimmia japonica*, *Alnus*-, *Euphorbia*- and

6.3 Pentacyclic Triterpenes, Baccharane Type

Rhododendron species. In providing the name of the 10 multifloranes known to date, *Gelonium multiflorum* contains (−)-7-multifloren-3β-ol (multiflorenol) in the leaves; the methylether is a constituent of the wax of the leaves of sugar cane *Saccharum officinarum* (Poaceae). The term of rarely abundant baueranes stems from *Achronychia baueri* with (−)-7-baueren-3β-ol as a constituent which also occurs in *Ilex* species (Aquifoliaceae), giving rise to the synonym ilexol.

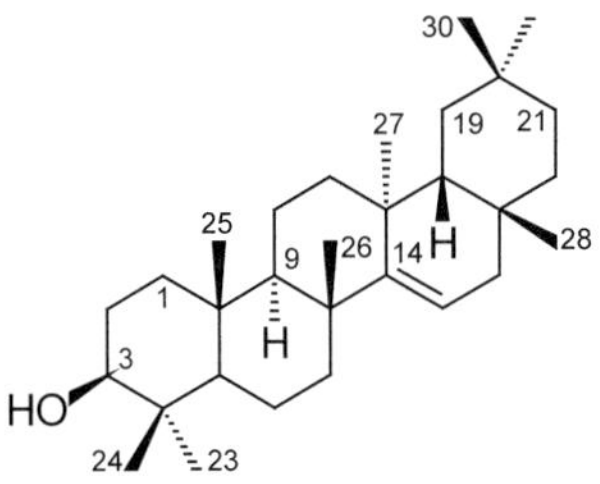

(+)-14-taraxeren-3β-ol
(taraxerol)

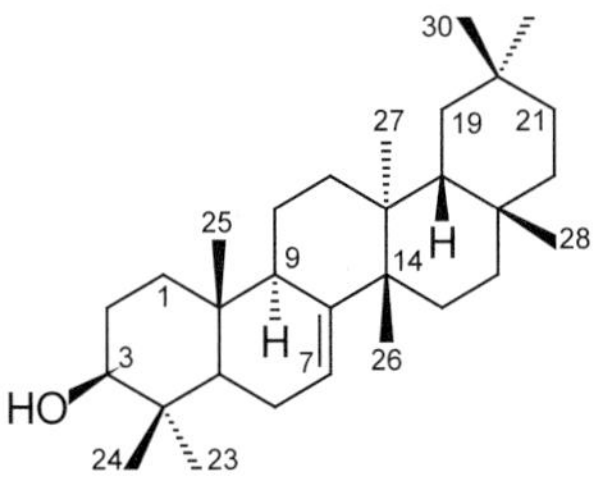

(−)-7-multifloren-3β-ol
(multiflorenol)

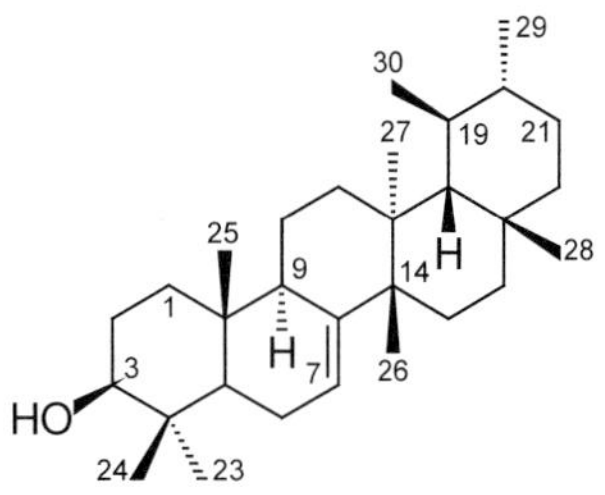

(−)-7-baueren-3β-ol
(ilexol)

6.3.5 Glutinanes, Friedelanes, Pachysananes

(+)-5-Glutinen-3β-ol, also named (+)-alnusenol, is one of the few naturally occurring glutinanes isolated from the black alder *Alnus glutinosa* (Betulaceae). Not more than five naturally abundant pachysananes are reported to exist; one of these, (+)-16,21-pachysanadiene-3β,28-diol is found in *Pachysandra terminalis* (Buxaceae). In contrast, about 50 friedelanes are known as constituents of higher plants; 3α- and 3β-friedelanol including their oxidation product 3-friedelanone, also known as friedelin, which is the major triterpene constituent of cork, are extracted from cork of the cork oak *Quercus suber* and from the bark of other *Quercus*- and *Castanopsis* species (Fagaceae).

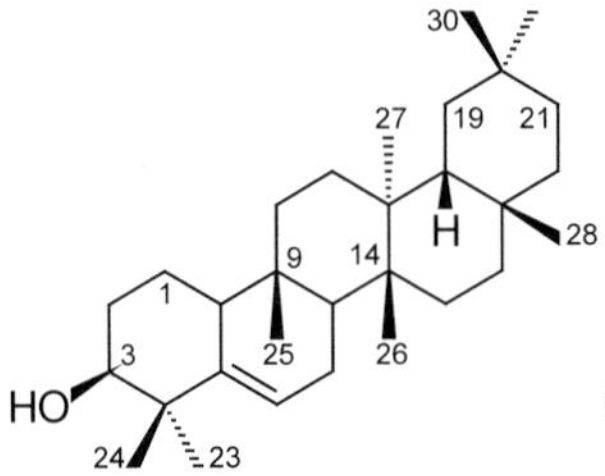

(+)-5-glutinen-3β-ol
(alnusenol)

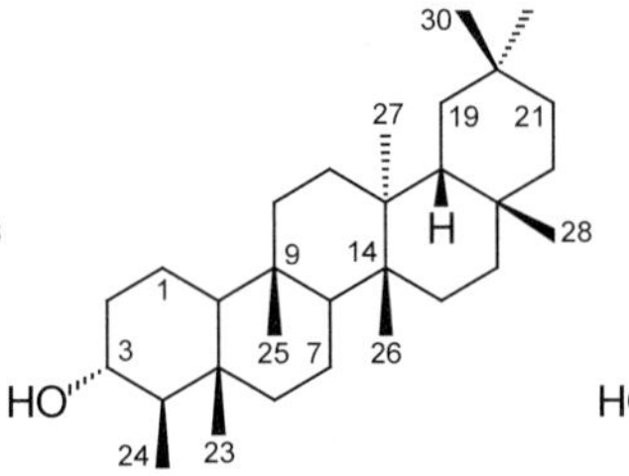

(+)-friedelan-3α-ol
(friedelinol)

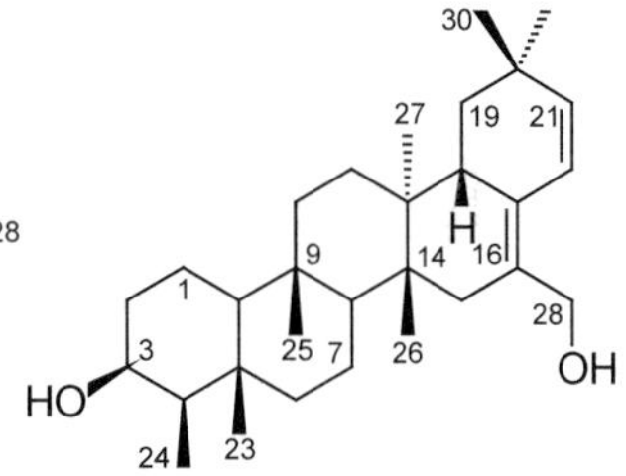

(+)-16,21-pachysanadiene-
3β,28-diol

6.3.6 Taraxastanes and Ursanes

To date, about 25 taraxastanes have been isolated, predominantly from Asteraceae [2]. Representatives include (+)-20-Taraxasten-3β-ol (ψ-taraxasterol) and (+)-20(30)-taraxasten-3β,16β-diol (arnidenediol) from lion's tooth *Taraxacum officinale* and *Arnica montana* as well as 20-taraxasten-3β,16β-diol (faradiol) from *Arnica montana*, *Tussilago farfara*, *Senecio alpinus* and gold-bloom *Calendula officinalis* (all Asteraceae).

(+)-20-taraxasten-3β-ol
(ψ-taraxasterol)

(+)-20-taraxastene-3β,16β-diol
(faradiol)

(+)-20(30)-taraxastene-
3β,16β-diol (arnidenediol)

More than 150 ursanes of plant origin are documented [2]. (+)-3β-Hydroxyursan-28-oic acid, for example, represents a saturated triterpene found in the leaves of oleander *Nerium oleander* (Apocynaceae). (+)-3β-Hydroxy-12-ursen-28-oic acid is the most prominent derivative; this was first isolated from the leaves and berries of bearberry *Arctostaphylos uva-ursi*, and therefore is commonly known as ursolic acid. It is also found in *Rhododendron* species, in cranberries *Vaccinum macrocarpon* (Ericaceae), and in the protective wax coating of apples, pears, prunes, and other fruits. Ursolic acid is reported to have antileukemic and cytotoxic activities; it is also used as an emulsifier in pharmaceuticals and foods, and is similar to (+)-3β,19α-dihydroxy-12-ursen-28-oic acid (known as pomolic acid) extracted from the wax coats of apples.

(+)-3β-hydroxyursan-28-oic acid

(+)-3β-hydroxy-12-ursen-
28-oic acid (ursolic acid)

(+)-3β,19α-dihydroxy-12-ursen-
28-oic acid (pomolic acid)

6.4 Pentacyclic Triterpenes, Hopane Type

Boswellic acids are constituents of incense (olibanum) from *Boswellia serrata* and *B. carterii* (Burseraceae); α-boswellic acid incorporates the 12-oleanene core structure (section 6.3.3), while 12-ursene is the parent hydrocarbon of β-boswellic acid and ketoboswellic acid. Various derivatives of boswellic acids are reported to have antiinflammatory properties and therefore are suggested to be potential cortisone substitutes.

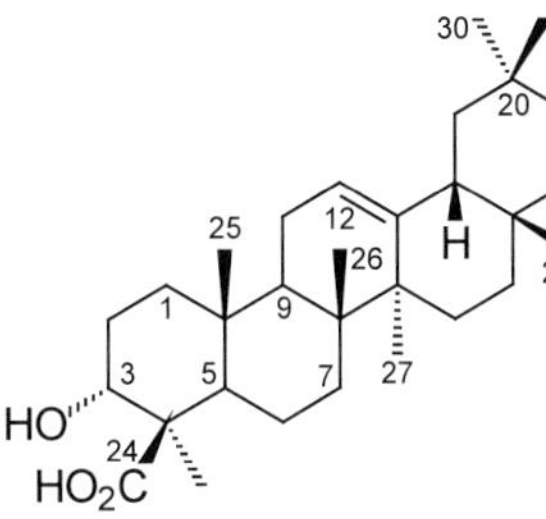

(+)-3α-hydroxy-12-oleanen-24-oic acid
(α-boswellic acid)

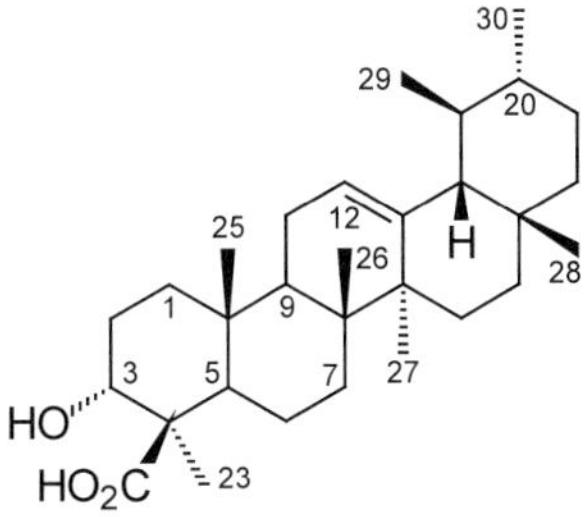

(+)-3α-hydroxy-12-ursen-24-oic acid
(β-boswellic acid)

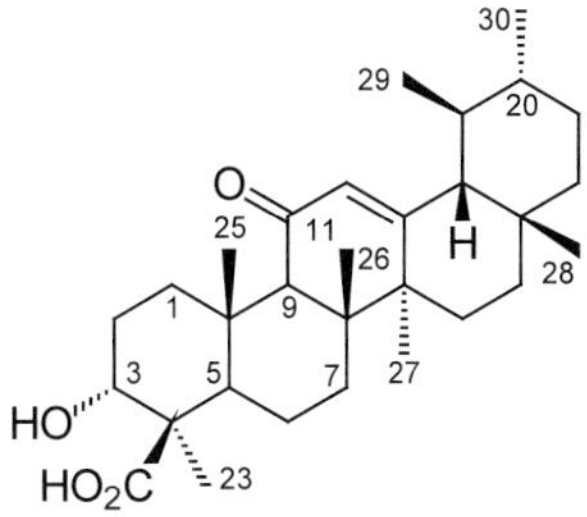

(+)-3α-hydroxy-11-oxo-12-ursen-24-oic acid
(ketoboswellic acid)

6.4 Pentacyclic Triterpenes, Hopane Type

6.4.1 Survey

The pentacyclic triterpene skeleton hopane is generated by 2,7-, 6,11-, 10,15-, 14,19-, and 18,22-cyclization of the carbenium ion in a fivefold chair conformation (as drawn) arising from regioselective protonation of the 2,3-double bond of squalene (but not of the 2,3-epoxide). This explains why hopanes are usually not hydroxylated in the 3-position.

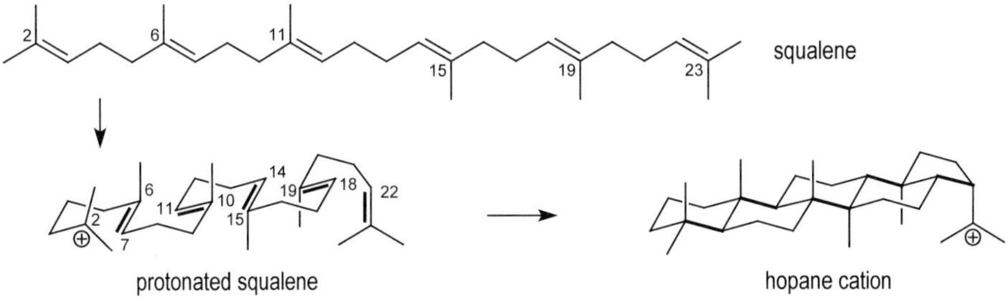

squalene

protonated squalene → hopane cation

Methyl shifts in hopane subsequently lead to neohopane (C-28 methyl from C-18 to C-17), fernane (C-27-methyl from C-14 to C-13; C-26 methyl from C-8 to C-14), adianane (C-25 methyl from C-10 to C-9) and, finally, filicane (C-24 methyl from C-4 to C-5). In another pathway, the cyclopentane ring of hopane expands to a cyclohexane substructure involving a shift of carbon atom C-17 from C-21 to C-22, thus generating gammacerane.

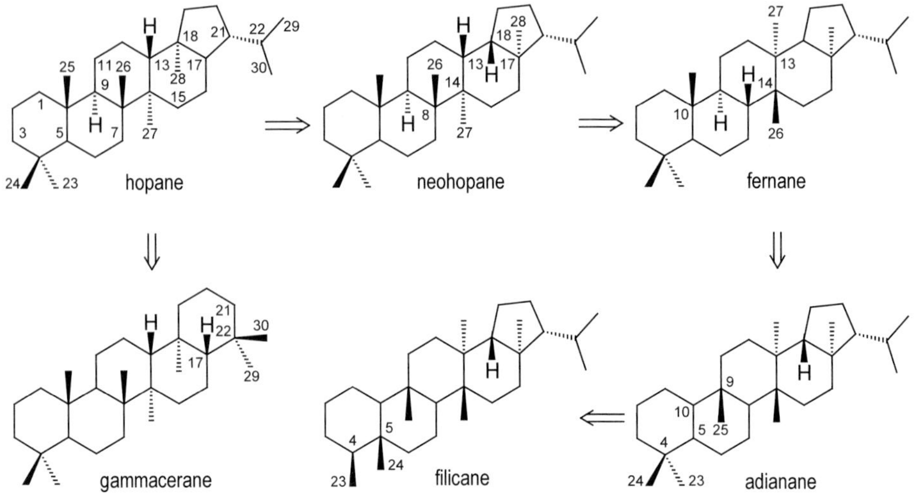

6.4.2 Hopanes and Neohopanes

Stereoisomers of hopane as the parent hydrocarbon of about 100 naturally abundant hopane triterpenes [2] are isolated in small amounts from oil slate and petroleum of various provenance. Hydroxylated hopanes such as (+)-6α,22-hopanediol are found to occur in some lichens and, associated with (+)-22(29)-hopen-6α,21β-diol, in the roots of *Iris missouriensis* (Liliaceae). Neohopanes exemplified by (+)-12-neohopen-3β-ol from *Rhododendron linearifolium* (Ericaceae) rarely occur as natural products.

Bacteriohopane-32,33,34,35-tetrol and other bio-hopanes partially substituting cholesterol in the cell walls of bacteria isolated from culture contain an unbranched polyhydroxylated C_5-C_6 alkyl chain attached to C-30 of the hopane skeleton. Following the death of the bacteria, the hydroxy functions of bio-hopanes are reduced; geo-hopanes which are isolated in large amounts from oil slate and other sediments were probably created *via* this pathway some 500 million years ago.

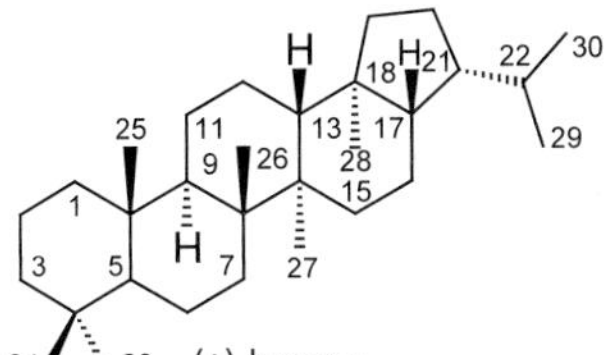

6.4 Pentacyclic Triterpenes, Hopane Type

(+)-hopane

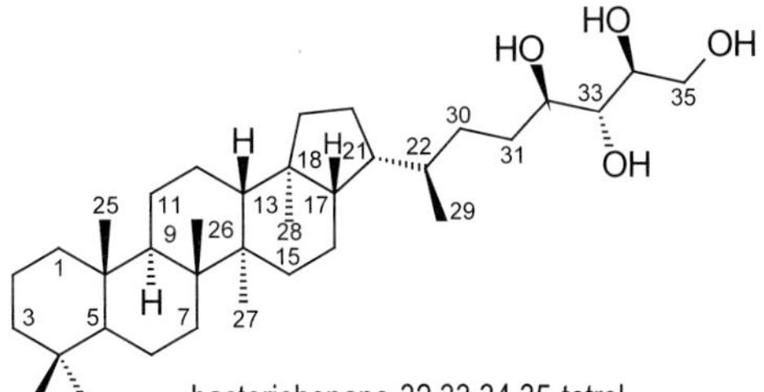

bacteriohopane-32,33,34,35-tetrol

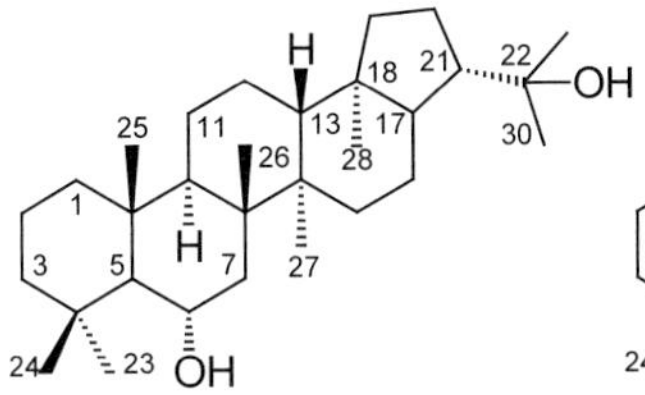

(+)-6α,22-hopanediol

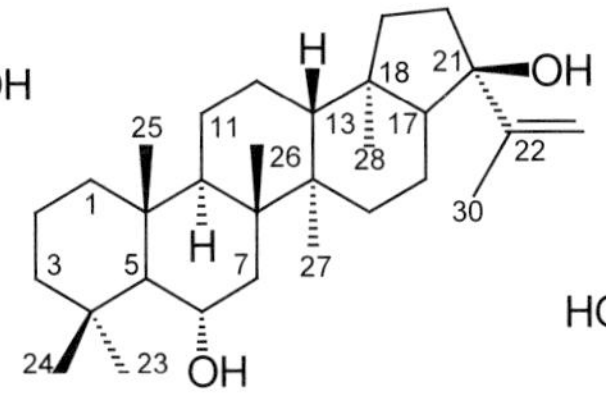

(+)-22(29)-hopene-6α,21β-diol

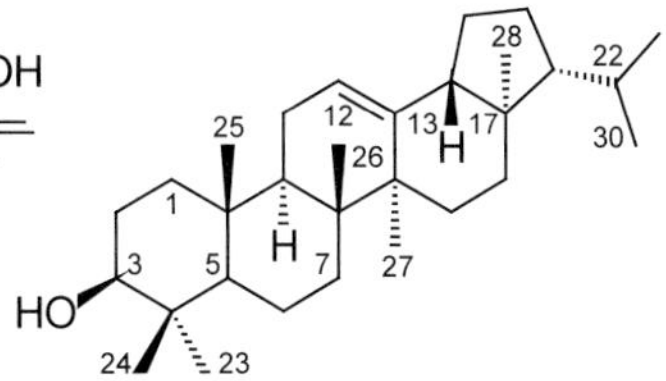

(+)-12-neohopen-3β-ol

6.4.3 Fernanes

Various fernanes are reported as constituents of some ferns, e.g. (+)-7- and (−)-8-fernene as well as (−)-7,9(11)-fernadiene from *Adiantum monochlamys* and *A. pedatum* (Pteridaceae). Hydroxylated derivatives are found in lichens such as retigeric acid A in *Lobaria retigera* (Sticataceae) and some higher plants, e.g. (−)-7-fernen-3β-ol in *Rhododendron linearifolium* (Ericaceae) and (−)-3β,11β-dihydroxy-8-fernen-7-one in *Euphorbia supina* (Euphorbiaceae).

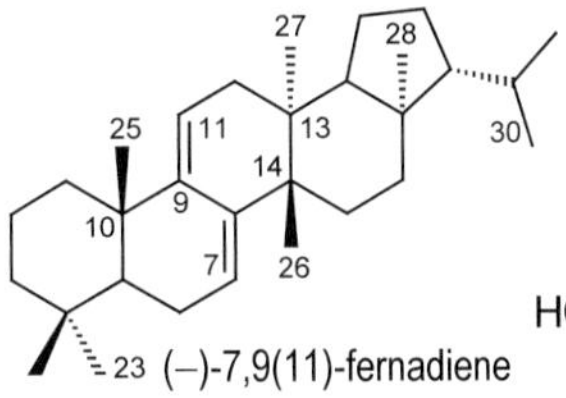

(−)-7,9(11)-fernadiene

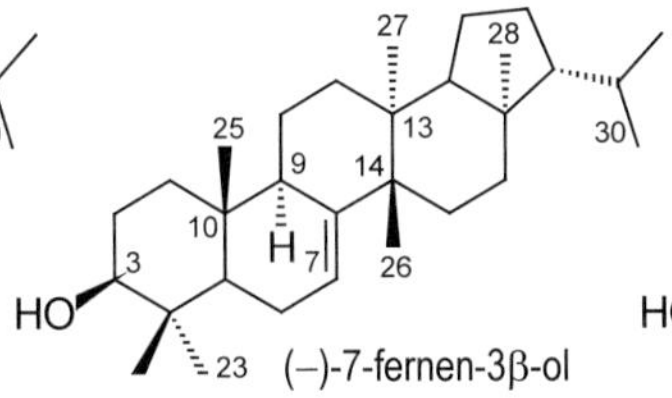

(−)-7-fernen-3β-ol

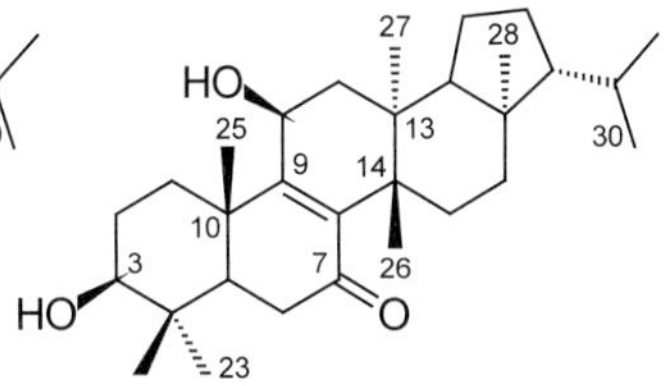

(−)-3β,11β-dihydroxy-8-fernen-7-one
(supinenolone B)

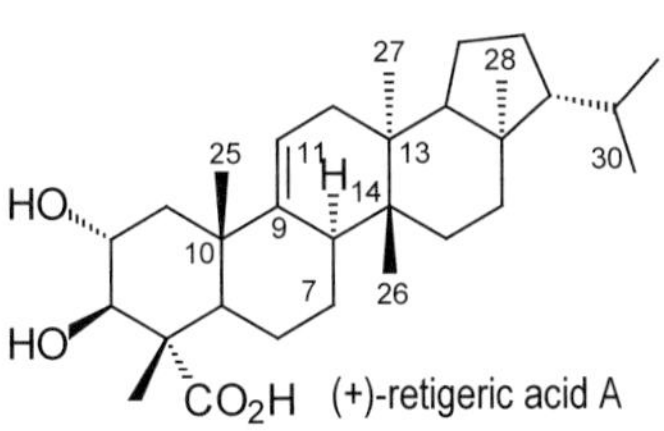

(+)-retigeric acid A

6.4.4 Adiananes and Filicanes

The term adiananes stems from the ferns of genus *Adiantum* (Pteridaceae) from which (−)-5-adianene was isolated. The leaves of *Rhododendron simiarum* (Ericaceae) were found to contain (+)-5-adianen-3β-ol (Simiarenol). Filicanes are also isolated from ferns; 3-filicen-23-al (filicenal) as an example is a constituent of maiden's hair fern *Adiantum pedatum*.

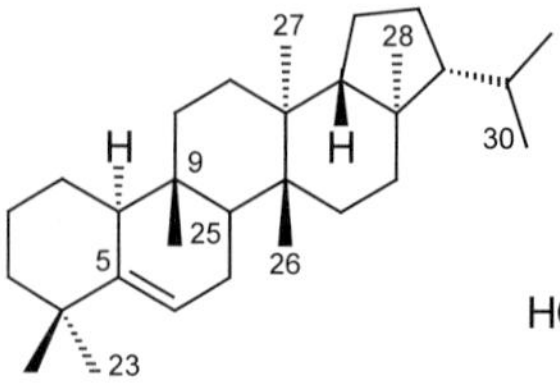

(−)-5-adianene 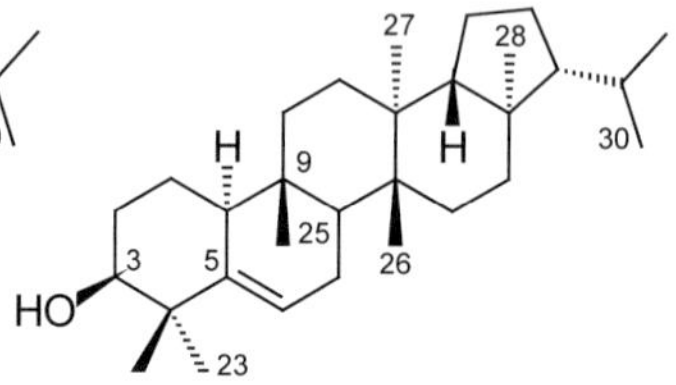   (−)-5-adianen-3β-ol (simiarenol) 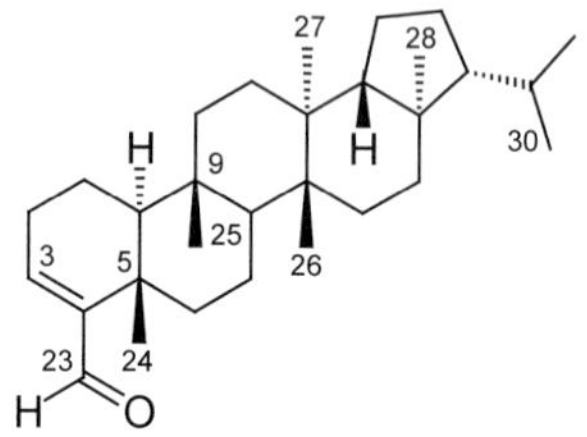   (−)-3-filicen-23-al (filicenal)

6.4.5 Gammaceranes

Small amounts of the dextrorotatory parent hydrocarbon of the ten naturally abundant gammaceranes have been extracted from oil slate. 3-Hydroxy-derivatives such as (+)-16-gammacerene-3β-ol are constituents of the roots of bitter herb *Picris hieracioides* (Asteraceae). 22β-Hydroxy-30-nor-gammaceran-21-one is found in the Japanese fern *Adiantum monochlamys*.

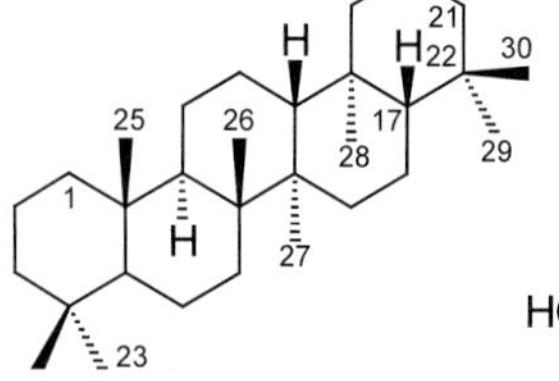

(+)-gammacerane 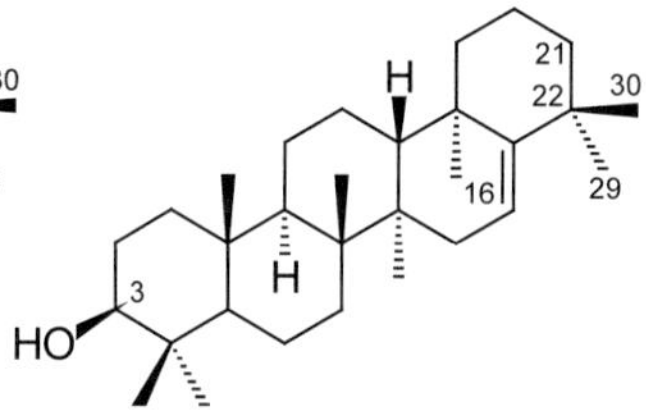   (+)-16-gammaceren-3β-ol 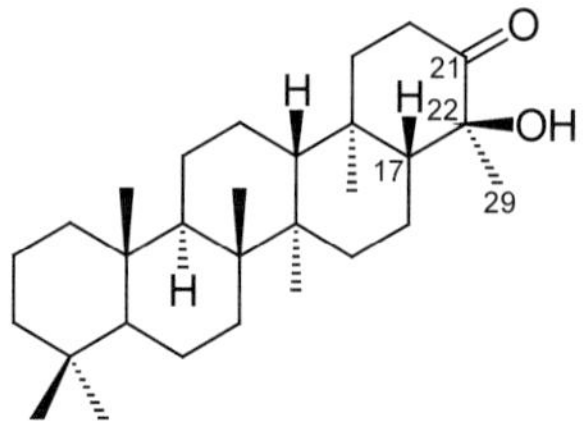   (+)-22β-hydroxy-30-nor-gammeran-21-one (ketohakonanol)

6.5 Other Pentacyclic Triterpenes

6.5.1 Survey

Cyclization of the carbenium ion (assumed to adopt a chair-boat-chair-chair-boat conformation) arising from protonation and ring opening of 2,3-epoxysqualene and subsequent ring expansion involving WAGNER-MEERWEIN rearrangement, leads to the stictane skeleton. Additional rearrangements contracting the terminal cyclohexane ring E convert stictane to arborinane.

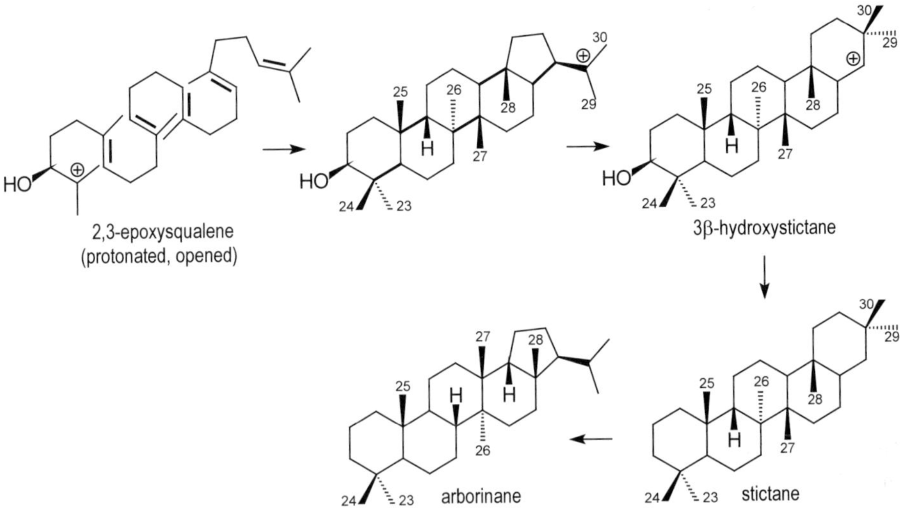

Another type of cyclization of the dication stemming from 2,3-/22,23-diepoxysqualene involving both ends initially generates the onocerane skeleton, finally resulting in the formation of serratane with an additional seven-membered ring.

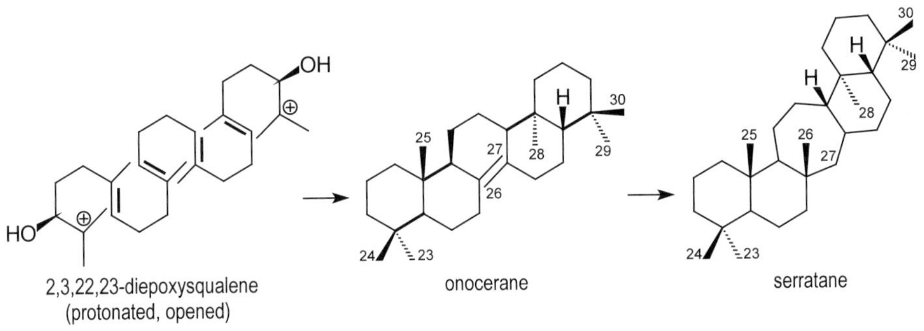

6.5.2 Stictanes and Arborinanes

The few known arborinanes exemplified by (+)-9(11)-arborinen-3α- and (+)-3β-ol (arborinols) are constituents of *Glycosmis arboreae* (Rutaceae). Lung lichens *Sticta pulmonaria* (Stictaceae) growing on the bark of old trees in woods and related species produce stictanes such as (+)-stictane-3β-22α-diol and (+)-stictane-2α,3β,22α-triol with six-membered ring *E*.

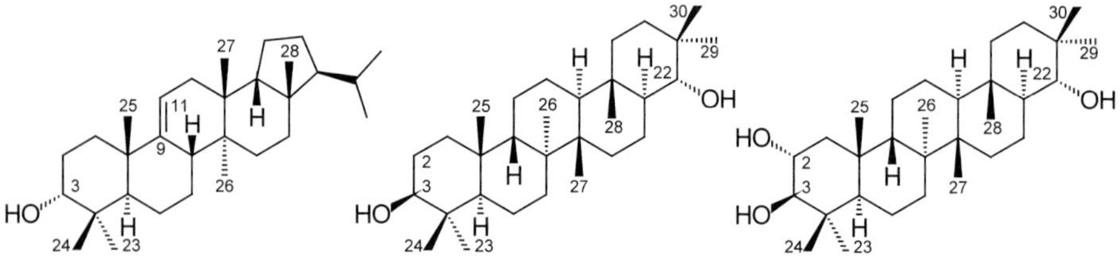

(+)-9(11)-arborinen-3α-ol   (+)-3β,22α-stictanediol   (+)-2α,3β,22α-stictanetriol

6.5.3 Onoceranes and Serratanes

About ten onoceranes named according to their abundance among *Ononis* species are reported [2], represented by (+)-8(26),14(27)-onoceradiene-3β,21α-diol from thorny *Ononis spinosa* (Leguminosae) and related species as well from club-moss spores (vegetable sulfur) *Lycopodium clavatum* (Lycopodiaceae), which also contains lyclavatol as a 26,27-di-nor-triterpene. The diketone 7,14(27)-onoceradiene-3,21-dione is isolated from *Lansium domesticum*.

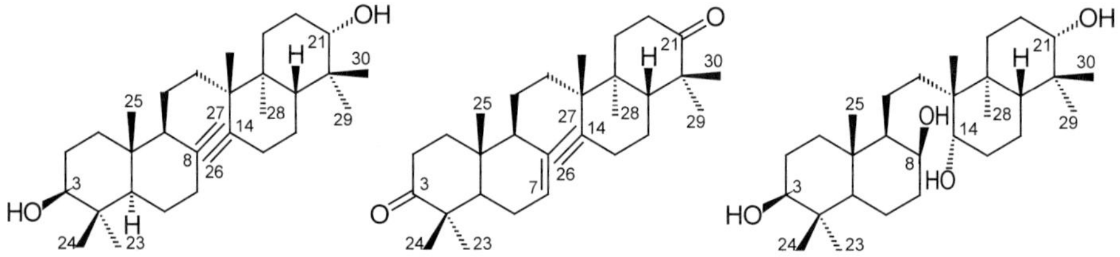

8(26),14(27)-onoceradiene-3β,21α-diol (onocerin)   7,14(27)-onoceradiene-3,21-dione   lyclavatol

(−)-14-Serratene from the rhizomes of European wood fern *Polypodium vulgare*, (+)-3α-methoxy-13-serraten-21β-ol from the bark of spruce *Picea sitchensis* (Pinaceae) and 3α,21β,24-trihydroxy-14-serraten-16-one from *Lycopodium clavatum* (Lycopodiaceae) represent a selection of about 20 serratane derivatives of natural origin.

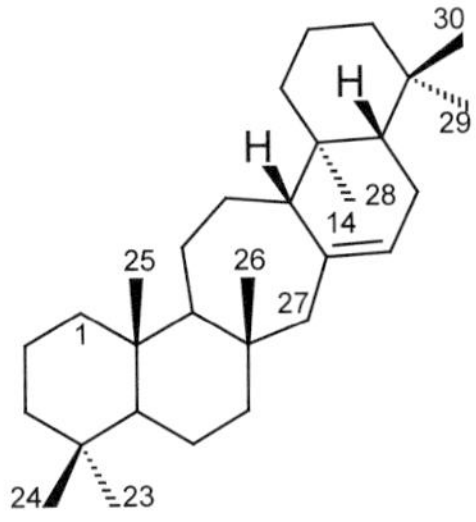

(−)-14-serratene

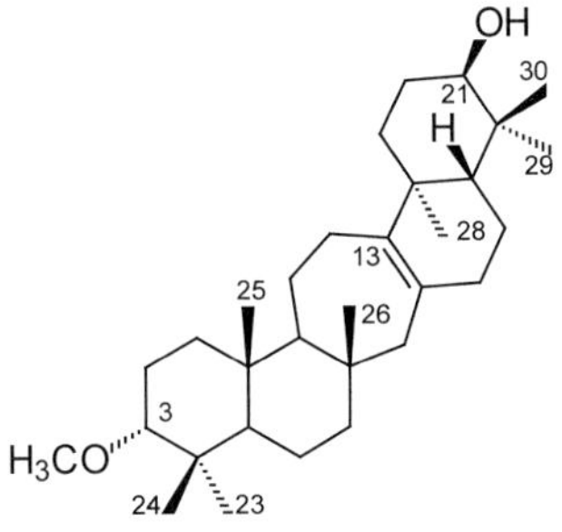

(+)-3α-methoxy-13-serraten-21β-ol

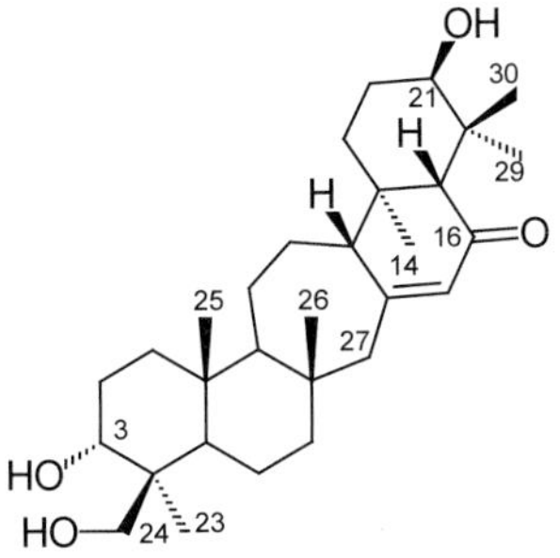

3α,21β,24-trihydroxy-14-serraten-16-one
(16-oxolycoclavanol)

6.6 Iridals

Iridals [2] represent a small group of unusual triterpenoid aldehydes including their degradation products, and homotriterpenes originating biogenetically from squalene otherwise difficult to classify. They occur predominantly in various *Iris* species (Liliaceae) exemplified by (+)-iridal and (+)-α-irigermanal from the Central European *Iris germanica*. The tricyclic, odorless triterpenoid alcohol (−)-ambrein is another prominent representative incorporating the iridal skeleton. It is the essential component of ambergris, the concretion from the intestinal tract of sperm whale *Physeter macrocephalus* and *P. catodon* (Physeterideae) found in tropical seas and seashores and chiefly used in perfumery as a tincture and essence for fixing delicate odors.

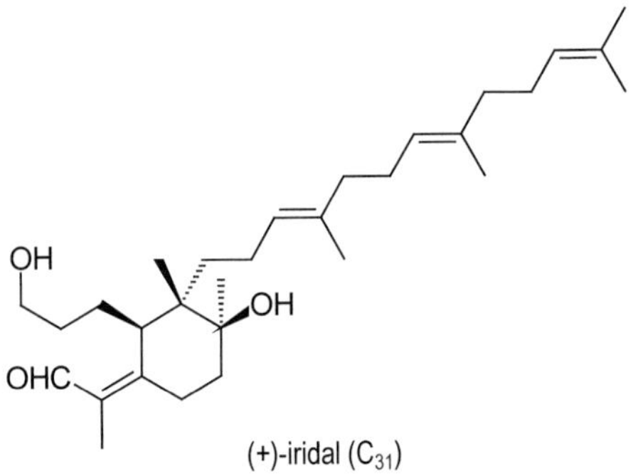

(+)-iridal (C_{31})

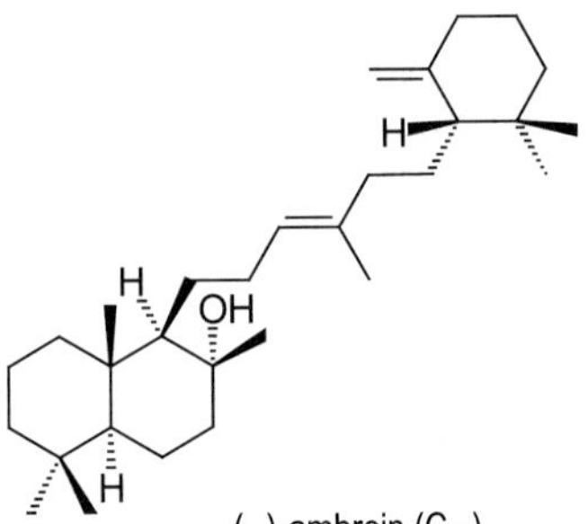

(−)-ambrein (C_{30})

Irones are fragrances in the oil of the dried rhizomes of various *Iris* species cultivated in Italy and Morocco (*Iris germanica, I. florentina*, *I. pallida*, Liliaceae) which is misleadingly referred to as oil of violet because of its pleasant violet-like odor. In fact, ionones (C_{13}, section 7.4) which belong to the class of megastigmanes and not irones (C_{14}) are the shaping fragrances of violets (*Viola odorata*, Violaceae).

Biogenetically, irones prove to be degradation products of iridals; their structural relation to the homotriterpenes (+)-α-irigermanal and (–)-iripallidal from *Iris* species (*Iris germanica* and *I. pallida*) indigenous to the center and south of Europe and northern parts of Africa is unmistakeable: Oxidative degradation of (–)-iripallidals yields *cis*-α-irone; correspondingly, (+)-irigermanal is degraded to dihydroirones.

(–)-*Trans*-α-irone emits a particularly pure and pleasant *Iris* odor, while (+)-β-irone is the most intensely smelling and shaping constituent [18] of the oil of *Iris* used in the production of fine perfumes.

(+)-*cis*-α-irone (–)-*trans*-α-irone (+)-β-irone (+)-*cis*-γ-irone

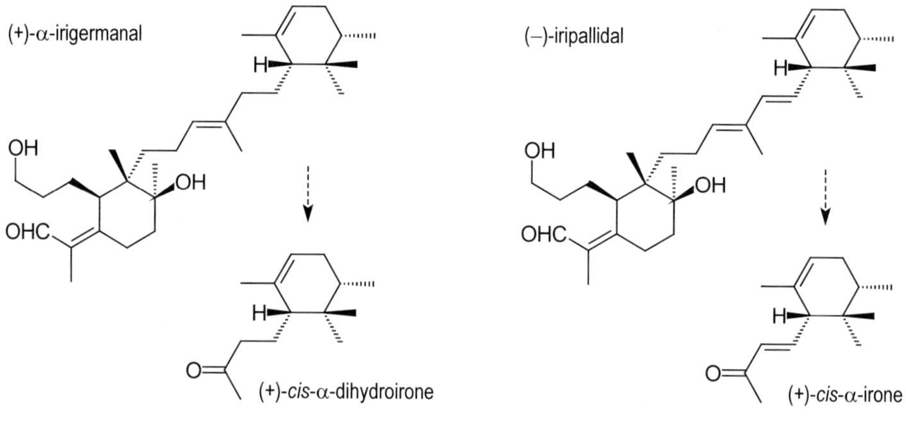

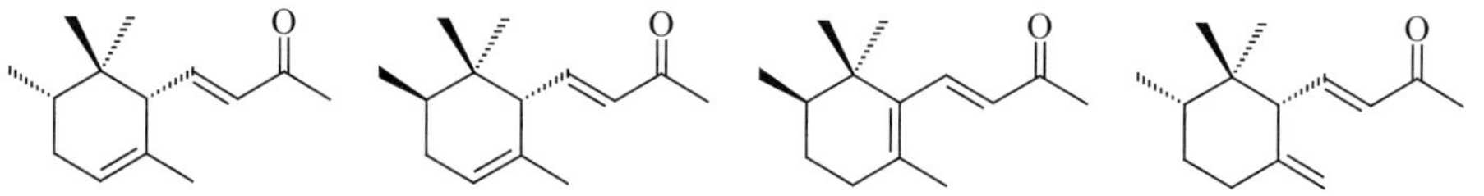

7 Tetraterpenes

7.1 Carotenoids

About 200 naturally abundant tetraterpenes are known to date and referred to as carotenoids [2], because all of them represent structural variants or degradation derivatives of β-carotene from the carrot *Daucus carota* (Umbelliferae) with 11 to 12 conjugated CC double bonds. The generally accepted parent name is "carotene"; two Greek letters (β, γ, ε, φ, κ, χ and ψ) define all seven of the known end groups.

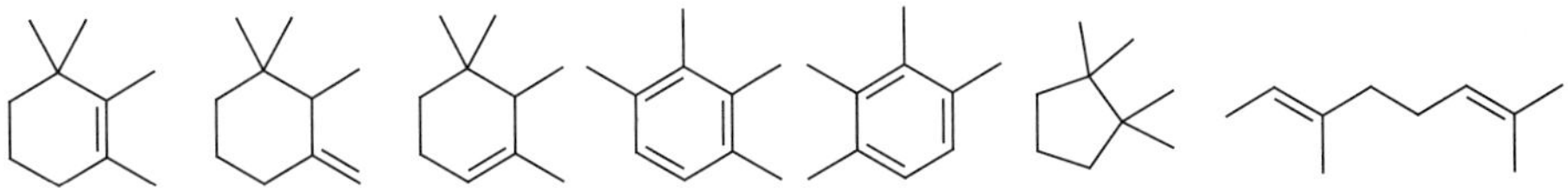

β-   γ-   ε-   φ-   χ-   κ-   ψ-

In keeping with this convention, the acyclic red tetraterpene lycopene occurring in tomatoes (*Lycopersicon esculantum*, Solanaceae), other fruits such as rose-hips, fungi and bacteriae, is systematically named ψ,ψ-carotene, and β,β-carotene is the correct synonym of orange-red β-carotene from carrots. Red and light-sensitive γ-carotene, a minor constituent of carrots and rare in other plants, most efficiently obtained from *Penicillium sclerotiorum*, is more precisely referred to as β,ψ-carotene, reflecting two different end groups in the name; the orange-yellow 7′,8′-dihydro-derivative shapes the color of corn (maize, *Zea mays*, Poaceae).

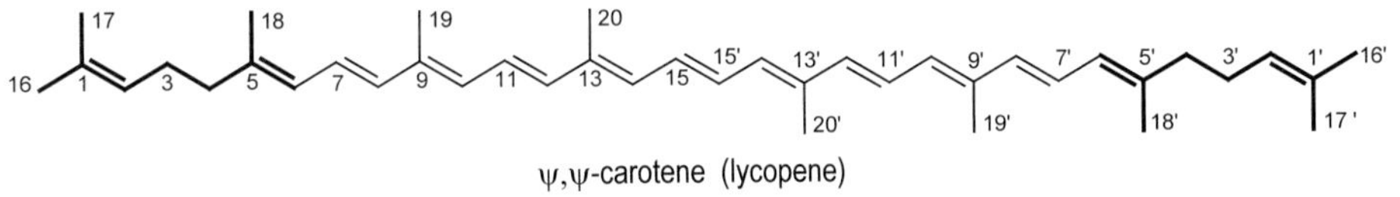

ψ,ψ-carotene (lycopene)

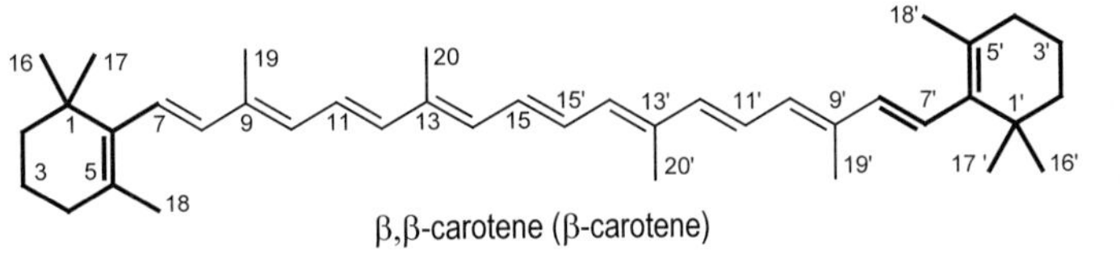

β,β-carotene (β-carotene)

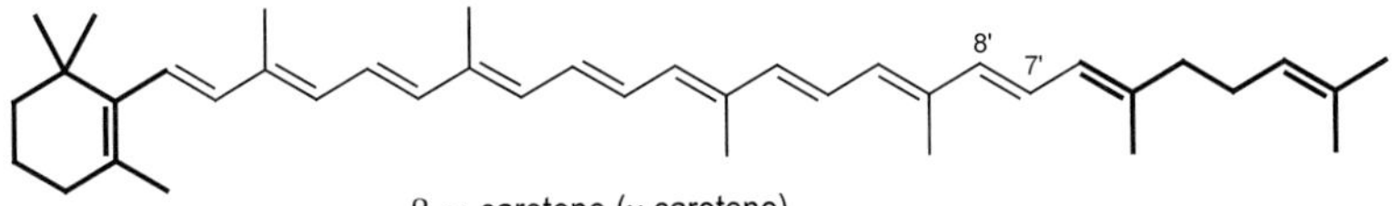

β,ψ-carotene (γ-carotene)

Table 8. Structure and occurence of selected carotenoids.

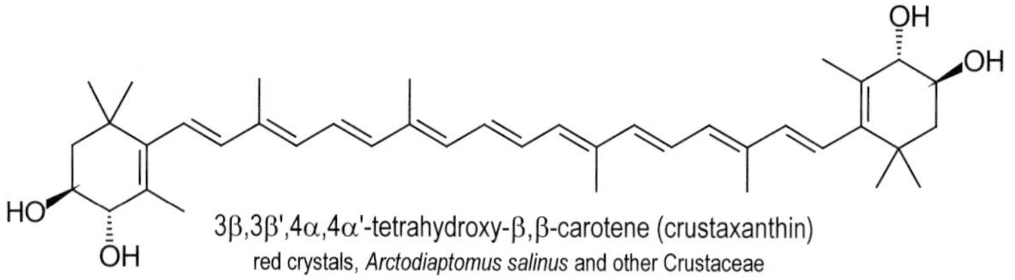

3β,3β',4α,4α'-tetrahydroxy-β,β-carotene (crustaxanthin)
red crystals, *Arctodiaptomus salinus* and other Crustaceae

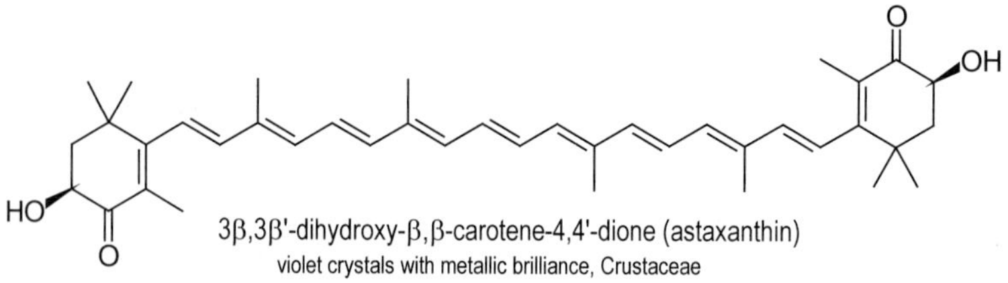

3β,3β'-dihydroxy-β,β-carotene-4,4'-dione (astaxanthin)
violet crystals with metallic brilliance, Crustaceae

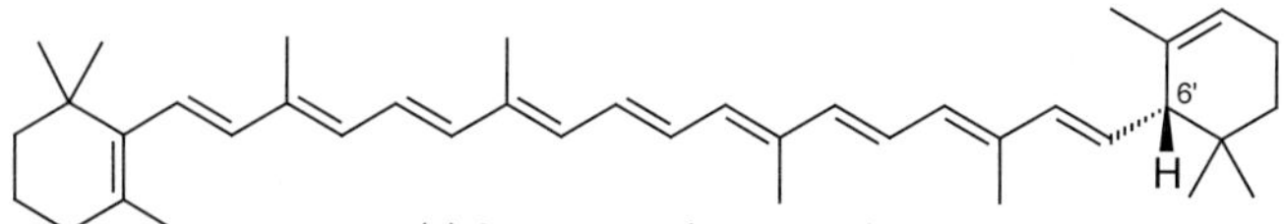

(+)-β,ε-carotene (α-carotene)
violet crystals, widely spread, vitamin A active

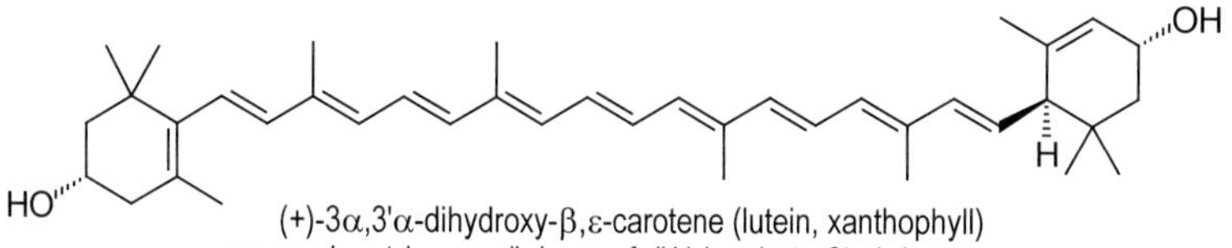

(+)-3α,3'α-dihydroxy-β,ε-carotene (lutein, xanthophyll)
copper red crystals, egg yolk, leaves of all higher plants, *Staphylococcus aureus*

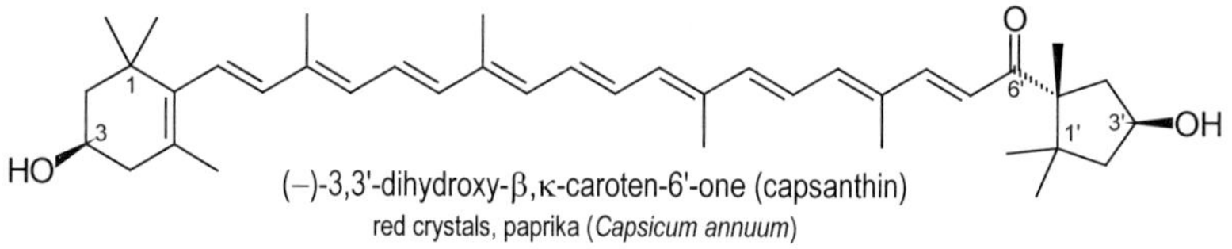

(−)-3,3'-dihydroxy-β,κ-caroten-6'-one (capsanthin)
red crystals, paprika (*Capsicum annuum*)

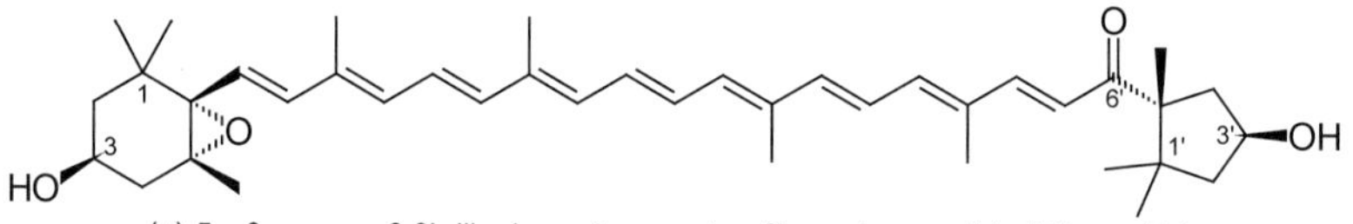

(+)-5α,6α-epoxy-3,3'-dihydroxy-β,κ-caroten-6'-one (capsanthin-5,6-epoxide)
red crystals, paprika (*Capsicum annuum*)

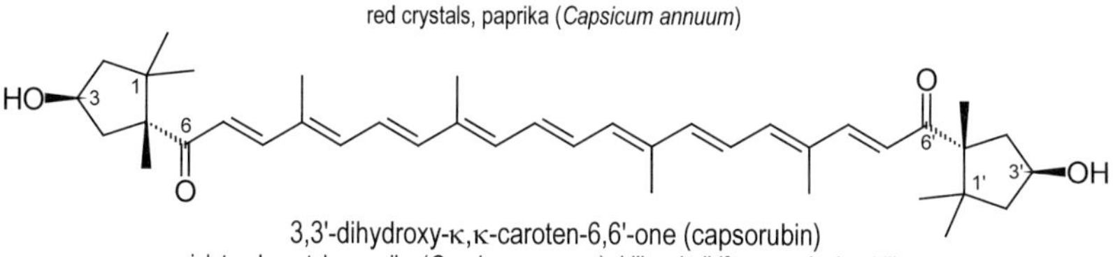

3,3'-dihydroxy-κ,κ-caroten-6,6'-one (capsorubin)
violet red crystals, paprika (*Capsicum annuum*), *Lilium bulbiferum* and other Liliaceae

7.2 Apocarotenoids

Carotenoids occur in the leaves, shoots and roots of all higher plants (content up to 0.1% of dried plant materials). They serve as color filters for photosynthesis in the leaves of plants, giving rise to the yellow and red color of the leaves during fall because they are more slowly degraded than the green chlorophyll. Many fruits such as paprika (*Capsicum annuum*, Solanaceae; Table 8) contain various carotenoids. As colors of flowers, carotenoids play a minor role when compared with anthocyanidines and flavonoids; nevertheless, they contribute to yellow and red shades in the blossoms and fruits of Rosaceae and Liliaceae.

The animal organism metabolizes carotenoids received with food, as it is unable to synthesize these compounds *de novo*. Ultimately, carotenoids and their metabolites are found as chromoproteins in blood plasma, egg yolk, in the feathers of some birds such as flamingos, in the skin of trout and in the meat of some fishes such as salmon and salmon trout, as well as in the shells of Crustaceae. Thus, the lobster changes color from greenish-brown to red upon cooking because the dark green chromoprotein in the shell is denatured in boiling water, thereby liberating the red β,β-carotenoid astaxanthin (Table 8). β,β-Carotene and some other carotenoids are vitamin A active as they are degraded to vitamin A aldehyde in the human and mammal organism; *trans-cis* isomerization of vitamin A aldehyde bound to the protein opsin in rhodopsin in the retina of the eyes is the key step of the visual process (section 4.2). As non-toxic natural compounds giving no cause for concern, β,β-carotene and some other carotenoids are used as coloring agents for foods and cosmetics, and/or as vitamin A precursors and antioxidants in medicines.

7.2 Apocarotenoids

Terpenoids formally arising from carotenoids by separation of terminal fragments are referred to as apocarotenoids [2]. The position of separation is indicated according to the numbering system of carotenoids (section 7.1). In keeping with this, β-carotenal isolated from orange peel and egg yolk is systematically referred to as 8′-apo-β-caroten-8′-al.

β,β-carotene

8′-apo-β-caroten-8′-al
(β-carotenal)

Correspondingly, neurosporaxanthin from the microorganisms *Neurospora crassa* and *N. sitophila* is denoted as 4´-apo-β,ψ-caroten-4´-oic acid.

β,ψ-carotene

4'-apo-β,ψ-caroten-4'-oic aci (neurosporaxanthin)

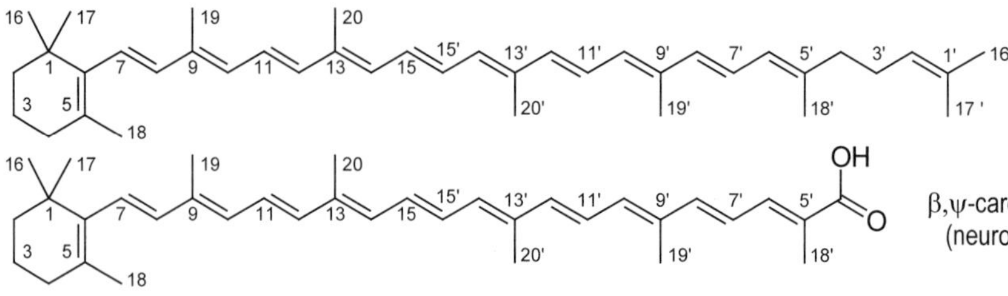

Other apocarotenoids such as the *E*- and *Z*-isomers of sinensiaxanthin and sinensiachrom occur in the flesh of various fruits. Persicachrom and its (3*S*,5*R*,8*S*)-diastereomer represent 12´-apocarotenoids from the flesh of peach *Prunus persica* (Rosaceae).

(3*S*,5*R*,8*R*,9*E*)-5,6-epoxy-10'-apo-β-carotene-3,10'-diol (sinensiaxanthin)

(3*S*,5*R*,8*R*,9*E*)-5,8-epoxy-5,6,7,8-tetrahydro-10'-apo-β-carotene-3,10'-diol (sinensiachrom)

(3*S*,5*R*,8*R*)-5,8-epoxy-5,8-dihydro-12'-apo-β-carotene-3,12'-diol (persicachrom)

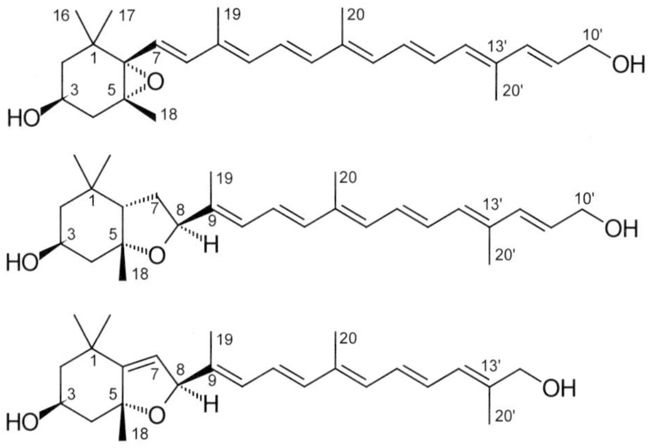

7.3 Diapocarotenoids

Well-known 8,8´-diapocarotenoids [2] include the orange-yellow crocetin in the blossoms of *Gardenia* species (Rubiaceae) and *Mimosa pudica* (Mimosaceae), the dimethylester γ-crocetin and the di-gentiobiose ester (+)-α-crocin in various *Crocus* species (Iridaceae). (+)-α-Crocin is one of the yellow-red constituents of saffron, the dried grains and pencils pulled from the blossoms of alpine *Crocus sativus*, used for coloring and flavoring of fine foods. Rosafluin, systematically named 10,10´-diapocaroten-10,10´-diol, occurs in yellow rose flowers.

8,8'-diapocarotene-8,8'-dioic acid (crocetin)

8,8'-diapocarotene-8,8'-dioic acid dimethyl ester (γ-crocetin)

8,8'-diapocarotene-8,8'-dioic acid di-gentiobiose ester [(+)-α-crocin]

10,10'-diapocarotene-10,10'-diol (rosafluin)

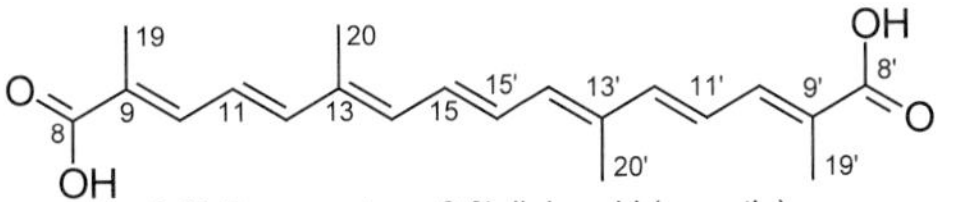

7.4 Megastigmanes

More than 150 C_{13}-isoprenoids, in which 2-butyl-1,1,3-trimethylcyclohexane as a partial structure of abscisic acid (section 3.2.1) and of β-carotene (section 7.1) forms the basic skeleton, are referred to as megastigmanes. Megastigmanes such as β-ionone belong to the most important pleasantly smelling degradation products of β-carotene in the flowers of many plants. Smaller metabolites of carotenoids, including 2,6,6-trimethyl-2-cyclohexenone, 2,4,4-trimethylcyclohexene-3-carbaldehyde and 5,5,9-trimethyl-1-oxabicyclo[4.3.0]-3-nonen-2-one, may also contribute to the fragrances of flowers.

β,β-carotene

2,6,6-trimethyl-2-cyclohexenone; 2,4,4-trimethyl-cyclohexene-3-carbaldehyde; 5,5,9-trimethyl-1-oxabicyclo[4.3.0]-3-nonen-2-one; β-ionone

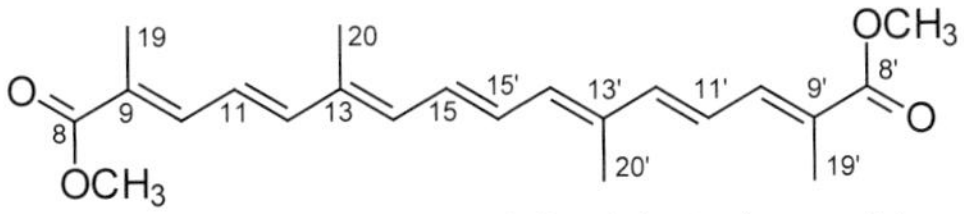

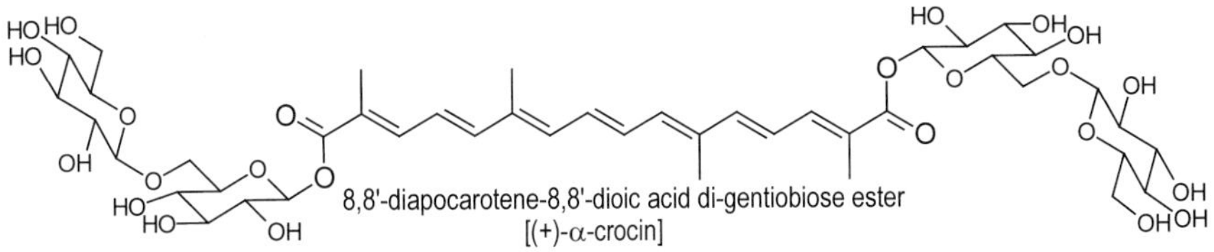

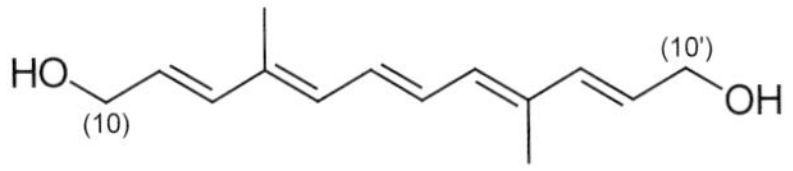

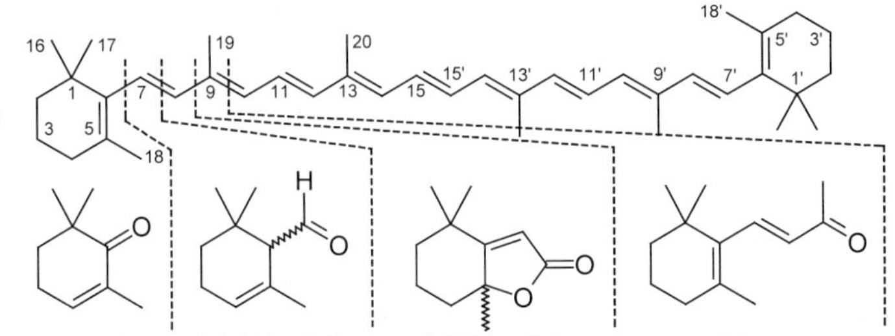

Ionones isolated from the ethereal oils of many flowers [18] are used in perfumery because of their intense and pleasant odor; however, they may cause allergic reactions. (R)-α- and β-ionone from violets *Viola odorata* (Violaceae), freesiae *Freesia refracta* and Australian boroniae *Boronia megastigma* (Rutaceae) as well as (S)-γ-ionone from *Tamarindus indica* (Leguminosae) are typical megastigmanes. (−)-5,6-Epoxy-7-megastigmen-9-one occurs in carrots (Umbelliferae), tomatoes, and tobacco (Solanaceae). Damascenone and damascones are fragrance-shaping constituents of the Bulgarian oil of rose from *Rosa damascena* (Rosaceae). 3β-Hydroxy-damascone is found in tobacco *Nicotiana tabacum* (Solanaceae). Edulanes shape the fragrance of the passion flower *Passiflora edulis* (Passifloraceae). Stereoisomeric theaspiranes and theaspirone essentially contribute to the flavor of black tea from *Camellia sinensis* (Theaceae) and the flowers of the oil tree *Osmanthus fragrans* (Oleaceae) indigenous to eastern Asia and used to perfume tea in China.

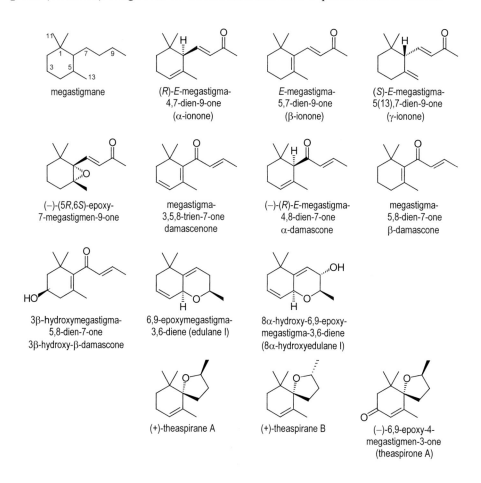

8 Polyterpenes and Prenylquinones

8.1 Polyterpenes

Isoprenoids with more than eight isoprene units are classified as polyterpenes [2]. Natural rubber (caoutchouc), formerly an important raw material for the rubber industry, is primarily obtained by coagulating the milk juice (latex) of *Hevea brasiliensis* (Euphorbiaceae) growing in the Amazonian area of Brazil and southeastern Asia. It consists essentially of *cis*-polyisoprene. The milky juice is an emulsion of this polyterpene in water stabilized by proteins as protecting colloids.

In former times, rubber was produced by vulcanization, the heating of sticky natural caoutchouc with up to 3 % of sulfur in the presence of catalysts. This process crosslinks the chains of *cis*-polyisoprene by addition of sulfur to the double bonds, resulting in the formation of disulfide bridges, thus giving rise to the three-dimensional network of rubber with high elasticity, strength and thermal stability. The large amounts of *cis*-polyisoprene required today for rubber production are made available almost exclusively by polymerization of isoprene synthesized by various large-scale procedures (synthetic caoutchouc).

Trans-polyisoprene is the main constituent of gutta-percha, the purified, coagulated, dried, milky exudate of gutta-percha trees *Palaquium gutta* and *P. oblongifolia* (Sapotaceae) growing in eastern India, Java, and Sumatra. Gutta-percha becomes pliable at 30 °C, plastic at 60 °C, and can be formed into vessels that resist aggressive chemicals such as hydrogen fluoride. It is also used as an insulator in electronics, as dental cement, for fracture splints, and for covering golf balls.

cis-polyisoprene (natural caoutchouc)

trans-polyisoprene (gutta-percha)

betulaprenols
(n = 6, 7, 8, 9, 10, 11, 12, 13)

dolichol

Various *trans*-oligoterpenols isolated from the birch *Betula verrucosa* (Betulaceae) are known as betulaprenols, labeled according to the number of isoprene units that their molecules contain. Betulaprenol-9 also occurs in tobacco (*Nicotiana tabacum*, Solanaceae). Betulaprenol-11 and -12 are found in the leaves of *Morus nigra* (Moraceae) and in the feces of silk-worms (*Bombyx mori*) eating these leaves.

Dolichols with 14 to 20 isoprene units occur as such or as phosphoric acid esters in lipid membranes, e.g., those of nerve cells or in the tissues of various endocrine glands of mammals. Their function is to carry and transfer oligosaccharides for the biosynthesis of glycolipids and glycoproteins. An increased content of dolichol in tissues indicates pathological changes, for example due to ALZHEIMER´s disease.

In bacteria, polyterpenols stabilize the cell walls and also perform other physiological functions. Violet bacterioruberin ($C_{50}H_{76}O_4$) and sarcinaxanthin ($C_{50}H_{72}O_2$) from *Flavobacterium dehydrogenatus* serving as sun protection for halophilic bacteria in salt lakes are examples.

bacterioruberin
halophilic bacteria

sarcinaxanthin
Flavobacterium dehydrogenatus

8.2 Prenylquinones

Prenylquinones contain terpenyl groups with up to ten isoprene units; they are capable of undergoing reductions to the corresponding hydroquinones and cyclizations to chromenols and chromanols.

Various lipid-soluble prenylbenzoquinones derived biogenetically from the amino acids phenylalanine and tyrosine occur in the cells of almost all aerobic organisms (bacteria, plants, animals), and are therefore referred to as ubiquinones (*lat.* ubique = everywhere). They are coenzymes involved in electron transport during the respiration processes in mitochondria, and cyclize to ubichromenols and ubichromanols when exposed to light [19]. For simplification, they are also denoted as ubiquinones UQ-n or coenzymes CoQ_n, the n accounting for the number of isoprene units they

8.2 Prenylquinones

contain as shown for the case of coenzyme Q_{10} which is also known as ubiquinone UQ-10. Some ubiquinones are used as cardiotonics.

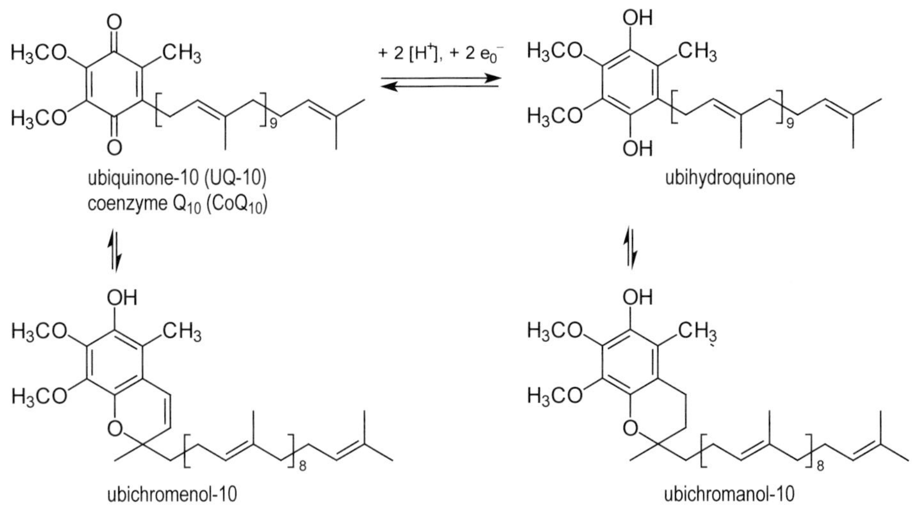

ubiquinone-10 (UQ-10)
coenzyme Q_{10} (CoQ_{10})

ubihydroquinone

ubichromenol-10

ubichromanol-10

In the chloroplasts of higher plants and algae, plastoquinones (PQs) structurally and functionally related to ubiquinones serve as electron-transfer agents in photosynthesis, as they are able to be reduced reversibly to the corresponding hydroquinones denoted as plastoquinols.

Archaebacteriae such as *Sulfolobus solfataricus* living in the solfatariae of southern Italy use sulfolobusquinone and various other prenylbenzothiophenequinones as electron-transfer agents for the oxidation of sulfur [2].

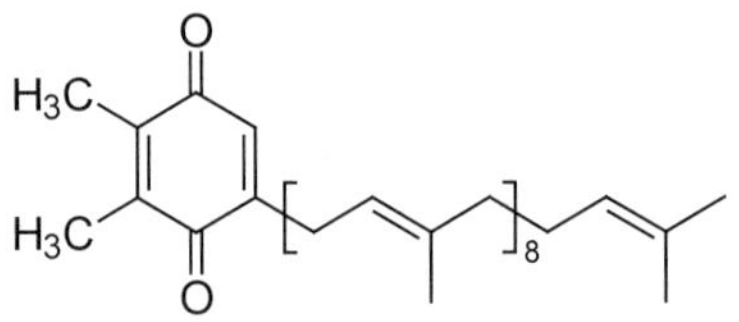

plastoquinone A

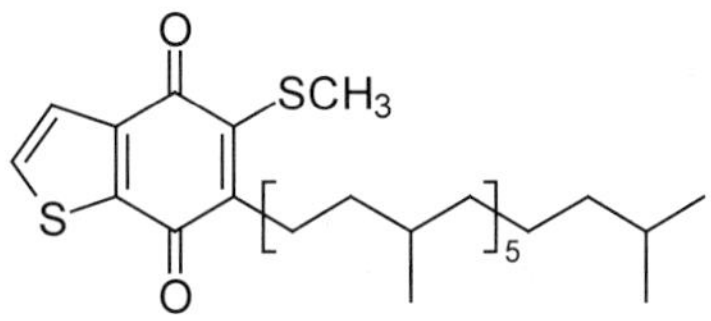

sulfolobusquinone

Vitamins of the K series ("*K*oagulation" vitamins) are chemically classified as prenyl-1,4-naphthoquinones. They are ingested with food originating from all green plants, are involved in oxidative phosphorylation during respiration processes and in the biosynthesis of glycoproteins in the liver, and are required as coagulation agents for blood.

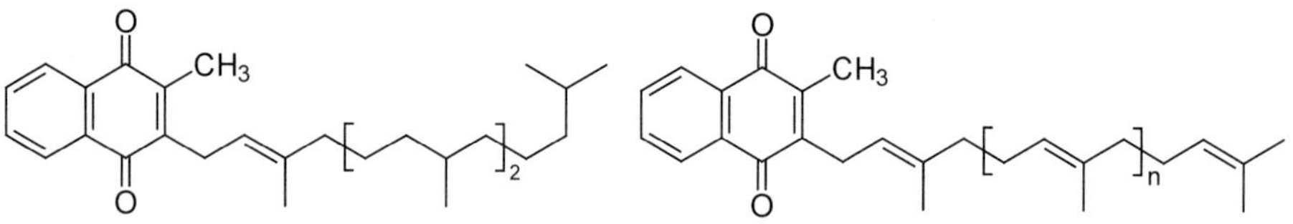

vitamin K_1
(2-methyl-3-phytyl-1,4-naphthoquinone)

vitamins $K_{2\,(35)}$ (n=5) and $K_{2\,(30)}$ (n=4)

Vitamin E also known as (+)-α-tocopherol or 2-prenyl-3,4-dihydro-2H-1-benzopyran-6-ol, represents a prenylchromanol. It occurs in fruits, vegetables and nuts, and is enriched in wheat germ and oils, particularly palm, soybean and sunflower. Tocopherol is oxidized to tocoquinone in the air (O_2 biradical, triplet oxygen) when exposed to light in the presence of chelated metal cations; therefore, it serves as a radical-capturing antioxidant which protects carotenoids and polyene lipids in biomembranes. It also protects the thiol groups of cysteine in enzyme proteins against oxidation by peroxides, and is added to foods as an antioxidant and vitamin E supplement.

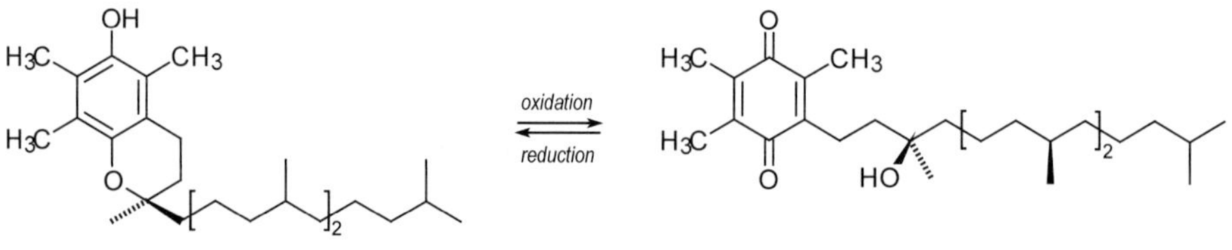

(+)-α-tocopherol (vitamin E)

tocoquinone

Vitamin E also has antiinflammatory and antirheumatic properties [19]. It influences fertility (fertility vitamin); for example, vitamin E deficient food causes sterility in rats and prevents honey-bees from metamorphosing to the queen.

9 Selected Syntheses of Terpenes

9.1 Monoterpenes

9.1.1 Concept of Industrial Syntheses of Monoterpenoid Fragrances

Many monoterpenes are desired fragrances in perfumery and flavors in food. They are produced on a larger scale from acetone (C_3) and ethyne (acetylene C_2) involving repetitive synthetic steps [23] (Fig. 5). Initially, acetone is ethynylated by acetylene in the presence of a base (sodium hydroxide, amines with sodium carbonate) yielding 3-butyn-2-ol (C_5) which is partially hydrogenated in the presence of deactivated catalysts (LINDLAR catalysts) to 2-methyl-3-buten-2-ol. This can be converted to the key intermediate 6-methyl-5-hepten-2-one (C_8) via two pathways, either by transetherification with methylpropenylether and subsequent oxa-COPE rearrangement, or by transesterification with methyl acetoacetate and subsequent CARROLL decarboxylation.

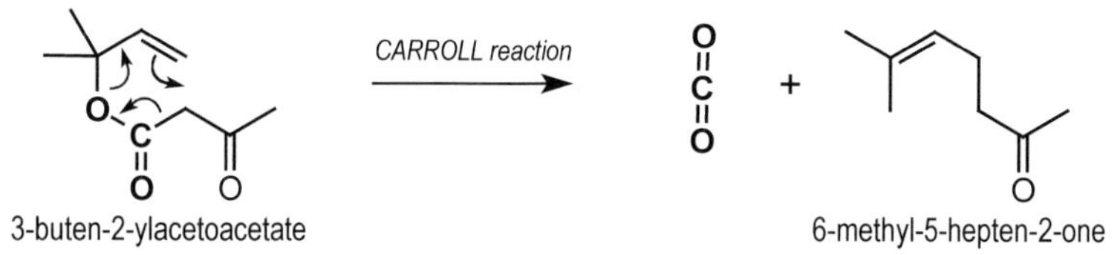

An additional ethinylation (C_2) of 6-methyl-5-hepten-2-one (C_8) leads to dehydrolinalool (C_{10}) as a monoterpene which is partially hydrogenated to linalool. The acid-catalyzed allyl rearrangement of linalool affords geraniol.

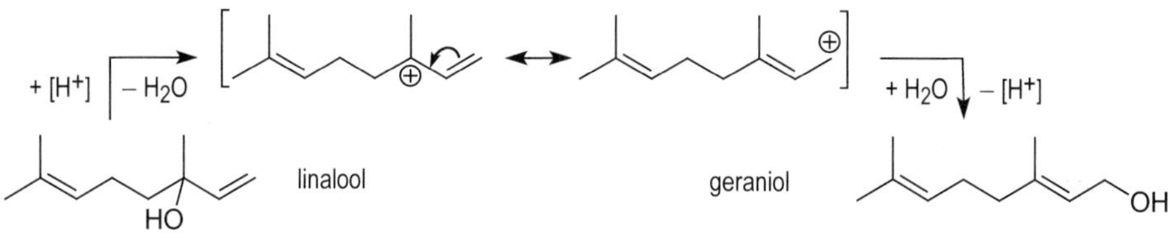

Transesterification of linalool with methyl acetoacetate followed by CARROLL decarboxylation provides access to 6,10-dimethylundeca-5,9-dien-2-one (C_{13}). This is ethynylated to the sesquiterpene dehydronerolidol (C_{15}), which is partially hydrogenated to nerolidol. Finally, nerolidol is subjected to an allyl rearrangement for the production of farnesol (Fig. 5).

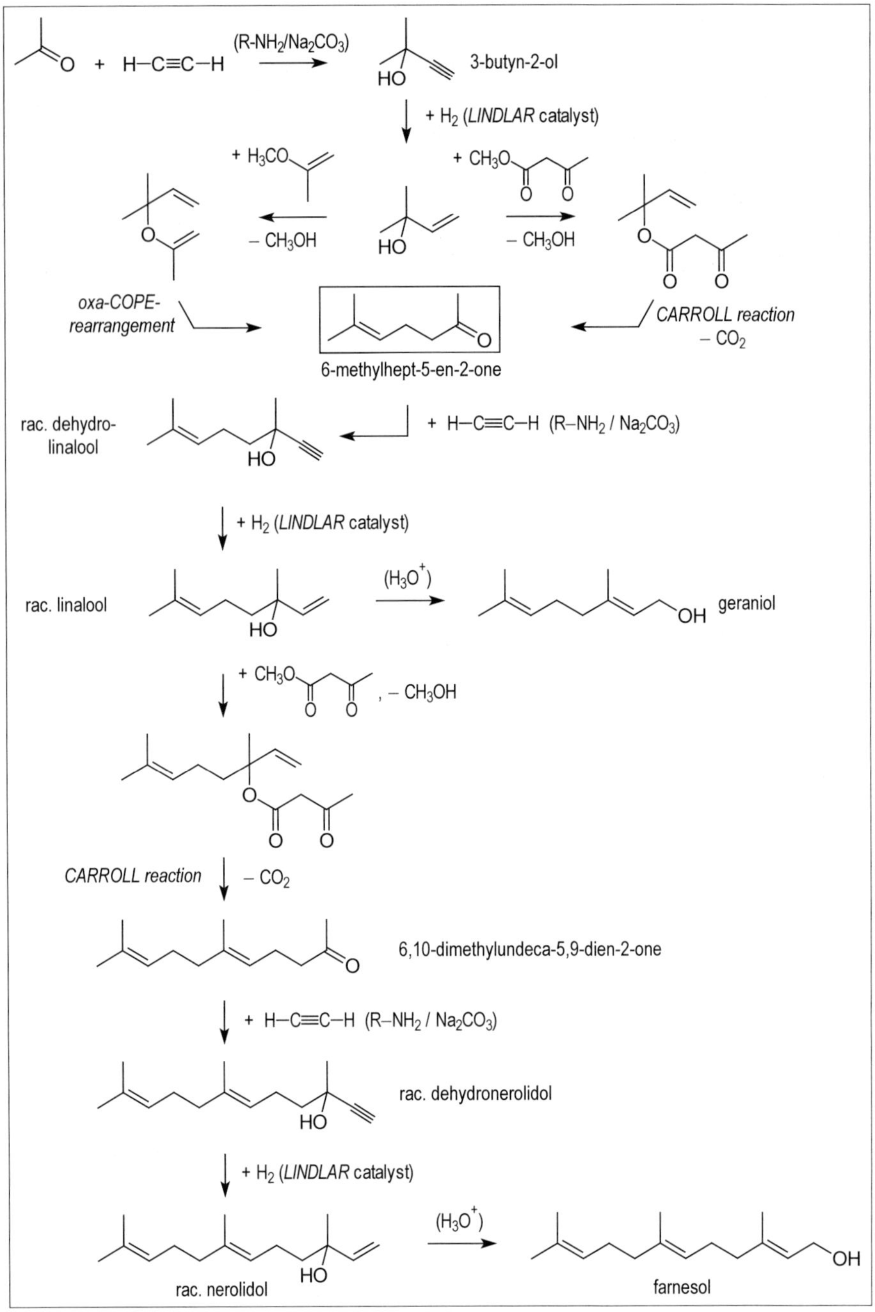

Figure 5. Industrial syntheses of acyclic mono- and sesquiterpenes.

9.1.2 (R)-(+)Citronellal

β-Pinene, a byproduct of the wood and paper industry, is the starting material for an enantioselective synthesis of (R)-citronellal [24]. Thermal cycloreversion leads to myrcene. The allylamine obtained by lithiation of myrcene with butyllithium and diethylamine involving an intermediate lithium chelate rearranges stereoselectively in the presence of a chiral catalyst containing the BINAP-ligand (2,2´-bis-(diphenylphosphino)-1,1´-binapthyl = BINAP) (telomerization) to the enamine, which then readily undergoes acid-catalyzed hydrolysis to (R)-(+)-citronellal with high enantiomeric excess.

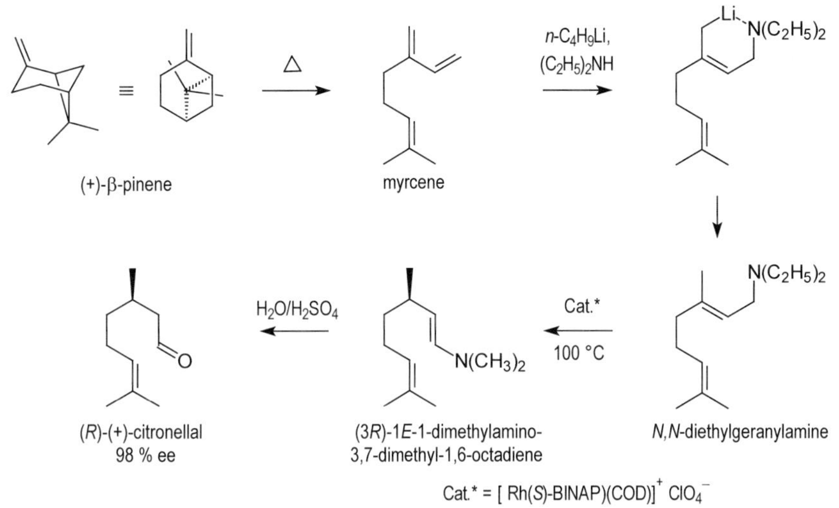

9.1.3 Rose Oxide

Reduction of (R)-(+)-citronellal by lithiumaluminumhydride yields (R)-(+)-citronellol which undergoes a photo ene reaction to the hydroperoxide in the presence of the xanthene dye rose Bengal as a photosensitizer of singlet-oxygen. Sodium sulfite reduces hydroperoxide to the diol which is, when catalyzed by acids, dehydrated to the *trans*-isomer of rose oxide [25].

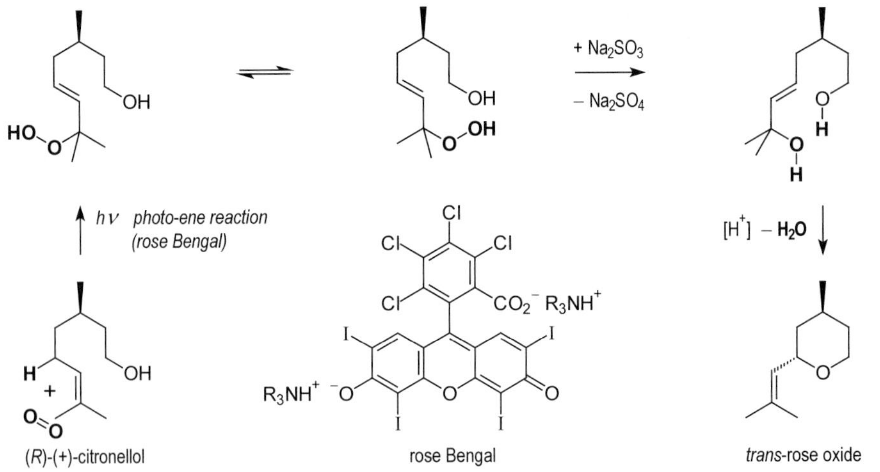

9.1.4 Chrysanthemic Acid Methyl Ester

Esters of chrysanthemic acid are rapidly acting insecticides with a comparatively low toxicity for human and mammalian organisms. Retrosynthetic disconnection of the cyclopropane ring following the path of a 1,3-elimination leads to a carbanion with the leaving group X in an allyl position. Provided that X stabilizes the carbanion by electron-withdrawing [(−)-M-effect], the intermediate on its part arises from a MICHAEL addition of the dimethylallyl-X-compound to the methyl ester of senecionic acid (3-methyl-2-butenoic acid).

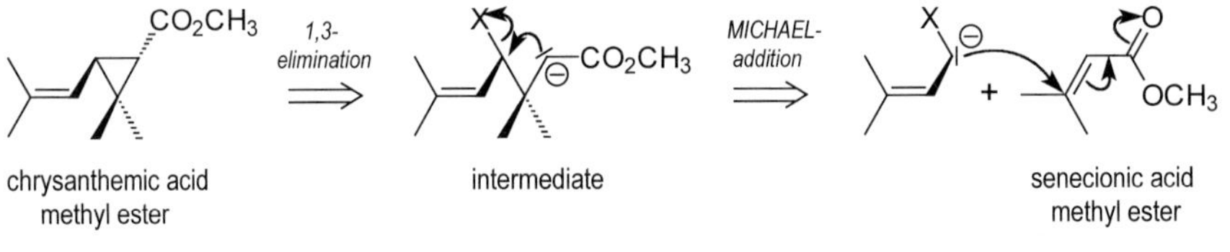

When carrying out the synthesis [26], 3,3,6-trimethyl-4-p-tolylsulfonyl-5-heptenoic acid methyl ester proves to be an appropriate synthetic equivalent of the intermediate. This precursor arises from MICHAEL addition of (3-methyl-2-buten-1-yl)-p-tolylsulfone to senecionic acid methyl ester. The p-tolylsulfone is obtained from sodium p-toluensulfinate and 1-bromo-3-methyl-2-butene by an S_N reaction and subsequent cationotropic 1,2-shift. 1-Bromo-3-methyl-2-butene is prepared by nucleophilic bromination of 3-methyl-2-buten-1-ol, which is produced by an allyl rearrangement of 2-methyl-3-buten-2-ol, described previously as a key intermediate for the synthesis of monoterpenes in section 9.1.1.

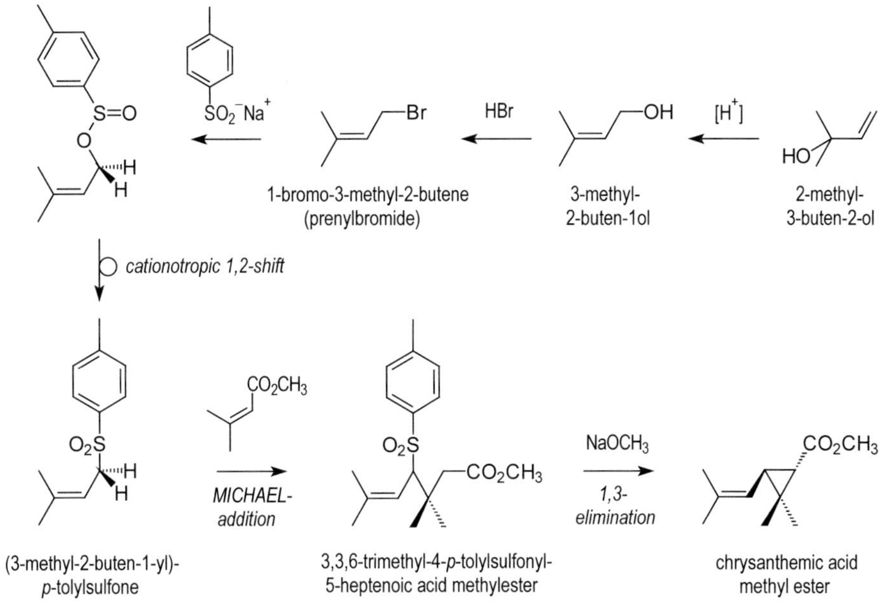

9.1.5 α-Terpineol

The cyclohexene ring of racemic α-terpineol, which smells like the blossoms of lilac and is used in perfumes and for denaturing fats in soap manufacture, is set up by a DIELS-ALDER reaction [27]. Isoprene is the diene, and methyl arylate the appropriate dienophile. The [4+2]-cycloaddition catalyzed by aluminum chloride as a LEWIS acid initially provides 1-methoxycarbonyl-4-methylcyclohex-3-ene. Transformation of the electrophilic methoxycarbonyl function to the tertiary alcohol α-terpineol is achieved by two equivalents of methylmagnesium bromide as carbon nucleophile.

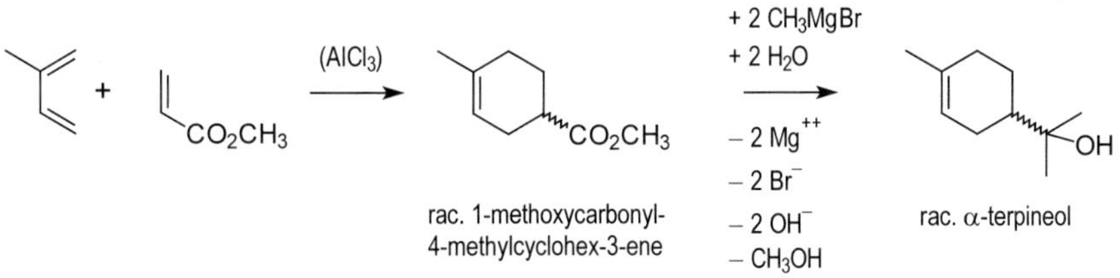

9.1.6 (1R,3R,4S)-(–)-Menthol

(1R,3R,4S)-(–)-Menthol, the major constituent of peppermint oils has a pleasant flavor and refreshing odor, and is used in cigarettes, cough-drops, nasal inhalors, ointments, as a mild antiseptic, local anesthetic, antipruritic, and internally as a carminative and gastric sedative. For these reasons, it is an important raw material for confectionery, cosmetics, perfumery, and pharmacy [18]. Its enantiospecific synthesis [24] is achieved in three steps from (R)-(+)-citronellal obtained by the procedure described in section 9.1.2, involving a carbonyl ene reaction catalyzed by zinc bromide as a LEWIS acid. The primarily formed (–)-isopulegol undergoes catalytic hydrogenation to (–)-menthol.

9.1.7 Camphor from α-Pinene

Camphor is used as flavorant, odorant, moth-repellant, plasticizer, preservative and as a starting reagent for syntheses. The industrial production of the racemate starts with α-pinene [18]. This is protonated at the CC double bond following MARKOWNIKOW'S rule to a carbenium ion which, on the path of WAGNER-MEERWEIN-rearrangement involving the intermediate non-classical bornyl cation (a carbonium ion), reacts with sodium acetate to isobornyl acetate. Hydrolysis yields isoborneol, and this is oxidized to camphor by means of various oxidation reagents. The intermediate symmetric carbonium ion causes racemic camphor to be formed *via* racemic isoborneol also from the pure enantiomers (+)- or (–)-α-pinene.

Oxidation of camphor with nitric acid opens the six-membered ring, affording *cis*- and *trans*-camphoric acid. Sulfonation of camphor with concentrated sulfuric acid and acetic anhydride selectively yields 10-camphorsulfonic acid.

9.1 Monoterpenes

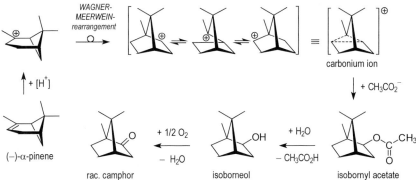

Several syntheses of chiral camphor derivatives make use of the CH acidity of the methylene group attached in α-position to the carbonyl function (C-3). Thus, iso-amyl nitrite converts camphor to 3-isonitrosocamphor which readily undergoes hydrolysis to the yellow camphorquinone. Bromination leads to 3-bromocamphor which is sulfonated to 3-bromocamphor-3-sulfonic acid with concentrated sulfuric acid. 3-Lithiated camphor obtained with phenyllithium is carboxylated to *endo-* and *exo-*isomers of camphor carboxylic acid. The CLAISEN condensation of camphor with esters of carboxylic acids provides enolized chiral 1,3-diketones, converting metal cations to chiral metal chelates.

The enantiomers of camphorcarboxylic and sulfonic acids are used for resolution of enantiomers from racemic chiral amines and alcohols *via* diastereomeric salts and esters, respectively. Europium(III)- and praseodymium(III)-chelates of hydroxy-methylenecamphor derivatives are suitable chiral shift reagents for the determination of enantiomeric purity by integration of NMR spectra, because they exchange ligands with enantiomeric substrates such as alcohols and amines, thus forming diastereomeric chelates characterized by different spectra.

9.1.8   α-Pinene and Derivatives for Stereospecific Syntheses of Chiral Monoterpenes

(+)-Epoxy-α-pinene, produced by epoxidation of (+)-α-pinene, rearranges in the presence of LEWIS acids to (−)-campholenealdehyde [28], which is not only one of the constituents of various juniper species, but also the starting reagent for the synthesis of fragrances with the odor of sandalwood. Sandalore, for example, is produced by KNOEVENAGEL alkenylation of campholenealdehyde with propanal followed by reduction of the aldehyde function with sodium borohydride [18].

The production of pure (*R*)-(−)-linalool [29] involves hydroperoxidation of (1*R*,2*S*,5*R*)-(+)-pinane by air oxygen; the pinane enantiomer is obtained by catalytic hydrogenation of (+)-α-pinene. Another catalytic hydrogenation converts the hydroperoxide to (1*R*,2*R*,5*R*)-(−)-2-pinanol which opens its cyclobutane ring by pyrolytic cycloreversion to the target compound with its pleasant flowery odor, widely used in perfumery instead of bergamot or French lavender.

(+)-α-pinene

(*R*)-(−)-linalool

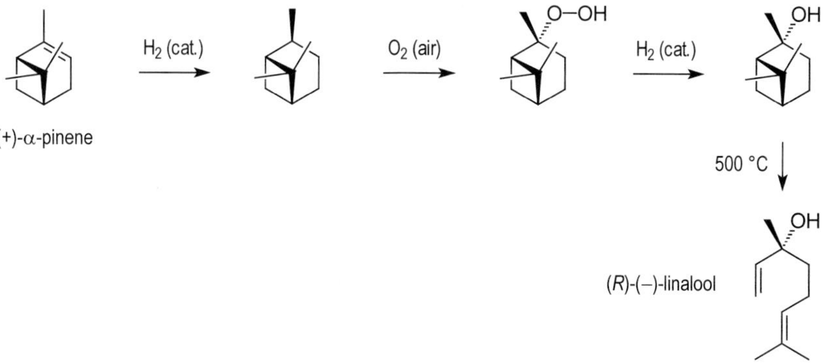

Bark beetles of the genus *Ips* are pests which attack pine and spruce trees. They use ipsdienols as aggregation pheromones, *Ips confusus* emitting the (*S*)-(+)-, and *Ips paraconfusus* the (*R*)-(−)-enantiomer [14-17]. The beetles receive the myrcenes (section 2.2) occurring in conifers with their food and metabolize them to ipsdienols; some evidence for *de-novo* synthesis by the bugs is also reported. In order to catch the beetles, pheromone traps are supplied with both enantiomers of ipsdienol which are produced from (+)-verbenone, a constituent of the Spanish verbena oil (section 2.4.3). This terpenone, also available by oxidation of α-pinene, is isomerized to the enantiomers of 2(10)-pinen-4-one *via* three steps (reduction, protonation, oxidation). After separation, both enantiomers are reduced by lithiumaluminumhydride. Pyrolytic cycloreversion of the resulting diastereomeric 2(10)-pinen-4-ols provides the enantiomers of ipsdienol [29].

(+)-verbenone

2(10)-pinen-4-one

(4R)-2(10)-pinen-4-ol

(4*R*)-(−)-ipsdienol

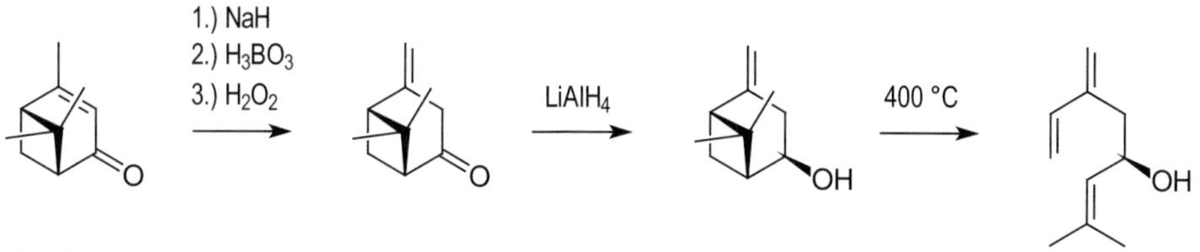

9.1.9 Hexahydrocannabinol

Retrosynthetic disconnection of hexahydrocannabinol **1** following the path of an intramolecular hetero DIELS-ALDER reaction leads to the *o*-quinonemethide **2** as an electron-deficient hetero-1,3-diene; **2** arises from a KNOEVENAGEL alkenylation of citronellal **3** with the carbanion resulting from deprotonation of the CH-acidic methylene group of the keto tautomer **4** of 5-pentylresorcinol known as olivetol.

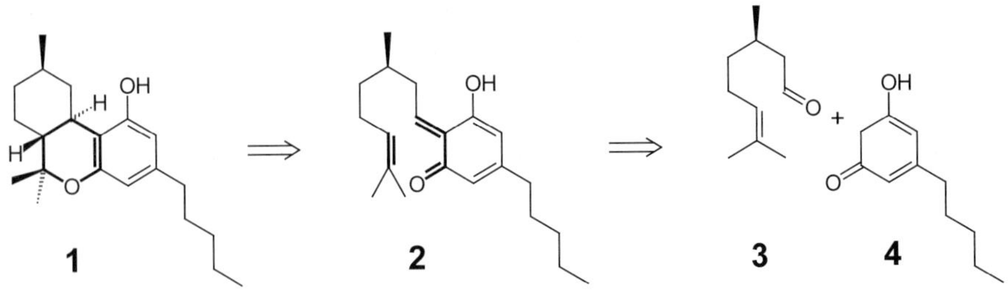

Following this concept, a synthesis of hexahydrocannabinol **1** [30] starts with the metallation of MOM-protected olivetol **5** (MOM = *methoxymethylether*) *ortho* to both *O*-alkyl groups with *n*-butyllithium. The intermediate aldol **6** obtained by aldol reaction with citronellal **3** is deprotected and dehydrated when refluxed in methanol solution in the presence of *p*-toluenesulfonic acid, and directly undergoes an intramolecular hetero DIELS-ALDER reaction to the target **1** on the expected pathway.

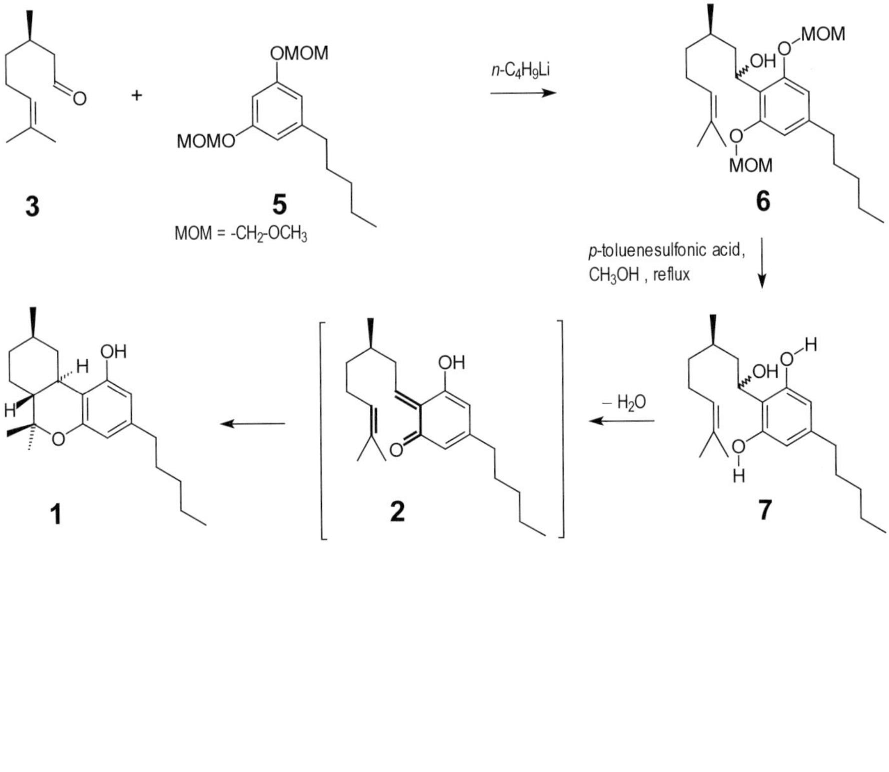

9.2 Sesquiterpenes

9.2.1 β-Selinene

The "right half" of the sesquiterpene (+)-β-selinene (as drawn below) includes (R)-(+)-limonene as a substructure. Retrosynthetic disconnection to (R)-(+)-limonene leads to the intermediate carbenium ions **1a** and **1b** *via* 15-nor-11-eudesmen-4-one (carbonyl alkenylation) and 15-nor-13-chloro-2-eudesmen-4-one (dehydrogenation, protective masking of the double bond in the side chain). These carbenium ions arise from (R)-9-chloro-p-menth-1-ene and the acylium ion **1c** (synthone) originating from 3-butenoic acid as reagent (synthetic equivalent). (R)-p-Menth-1-en-9-ol, on its part obtained by hydroboration and oxidation of (R)-(+)-limonene, turns out to be the precursor of the chloromenthene.

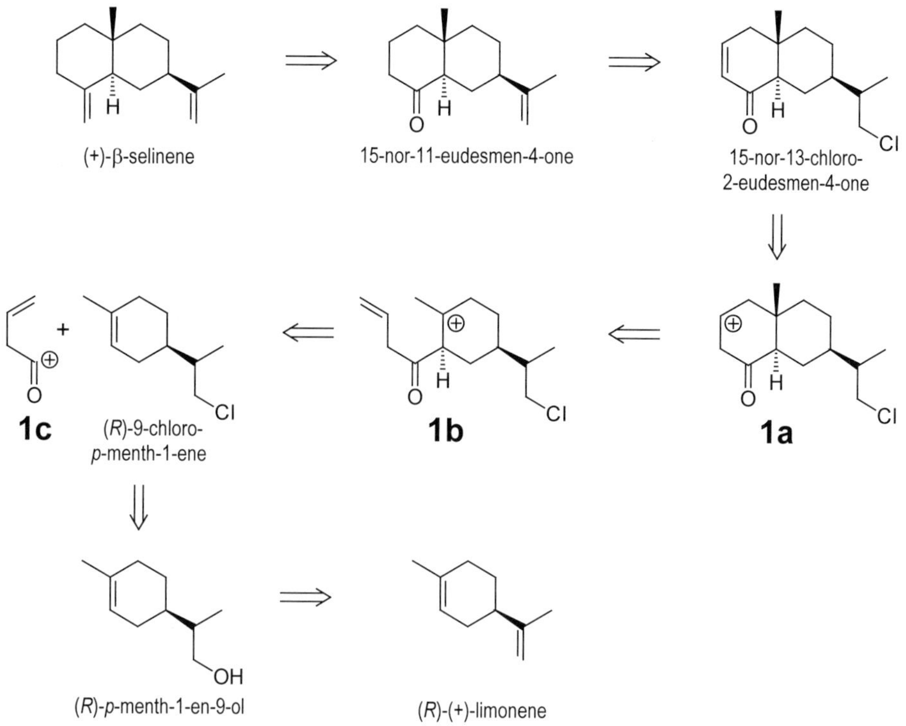

Thus, the first step of a stereoselective synthesis of β-selinene reported by MACKENZIE, ANGELO and WOLINSKI [31] involves hydroxylation of the C-8–C-9-double bond of (R)-(+)-limonene by hydroboration with diborane and subsequent oxidation with hydrogen peroxide. The nucleophilic chlorination to (R)-9-chloro-p-menth-1-ene is achieved with tetrachloromethane and triphenylphosphane. 3-Butenoic acid chloride in the presence of aluminum chloride closes the ring to 15-nor-13-chloro-2-eudesmen-4-one. Catalytic hydrogenation and simultaneous dehydrochlorination in glacial acetic acid affords 15-nor-11-eudesmen-4-one as precursor, which is finally subjected to WITTIG methylenation to (+)-β-selinene.

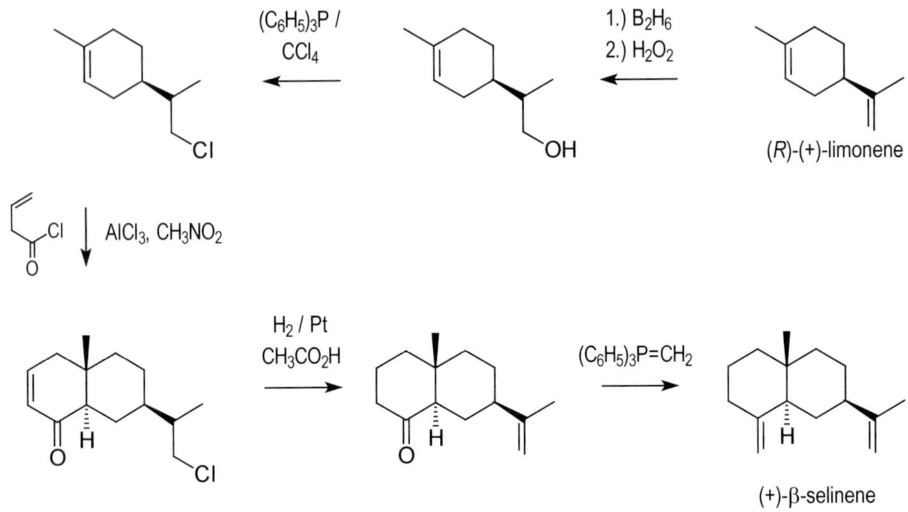

9.2.2 Isocomene

The retrosynthetic disconnection of isocomene leads primarily to the intermediate tertiary carbenium ion **1**, which may arise from the intermediate carbenium ion **2** by anionotropic 1,2-alkyl shift. The latter turns out to be the protonation product of the tricycle **3** containing an exocyclic CC-double bond which is generated by a WITTIG-methylenation of the tricyclic ketone **4**. The concept behind this is formation of the cyclobutane ring in **4** by means of an intramolecular [2+2]-photocycloaddition of the 1,6-diene **5**. The enone substructure in **5** results from hydrolysis of the enolether and dehydration of the tertiary alcohol function in (6S)-1-alkoxy-2,4-dimethyl-3-(2-methyl-1-penten-5-yl)cyclohexene **6**. The tertiary alcohol **6** emerges from a nucleophilic alkylation of (6S)-3-alkoxy-2,6-dimethyl-2-cyclohexen-1-one **7** with metallated 5-halo-2-methyl-1-pentene obtained by GRIGNARD reaction or

9.2 Sesquiterpenes

lithiation. The desired enantiomer **7** becomes feasible by methylation of the methylene group α to the carbonyl function of 3-alkoxy-2-methyl-2-cyclohexen-1-one, followed by resolution of the racemic mixture.

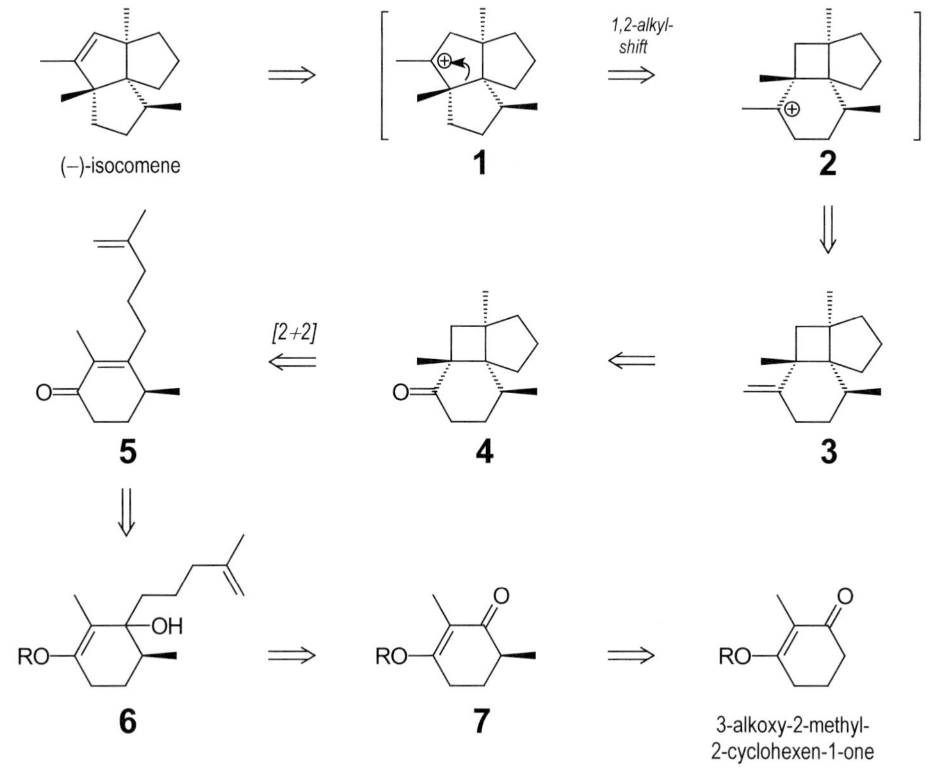

A synthesis reported by PIRRUNG [32] follows this concept. 3-Ethoxy-2-methyl-2-cyclohexen-1-one obtained from 1,3-cyclohexanedione in two steps involving C- and O-alkylation is lithiated with lithiumdiisopropylamide (LDA) and methylated with iodomethane to racemic **7**. The GRIGNARD-reagent **8** prepared from 5-bromo-2-methyl-1-pentene alkylates **7** to the unstable tertiary alcohol **6** which is not isolated, reacting during workup of the reaction mixture with aqueous hydrochloric acid directly to the enone **5**. This undergoes an intramolecular [2+2]-photocycloaddition to the tricyclic ketone **4** upon irradiation with UV light (350 nm) in *n*-hexane solution. Sterically forced by the methyl groups at the six-membered ring, the cyclobutane ring closes opposite to these methyl groups. A crystal structure confirms the relative configuration of the tricycle **4** [32].

Carbonyl methylenation with methylentriphenylphosphorane, generated *in situ* from methyltriphenylphosphonium iodide in dimethylsulfoxide at 70 °C, introduces the exocyclic double bond. The resulting alkene **3** is protonated with *p*-toluenesulfonic acid in refluxing benzene solution to the intermediate carbenium ion **2**. This undergoes a ring-expanding 1,2-alkyl shift to the carbenium ion **1**, which deprotonates to racemic isocomene, as expected.

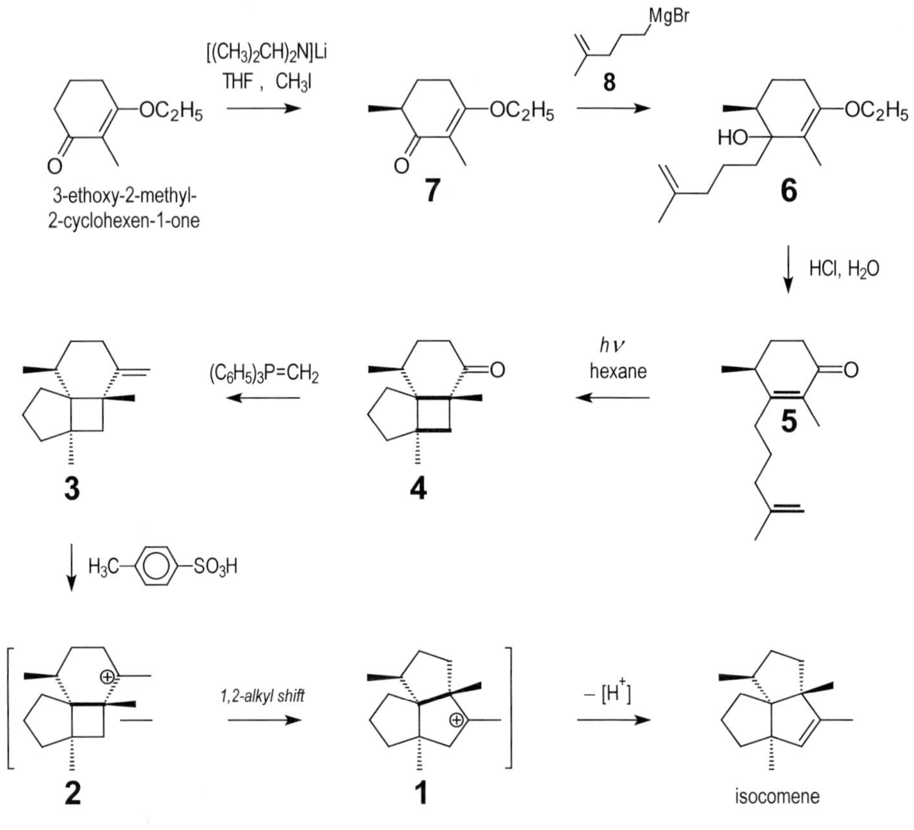

9.2.3 Cedrene

Cedrene and cedrol are used as fragrances in perfumery and as insect repellants. Acid-catalyzed dehydration of cedrol, the product of a nucleophilic addition of

9.2 Sesquiterpenes

methyllithium to the carbonyl function of 15-nor-3-cedrone, is a straightforward method to introduce the exocyclic 3(15)-double bond. An intramolecular opening of the cyclopropane ring in the tricyclic ketone **1** by the (nucleophilic) CC double bond permits formation of the five-membered ring *B* in 15-nor-3-cedrone. [2+1]-Cycloaddition with diazomethane as carbene generator or the SIMMONS-SMITH reaction are suitable methods for cyclopropanation of the corresponding dienone precursor. In order to avoid the competing ring homologization of the cycloalkanone, the carbonyl function in **1** must be masked before cyclopropanation. The obvious way to do so is to use the secondary alcohol 2-cyclopentenol **3** as the precursor. Thus, the retrosynthesis leads from **1** via **2** to **3**. The latter turns out to be the reduction product of the cyclopentenone **4**, which is feasible by nucleophilic addition of lithiated 2-halo-6-methyl-5-heptene **5** with the enolether **6** of 1,3-cyclopentanedione.

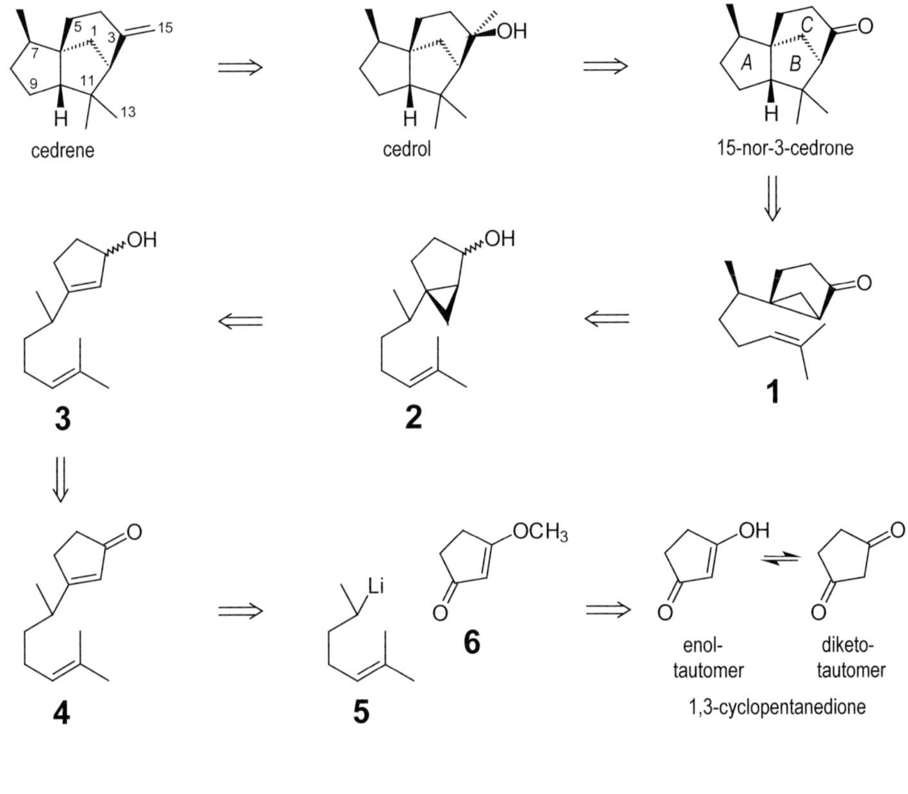

This is the concept of a synthesis of racemic cedrene reported by E.J. COREY and R.D. BALANSON [33]. Lithiated 2-chloro-6-methyl-5-heptene **5** as carbon nucleophile alkylates 3-methoxy-2-cyclopentenone **6**; work-up of the reaction mixture in aqueous acidic solution converts the primarily formed alkylated 3-methoxy-2-cyclopentenol directly to the corresponding 2-cyclopentenone **4**, which is reduced to the secondary allyl alcohol **3** by diisobutylaluminumhydride. Regioselective cyclopropanation using the SIMMONS-SMITH reaction gives the diastereomeric bicyclo[3.1.0]hexanols **2**, which are oxidized to the ketones by chromium(VI)oxide in pyridine. The desired diastereomer **1** rearranges to 15-nor-3-cedrone in the presence of the mixed anhydride from methanesulfonic and acetic acid as catalyst. Nucleophilic addition of methyllithium to the carbonyl function of **1** affords cedrol, which dehydrates in the presence of formic acid to racemic cedrene.

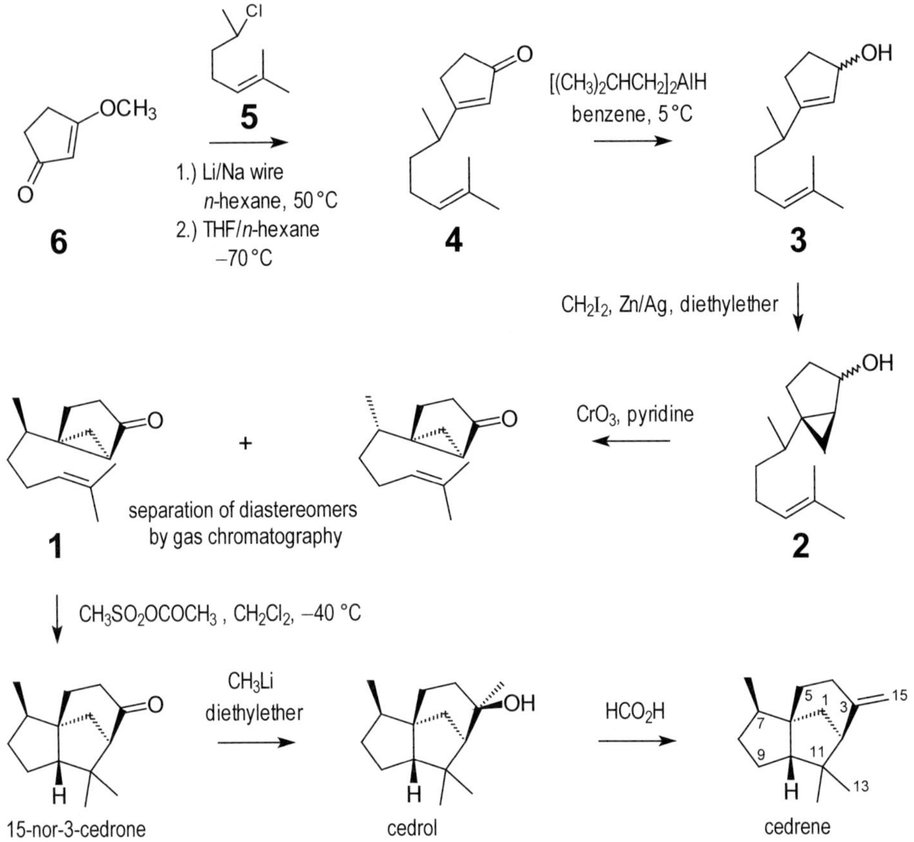

9.2.4 Periplanone B

Periplanone B is the most active sex pheromone found in the alimentary tract and excreta of the American cockroach *Periplaneta americana*. An elegant total synthesis of this germacrane sesquiterpene was achieved by SCHREIBER and SANTINI [34]. Cyclodecatrienone **1** is an obvious precursor. One of the oxirane rings arises from epoxidation of the enone CC double bond, the other from [2+1]-cycloaddition of a carbene to the carbonyl bond of the enone. Oxidation of the methylene group introduces the additional carbonyl double bond. The CC double bond of the enone results from an elimination of HX in the α-X-substituted cyclodecadienone **2**, which, on its part, is feasible by substitution of cyclodecadienone **3**. An electrocyclic opening of the cyclobutene ring in **4** provides the 1,3-diene substructure in **3**.

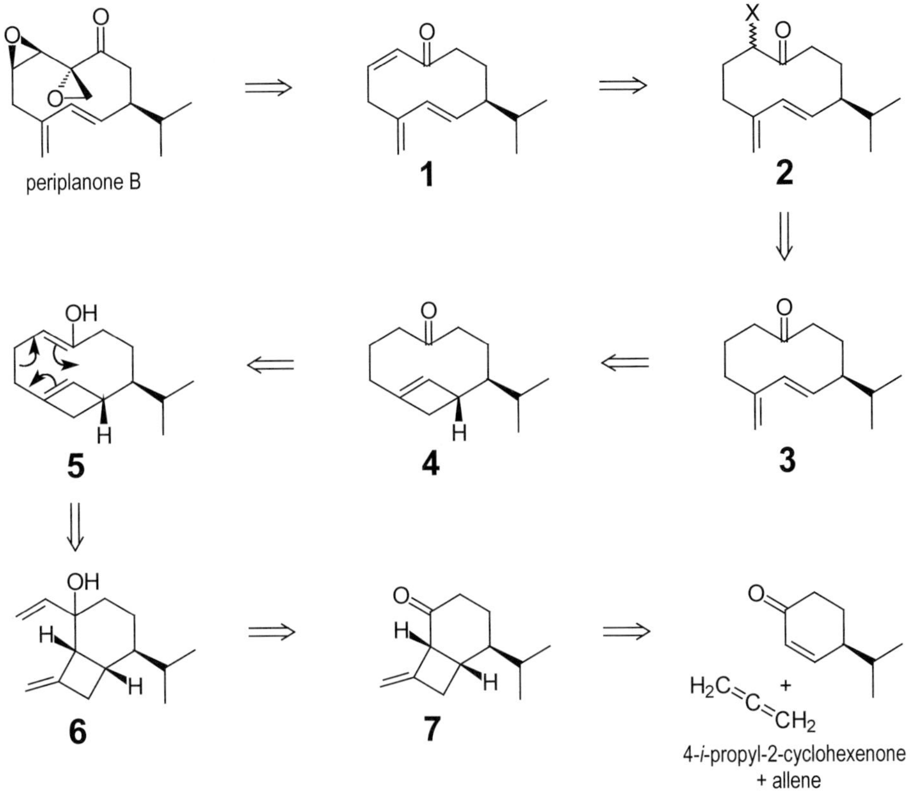

Enolization of the carbonyl function in **4** gives rise to a 1,5-dien-1-ol **5** which is nothing but the product of a COPE-rearrangement of the bicyclo[4.2.0]octanol **6**. The vinyl group in **6** is introduced by 1,2-addition of vinylmagnesiumhalide to the carbonyl function of bicyclo[4.2.0]octanone **7**; the precursor **7** turns out to be the product of a [2+2]-photocycloaddition of allene to the CC double bond of 4-*i*-propyl-2-cyclohexenone.

In the first step of this synthesis [34], [2+2]-photocycloaddition of allene to racemic 4-*i*-propyl-2-cyclohexenone yields a mixture of the *anti*- and *syn*-cycloadducts **7a** and **7b**, which do not have to be separated because the cyclodecadienone **3** is built from both isomers in the course of subsequent procedures. The 1,2-addition of vinylmagnesiumbromide to the ketones **7a** and **7b** gives the diastereomeric alcohols **6a** which are deprotonated with potassium hydride and 18-crown-6 to the alcoholates **6b**, and these undergo oxa-COPE rearrangement to the bicyclic ketone **5** with the expected 2:1 ratio of isomers. Thermally induced electrocyclic ring opening yields the desired *trans*-cyclohexadienone **3** as a major product. Lithium-bis(trimethylsilyl)amide (LBTMSA) activates **3** to the enolate which is sulfenylated by diphenyldisulfide dioxide (TROST's reagent), predominantly to the desired regioisomer **2**. Subsequent oxidation of the thioether **2** with sodium periodate affords the intermediate phenylsulfone, which undergoes elimination to the cyclodecatrienone **1** upon heating in toluene solution.

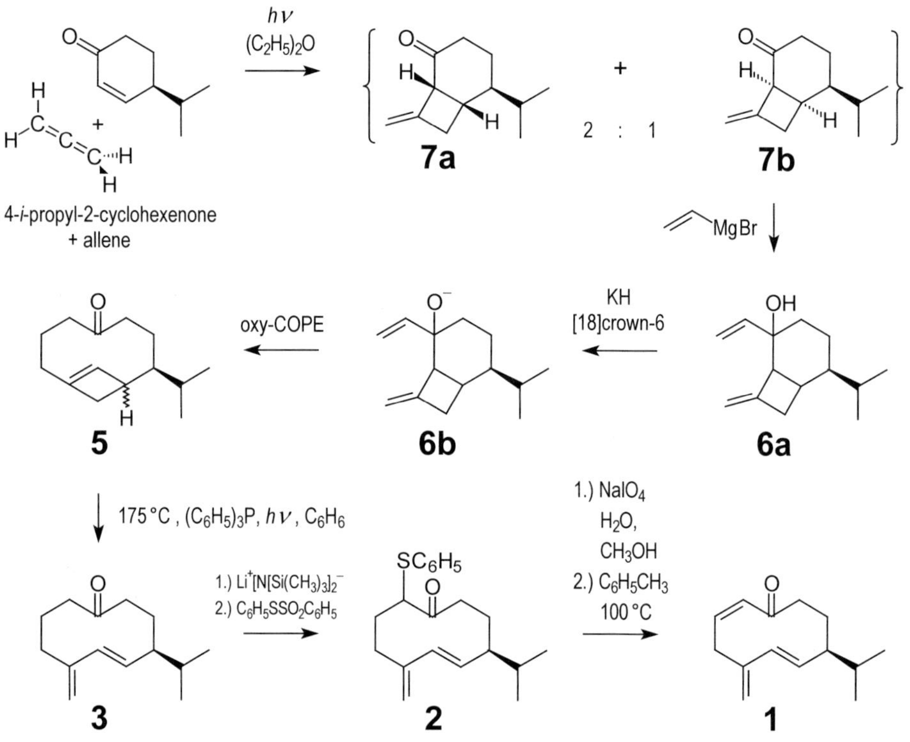

9.2 Sesquiterpenes

Insertion of the oxirane ring in the 1,2-position is achieved by epoxidation of the electron-deficient enone CC double bond in **1** with *t*-butylhydroperoxide. In order to introduce the carbonyl function at C-10, the 1,2-epoxycyclodecadienone **8** is activated once again with LBTMSA to the enolate, which is converted by electrophilic addition of phenylselenylbromide to the phenylselenide **9** as a masked ketone. Hydrogen peroxide oxidizes the selenide **9** to the selenoxide **10** which, upon acylation with acetic anhydride and sodium acetate in tetrahydrofuran, undergoes a selena-PUMMERER rearrangement [34] to the epoxycyclodecadienedione **14** involving the intermediates **11-13** which are not isolated. A regioselective cycloaddition of dimethylsulfoniummethylide to the carbonyl double bond next to first oxirane ring completes the last synthetic step to racemic periplanone B.

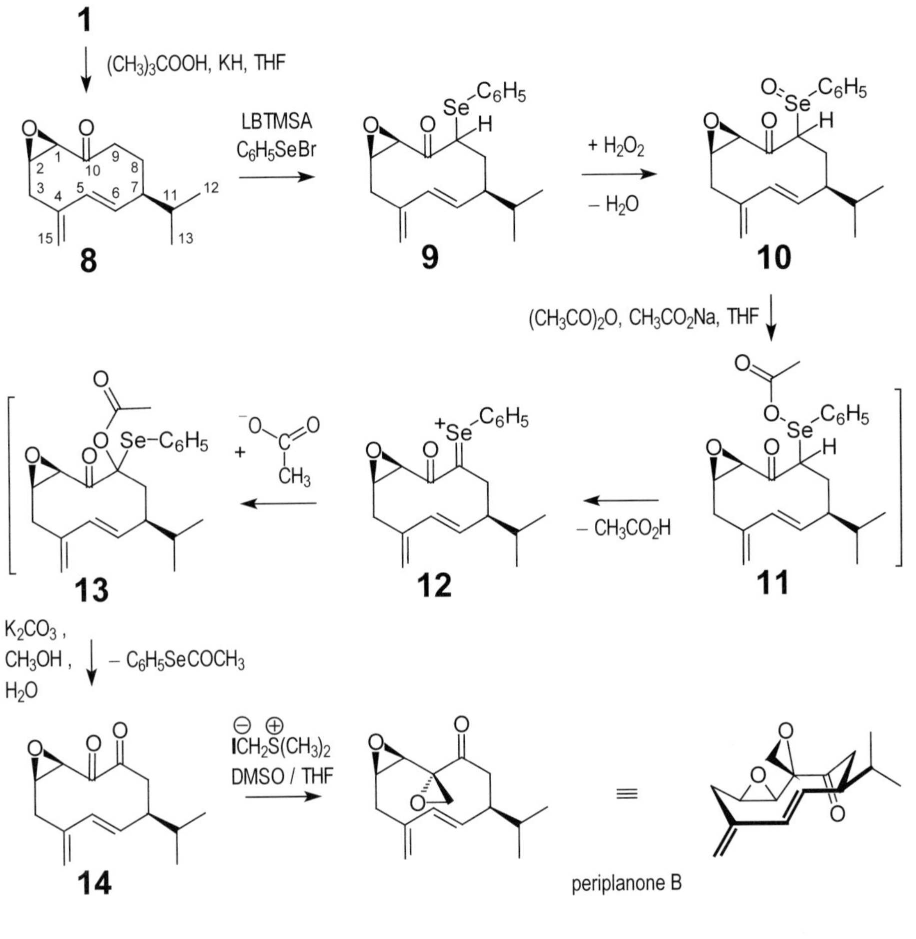

9.3 Diterpenes

9.3.1 Vitamin A (Retinol Acetate)

The C-11–C-12 double bond of the diterpene retinol acetate is obviously disconnected on the path of a WITTIG reaction to C_{15} WITTIG salt and C_5 acetate. Retrosynthesis of the C_{15} salt leads to dehydrolinalool as the starting reagent *via* β-ionylidenethanol, vinylionol, ethynylionol, β-ionone, and pseudoionone involving procedures as outlined for industrial syntheses of monoterpenes (section 9.1.1). The C_5 acetate arises from an allyl rearrangement of 4,4-dialkoxy-3-methyl-1-buten-3-ol. The latter is, of course, the product of hydrogenation of 4,4-dialkoxy-3-methyl-1-butin-3-ol feasible by ethynylation of dialkoxyacetone which finally emerges from dialkoxyacetone obtained by oxidation of acetoneketal. This is the concept of a convergent industrial synthesis of retinol acetate elaborated by POMMER [35].

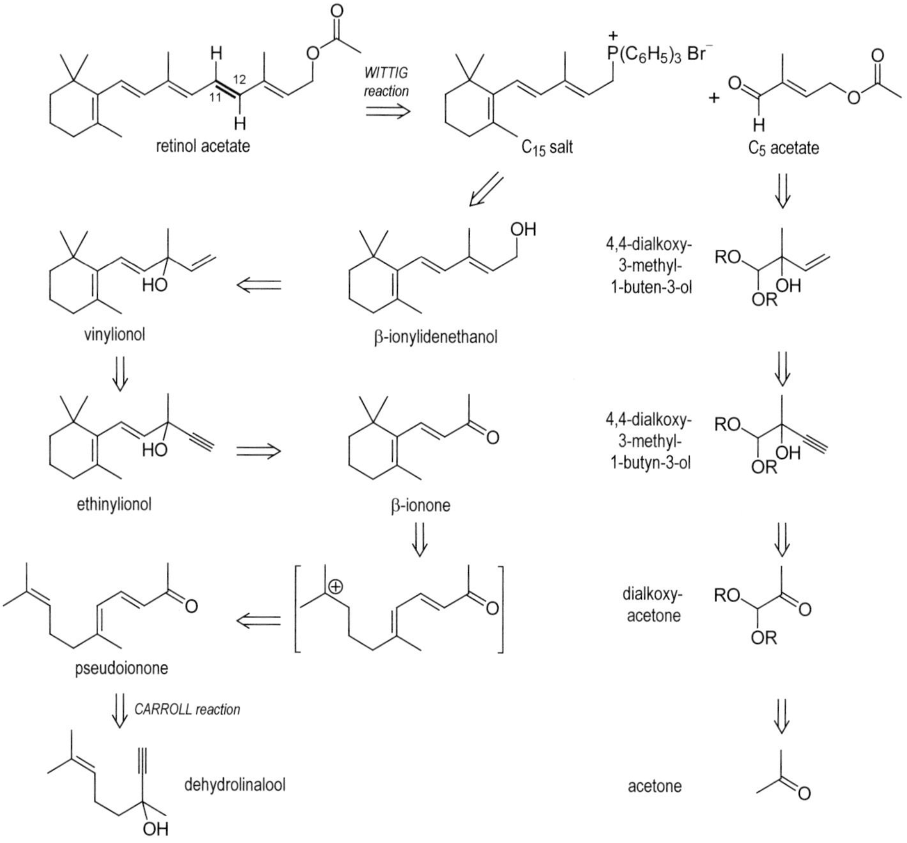

9.3 Diterpenes

In order to accomplish the synthesis, methylacetoacetate is transesterified with racemic dehydrolinalool and the resulting dehydrolinaloylacetoacetate subjected to CARROLL decarboxylation to pseudoionone. This undergoes acid-catalyzed cyclization to β-ionone which is ethynylated to ethynylionone by ethyne in the presence of sodium hydroxide as the base. Ethynylionone undergoes partial catalytic hydrogenation to vinylionol using a deactivated catalyst. In the presence of hydrobromic acid, allyl rearrangement of vinylionol directly yields β-ionylidenebromoethane which reacts with triphenylphosphane to the crystallizing C_{15}-WITTIG salt.

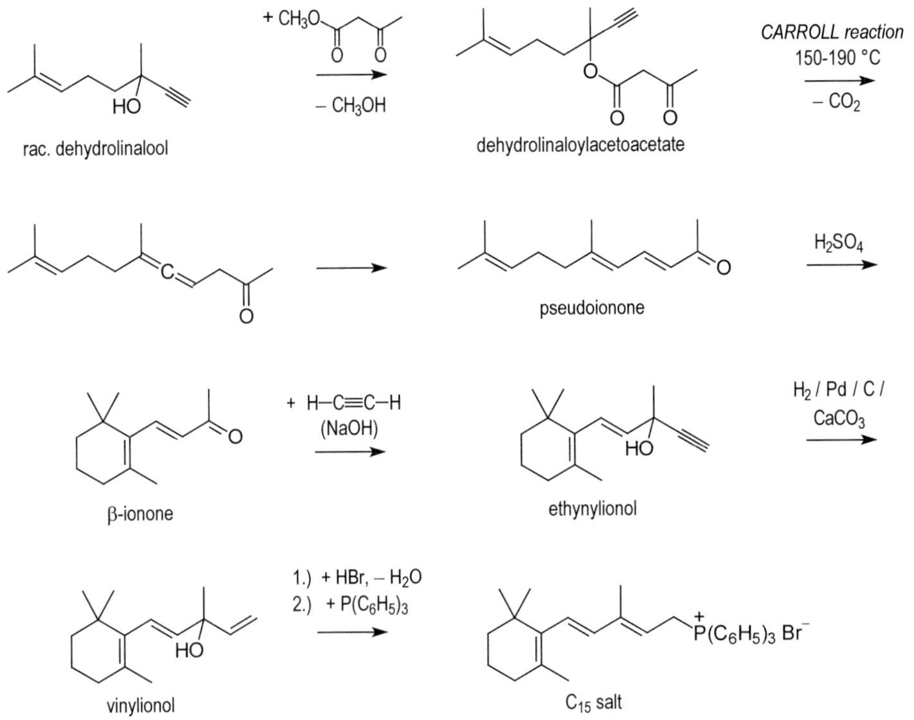

In order to synthesize the C_5 acetate, dimethoxyacetone obtained by oxidation of acetone is ethynylated to 4,4-dimethoxy-3-methyl-1-butyn-3-ol in the presence of sodium hydroxide. Partial catalytic hydrogenation of the alkynol leads to 4,4-dimethoxy-3-methyl-1-buten-3-ol as the C_5 alcohol which rearranges in acetic anhydride to the C_5 acetal ester. Deprotection of the aldehyde function necessary before the WITTIG alkenylation is achieved thermally in the presence of copper(II)-salt as catalyst.

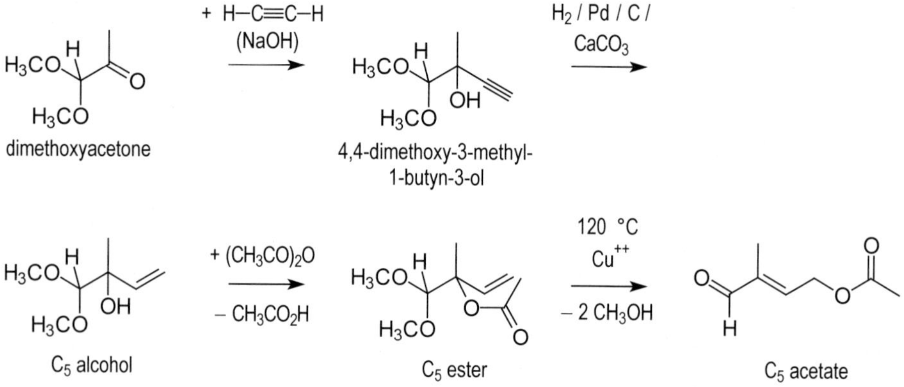

9 Selected Syntheses of Terpenes

[Reaction scheme: dimethoxyacetone + H–C≡C–H (NaOH) → 4,4-dimethoxy-3-methyl-1-butyn-3-ol → (H₂/Pd/C/CaCO₃)]

[Reaction scheme: C_5 alcohol + $(CH_3CO)_2O$, $-CH_3CO_2H$ → C_5 ester → (120 °C, Cu^{++}, $-2\,CH_3OH$) → C_5 acetate]

WITTIG alkenylation of the C_{15} salt with the C_5 acetate proceeds almost quantitatively. Retinol acetate which is more stable than vitamin A (retinol) is purified by recrystallization from *n*-hexane.

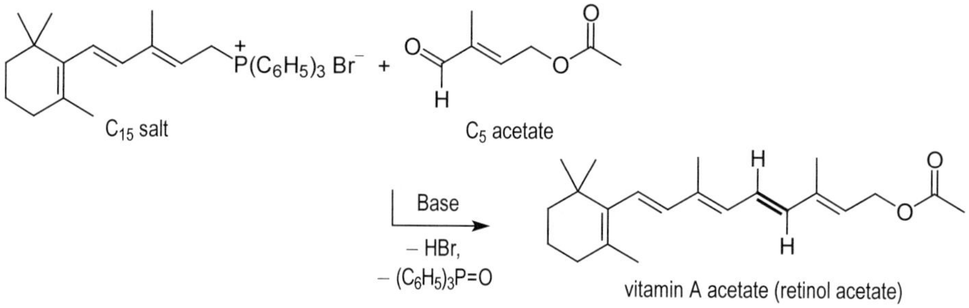

[Reaction scheme: C_{15} salt $\overset{+}{P}(C_6H_5)_3\,Br^-$ + C_5 acetate → (Base, –HBr, –$(C_6H_5)_3P=O$) → vitamin A acetate (retinol acetate)]

An intermolecular MCMURRY deoxygenative coupling of retinal (vitamin A aldehyde) finally yields β-carotene [36].

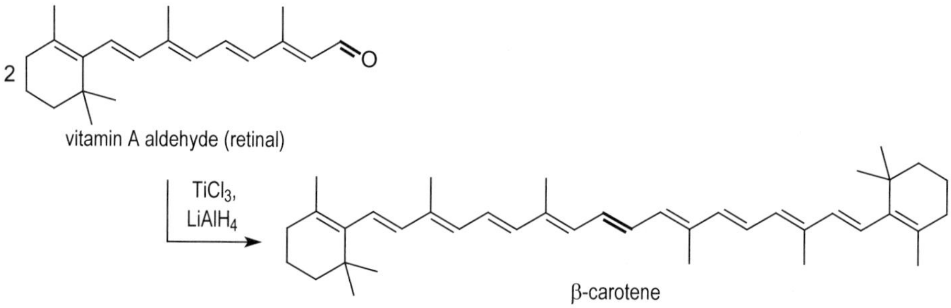

[Reaction scheme: 2 vitamin A aldehyde (retinal) → (TiCl₃, LiAlH₄) → β-carotene]

9.3.2 Cafestol

(−)-Cafestol, an antiinflammatory diterpenoid found in coffee belongs to the class of rearranged kauranes (section 4.5.3). An obvious design of its synthesis suggests introducing the 1,2-diol substructure of cafestol by dihydroxylation of the exocyclic C-16–C-17 double bond in the precursor **1**. Additional double bonds in the positions 5,6 and 11,12 of the precursor **2**, the purposes of which are less obvious at first glance, arise from the idea of COREY and coworkers [37] to use the ring strain of the cyclopropylmethylium ion **4** as a vehicle to build up the kaurane pentacycle: thus, **4** is expected to generate the precursor **2** via carbenium ion **3**. Protonation of the primary alcohol **5** obtained by reduction of the carboxylic acid ester function in **6** should give rise to the intermediate carbenium ion **4**. The additional keto function in **6** should enable preparation of the carbene intermediate **7** required for cyclopropanation by an intramolecular [2+1]-cycloaddition on one hand, and of the β-ketoester **8** by CLAISEN condensation with the ester **9** on the other hand. The ester **9** on its part could be made available by alkylation of 1,3-cyclohexadien-5-carboxylate **10** with the halide **11** as electrophile. The tertiary alcohol **12** is a clear precursor of the dihydrobenzofuran **11** obtainable by alkylation of the tetrahydrobenzofuranone **14**, once in α-position to the carbonyl carbon, giving **13**, and then exactly there, giving **12**. Ultimately, tetrahydrobenzofuranone **14** arises from a FEIST-BENARY furan synthesis reacting 1,3-cyclohexanedione **15** and chloroacetaldehyde **16**.

The synthesis of racemic cafestol reported by COREY and coworkers [37] begins with the predominant enol tautomer **15** of 1,3-cyclohexanedione which cyclizes with chloroacetaldehyde **16** in a solution of sodium hydroxyide in ethanol to tetrahydrobenzofuranone **14**. This is, after deprotonation with lithiumdiisopropylamide (LDA), methylated by iodomethane in α-position to the carbonyl group, yielding **13**. In a variation of the synthetic concept, the carbonyl function is alkynylated by lithiated trimethylsilylethyne which directly undergoes dehydration to the trimethylsilylenyne **17** in the presence of pyridinium p-toluenesulfonate (PPTS) and magnesium sulfate in benzene. Desilylation with potassium fluoride in dimethylsulfoxide yields the ethyne **18** which, upon hydroboration with diisoamylborane (iA$_2$BH) and subsequent oxidation by hydrogen peroxide, is converted to the aldehyde **19**. Sodium borohydride in methanol reduces **19** to the primary alcohol which is reacted with iodine in the presence of triphenylphosphane and imidazole to the iodethyl compound **11**. The latter alkylates methylcyclohexa-1,3-diene-5-carboxylic acid methyl ester **10** after deprotonation with LDA to the tricyclic ester **9**. In order to prepare the β-ketoester **8**, the ester **9** is hydrolyzed with sodium hydroxide; 1,1′-carbonyldiimidazole converts the resulting carboxylate to the imidazolide which is reacted with α-lithiated t-butylacetate. Subsequent reaction of the β-ketoester **8** with p-toluenesulfonylazide in a solution of diazabicyclo[5.4.0]undec-7-ene (DBU) in dichloromethane yields the diazoketoester **20**.

9.3 Diterpenes

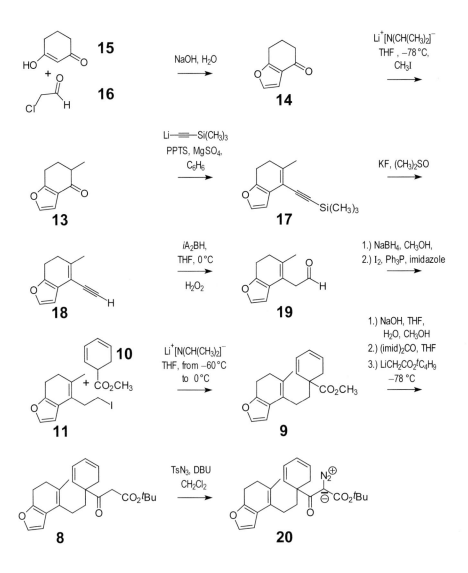

The diazoketoester **20** undergoes cyclopropanation to the unstable pentacycle **6** (major product) when added dropwise to a solution of copper(II)bis-salicylaldehyde-*t*-butylimine in toluene, involving the intermediate carbene **7**. Sodium borohydride reduces **6** in methanol solution to the secondary alcohol which is, for protection, converted to the benzylether **21** by WILLIAMSON synthesis involving deprotonation to the alcoholate by sodium hydride and O-alkylation by benzylbromide so that diisobutylaluminumhydride (DIBAH) reduces the *t*-butylester in **21** to the primary alcohol **22** in dichloromethane.

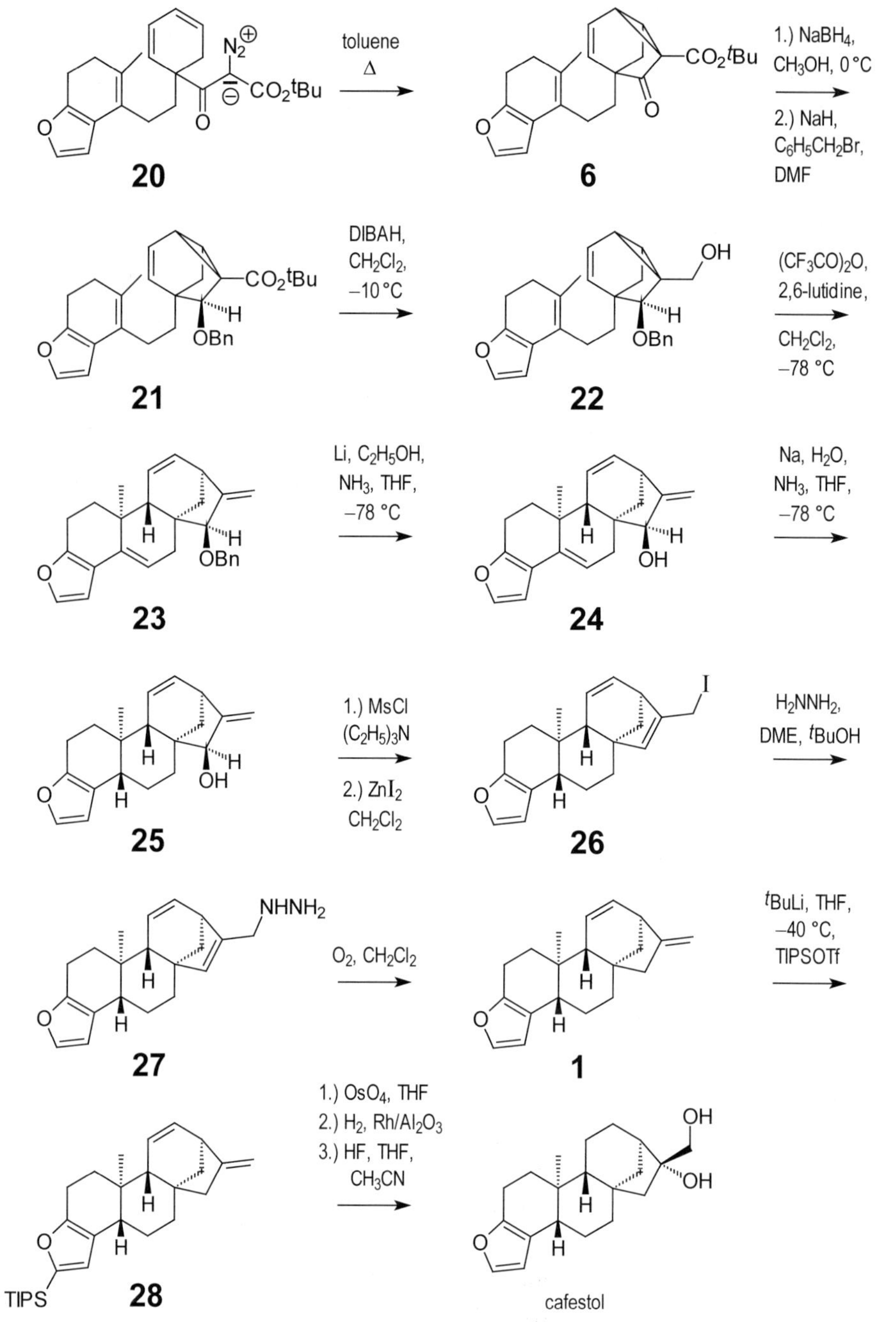

9 Selected Syntheses of Terpenes

20 → (toluene, Δ) → **6** → (1.) NaBH₄, CH₃OH, 0 °C; 2.) NaH, C₆H₅CH₂Br, DMF) → **21**

21 → (DIBAH, CH₂Cl₂, −10 °C) → **22** → ((CF₃CO)₂O, 2,6-lutidine, CH₂Cl₂, −78 °C) → **23**

23 → (Li, C₂H₅OH, NH₃, THF, −78 °C) → **24** → (Na, H₂O, NH₃, THF, −78 °C) → **25**

25 → (1.) MsCl, (C₂H₅)₃N; 2.) ZnI₂, CH₂Cl₂) → **26** → (H₂NNH₂, DME, tBuOH) → **27**

27 → (O₂, CH₂Cl₂) → **1** → (tBuLi, THF, −40 °C, TIPSOTf) → **28**

28 → (1.) OsO₄, THF; 2.) H₂, Rh/Al₂O₃; 3.) HF, THF, CH₃CN) → cafestol

9.3 Diterpenes

The key reaction to the pentacycle **23**, unstable towards acids, is performed by refluxing **22** with trifluoroacetic anhydride in 2,6-lutidine. After deprotection of **23** to the secondary alcohol **24** by lithium in ethanol, the CC double bond in conjugation with the furan ring is reduced by sodium in liquid ammonia. Work-up in tetrahydrofuran (THF) and water predominantly yields the desired *trans*-isomer **25**.

To remove the alcohol function no longer required, **25** is prepared to undergo an allyl rearrangement by reacting it with methanesulfonyl chloride in the presence of triethylamine, so that the allyl iodide **26** is formed upon addition of zinc iodide in dichloromethane, and the exocyclic CC double bond can be regenerated in two steps: Hydrazinolysis of **26** in a mixture of dimethoxyethane and *t*-butylalcohol affords the allylhydrazine **27** which, upon oxidation to the diimine involving a 1,5-sigmatropic hydrogen shift followed by elimination of nitrogen, provides the cafestol precursor **1**. Prior to catalytic hydrogenation, the furan ring must be protected by α-lithiation and subsequent reaction with triisopropylsilyltriflate (TIPSOTf) to the α-triisopropylsilylfuran **28**. The last steps to racemic cafestol are accomplished by dihydroxylation of the exocyclic CC double bond with osmium tetroxide in THF, catalytic hydrogenation of the C-11–C-12 double bond in the presence of rhodium and aluminum oxide and, finally, desilylation of the furan ring with hydrogen fluoride in a mixture of THF and acetonitrile.

9.3.3 Baccatin III as the Precursor of Taxol

Baccatin III occurs in the European yew tree *Taxus baccata*. It is an attractive precursor in the partial syntheses of (–)-taxol isolated from the pacific relative *Taxus brevifolius* which is used in the chemotherapy of leukemia and various types of cancer. In competition with various other groups, a total synthesis of baccatin III was achieved by NICOLAOU and coworkers [38]. When designing the synthesis, they supposed protection of the 1,2-diol by cyclocarbonate to be valid after deoxygenation of baccatin III in allyl position (C-13) to the precursor **1**. Retrosynthetic disconnection of the oxetane ring in the cyclocarbonate **2** leads to the alkene **3** in which the *vicinal* alcohol functions are masked as acetone ketal; the oxetane ring is then expected to close by addition of the primary alcohol group to the C-5–C-6 double bond in **3**. The introduction of functional groups in 9,10-position may then be prepared by a double bond in **4**, which is found to result from an intramolecular MCMURRY deoxygenative coupling of the dialdehyde **5** which, in its part, is the product of oxidation of the (protected) primary alcohol functions in **6**.

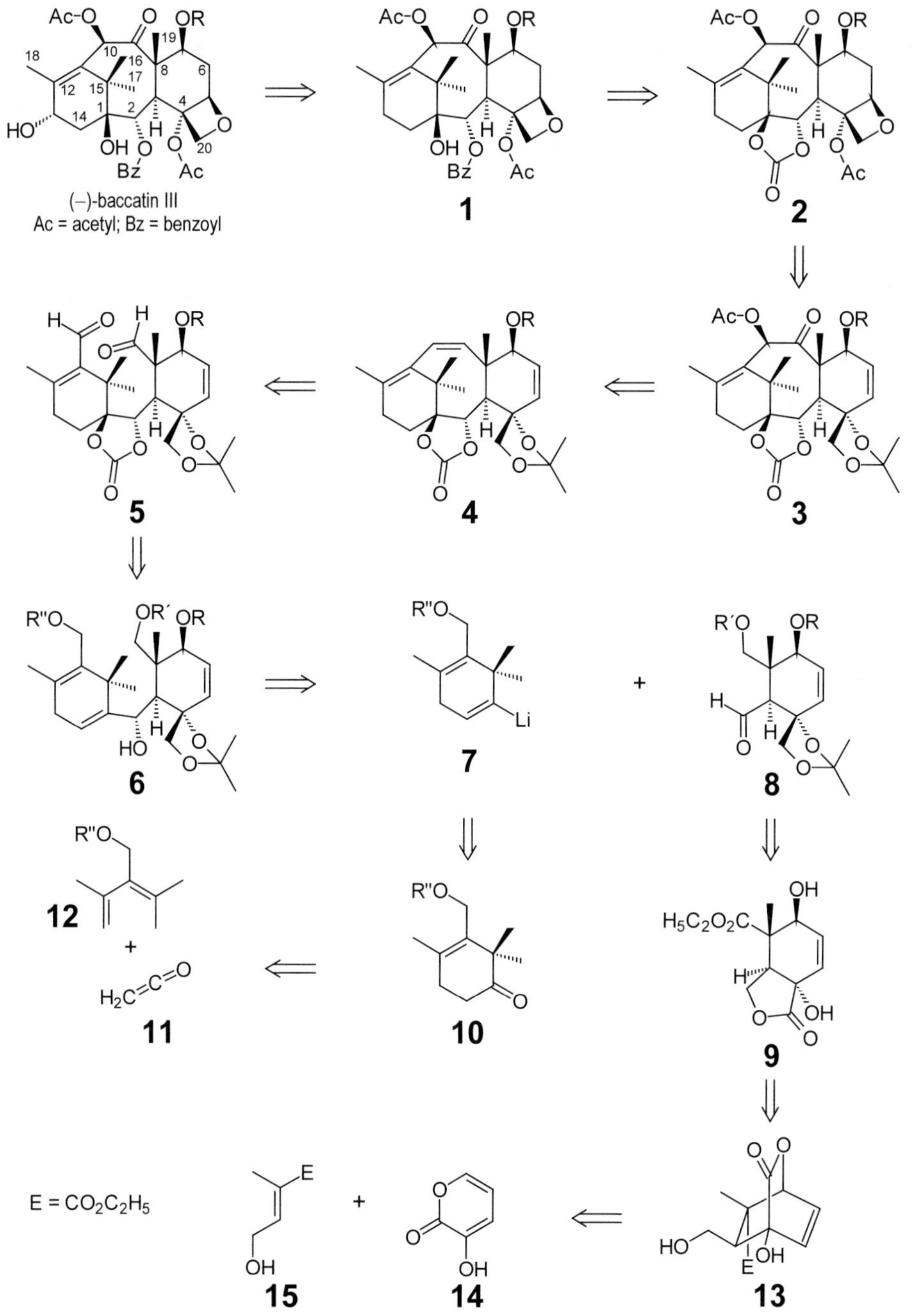

146   9 Selected Syntheses of Terpenes

(−)-baccatin III
Ac = acetyl; Bz = benzoyl

E = CO₂C₂H₅

9.3 Diterpenes

The allylalkohol **6** turns out to be the obvious precursor of the protected diol **5**, and **6** reasonably arises from a SHAPIRO coupling of cyclohexadienyllithium **7** with the cyclohexene-4-aldehyde **8** which is the product of oxidation and subsequent diol protection of the bicyclic hydroxylactone **9**; the latter emerges from rearrangement of the DIELS-ALDER cycloadduct **13** of 3-hydroxy-2-pyrone **14** as the diene and 4-hydroxy-2-methyl-2-butenoate **15** as the electron-deficient dienophile. The cyclohexadienyllithium **7** originates from the sulfonylhydrazone of the ketone **10** which is, once again, a DIELS-ALDER cycloadduct of the protected 3-hydroxymethyl-2,4-dimethyl-1,3-pentadiene **12** and ketene **11** as the dienophile.

The arylsulfonylhydrazone **18** is a stable precursor of the cyclohexadienyllithium **7**. In order to prepare this starting reagent, 3-acetoxymethyl-2,4-dimethyl-1,3-pentadiene **12** is subjected to a DIELS-ALDER reaction with the ketene equivalent chloroacrylnitrile **11a**. The cycloadduct **16** primarily obtained is hydrolyzed to the hydroxyketone **17** in *t*-butyl alcohol. Subsequent reaction with *t*-butyldimethylsilylchloride (TBSCl) and imidazole in dichloromethane serves to protect the primary alcohol function of the intermediate **10** in which the keto carbonyl group is derivatized to the required arylsulfonylhydrazone **18** with 2,4,6-triisopropylphenylsulfonylhydrazide in tetrahydrofuran (THF) as solvent.

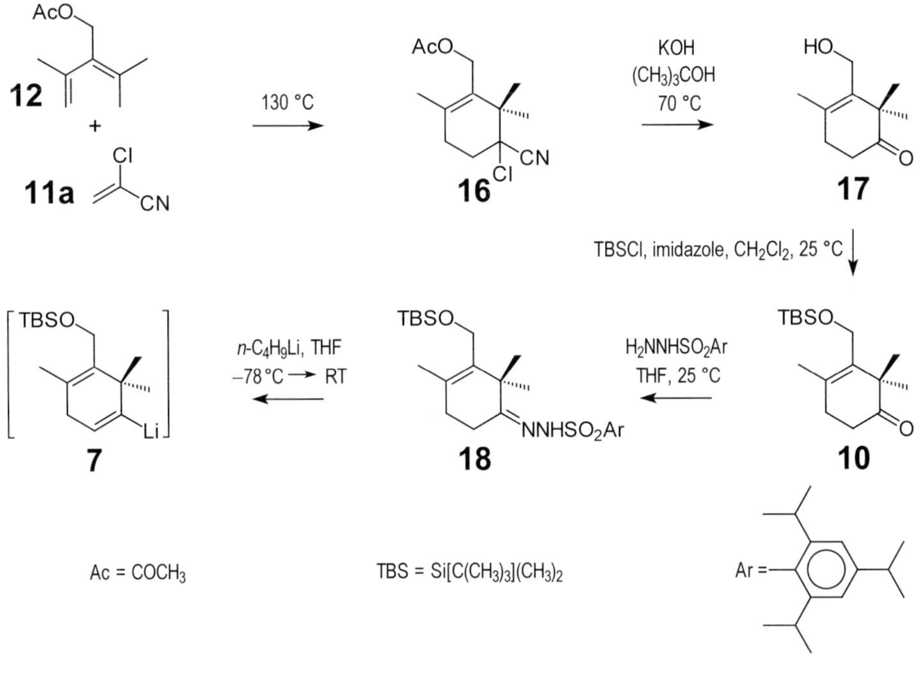

In order to prepare the cyclohexenaldehyde **8**, 3-hydroxy-2-pyrone **14** and ethyl 4-hydroxy-2-methyl-2-butenoate **15** are subjected to a DIELS-ALDER reaction in the presence of phenylboronic acid which arranges both reactants to the mixed boronate ester **19** as a template to enable a more efficient intramolecular DIELS-ALDER reaction with optimal control of the regiochemical course of the reaction. Refluxing in benzene affords the tricyclic boronate **20** as primary product. This liberates the intermediate cycloadduct **21** upon transesterification with 2,2-dimethylpropane-1,3-diol which, on its part, relaxes to the lactone **22**. Excessive *t*-butyldimethyl-silyltriflate (TBSTf) in dichloromethane with 2,6-lutidine and 4-*N,N*-dimethyl-aminopyridine (DMAP) as acylation catalysts protects both OH goups so that the primary alcohol **23** is obtained by subsequent reduction with lithiumaluminum-hydride in ether.

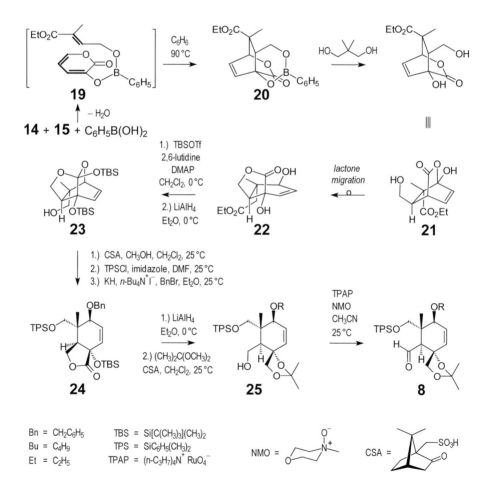

9.3 Diterpenes

Catalytic amounts of camphor-10-sulfonic acid (CSA) in methanol and dichloromethane smoothly cleave the orthoester function in **23**, giving the intermediate γ-lactone-1,3-diol. Subsequent protection of the primary alcohol function with dimethylphenylchlorosilane (TPSCl) in dimethylformamide with imidazole as the base and of the secondary alcohol function *via* alcoholate with benzylbromide (BnBr) following WILLIAMSON's ether synthesis yields the γ-lactone **24**. Its reduction with lithiumaluminumhydride leads to two *vicinal* primary alcohol groups. Thereafter, camphor-10-sulfonic acid (CSA) in dichloromethane selectively cleaves the TBS ether and catalyzes the transketalization with acetone dimethylketal to the precursor **25** of the aldehyde **8**. Smooth oxidation of the primary alcohol function in **25** is achieved with tetrapropylammoniumperruthenate (TPAP) and *N*-methylmorpholine-*N*-oxide (NMO) in acetonitrile.

SHAPIRO coupling of cyclohexadienyllithium **7**, prepared by reacting the sulfonylhydrazone **18** with butyllithium in THF, with the cyclohexenaldehyde **8**, leads exclusively to the desired stereoisomer **6**. The unexpected selectivity probably arises from steric overcrowding of the *Si* face of the prochiral chelated aldehyde carbonyl in **8** [38], thus enabling the nucleophilic alkenyllithium to approach predominantly from the less-hindered *Re* face.

The coupling product **6** incorporates an allylic alcohol substructure which is epoxidized by *t*-butylhydroperoxide in benzene catalyzed by small amounts of vanadyl(IV)acetylacetonate following the SHARPLESS epoxidation. Lithiumaluminumhydride in ether reductively opens the oxirane ring, giving the *trans*-diol **27** which is, after deprotonation to the *trans*-diolate by potassium hydride in hexamethylphosphoric acid triamide (HMPA), reacted for protection at room temperature with phosgene to the cyclocarbonate **28**. Tetra-*n*-butylammoniumfluoride (TBAF) in THF at room temperature cleaves both silyl protective groups thus liberating the primary alcohol groups for oxidation to the dialdehyde **5** required for intramolecular MCMURRY coupling by tetrapropylammoniumperruthenate (TPAP) and *N*-methylmorpholine-*N*-oxide (NMO) in dichloromethane. The conditions of the MCMURRY reaction are modified in order to prepare the 9,10-diol **29** more closely related to the baccatin structure than to the alkene. To do so, excessive titanium(III)chloride in dimethoxyethane (DME) with excessive zinc-copper couple as reducing agent is applied. This step leads in moderate yields to the racemic *cis*-diol **29** which is reacted with (1*S*)-(–)-camphanic chloride in dichloromethane and triethylamine as the base to the diastereomeric camphanates **30**. Diastereomers are separated chromatographically and identified by X-ray crystallography. The desired dextrarotatory diol **29** obtained by hydrolysis of its camphanate is used for the remaining steps of the synthesis.

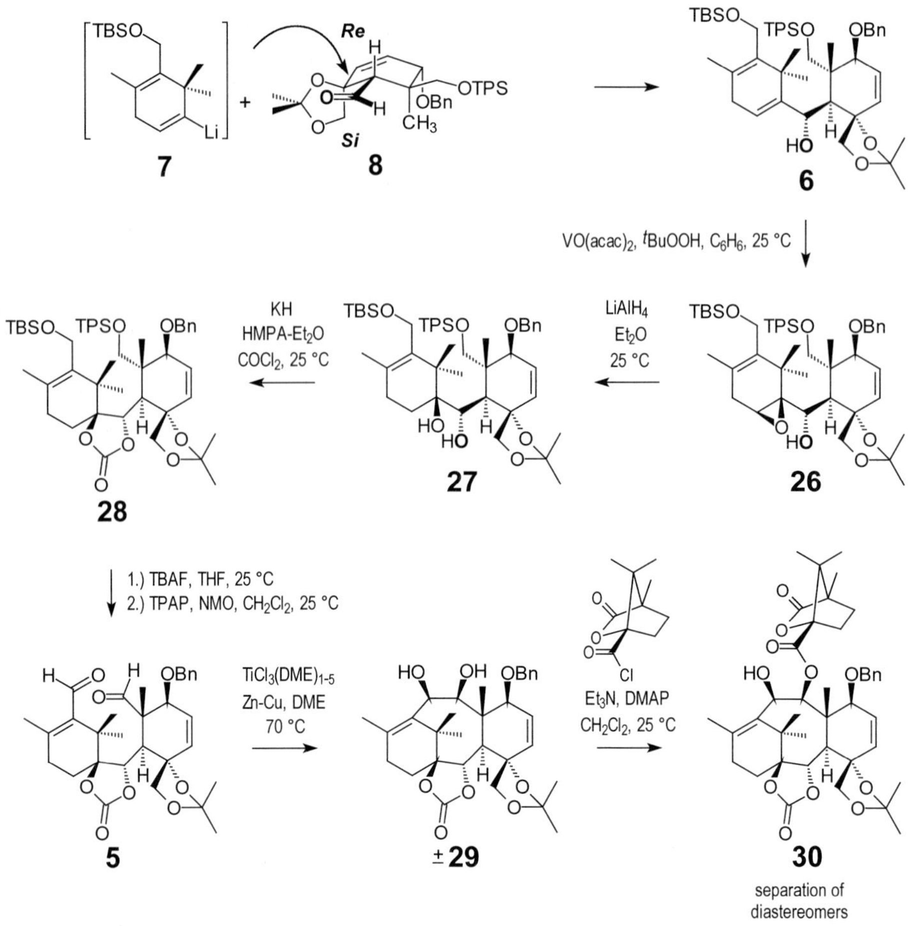

The alcohol functions of the *cis*-diol **29** clearly exhibit different reactivities: camphanic chloride selectively acylates the 9-OH group to **30**, while acetic anhydride in dichloromethane acylates the allyl alcohol function 10-OH at room temperature in the presence of DMAP as acylation catalyst. There is no straightforward explanation of this advantageous selectivity of acylation [38], which enables oxidation of the free 9-OH group with TPAP and NMO in dichloromethane to the α-acetoxyketone **3**. In contrast, the selectivity of the subsequent hydroboration (BH$_3$/THF) and oxidation (H$_2$O$_2$, NaHCO$_3$) at the sterically less hindered C-5–C-6 double bond in favor of the desired alcohol **31** is reasonable. The undesirable regioisomeric C-6-alcohol is the minor product. From the hydroxy functions deprotected by hydrolysis of the ketal with hydrogen chloride in methanol and water at room temperature, the sterically less-hindered primary alcohol group is acetylated selectively. After re-

9.3 Diterpenes

moval of the benzyl protective group by hydrogenation in ethyl acetate, 7-OH is protected once again by triethylchlorosilane, while 5-OH is activated with methanesulfonyl chloride. This enables C-20-OH, deprotected by potassium carbonate in methanol and water, to substitute intramolecularly the C-5-mesylate in butanone containing tetra-*n*-butylammonium acetate, thus closing the oxetane ring in **34**. Acetylation of the tertiary 4-OH catalyzed by DMAP is achieved in spite of steric hindrance, and phenyllithium in THF opens the cyclocarbonate ring at −78 °C to the benzoate **35**. In order to introduce the allyl alcohol function C-13-OH, pyridiniumchlorochromate (PCC) oxidizes in the allyl position in benzene solution, yielding the enone precursor. Sodium borohydride in methanol solution finally reduces the enone to the target 7-*O*-triethylsilylbaccatin III.

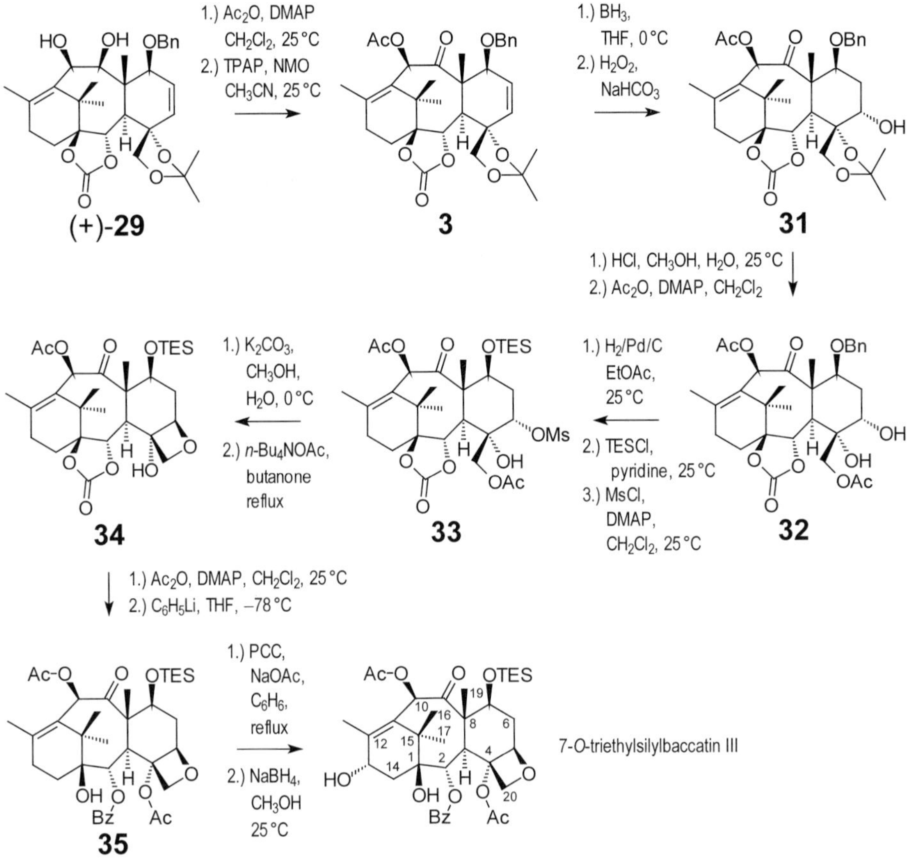

9.4 Triterpenes

9.4.1 Lupeol

Well-documented total syntheses of triterpenes appear rather sparingly in the literature. The synthesis of racemic lupeol reported by G. STORK and coworkers [39] is an example. (+)-Lupeol represents a pentacyclic triterpene most frequently occurring in plants. It was useful for the design of the synthesis to know that genuine (+)-lupeol can be degraded to the pentacyclic ketal ester **1** [39]. Excessive methyllithium adds to **1** in refluxing 1,4-dioxane, yielding the ethyleneketal of lupan-20-ol-3-one. Dehydration with phosphorylchloride in pyridine regenerates the isopropenyl group. After hydrolysis of the ketal, sodium borohydride in methanol reduces the ketone to authentic (+)-lupeol. To conclude, the ketalester **1** as a precursor of lupeol will be an attractive target for a total synthesis.

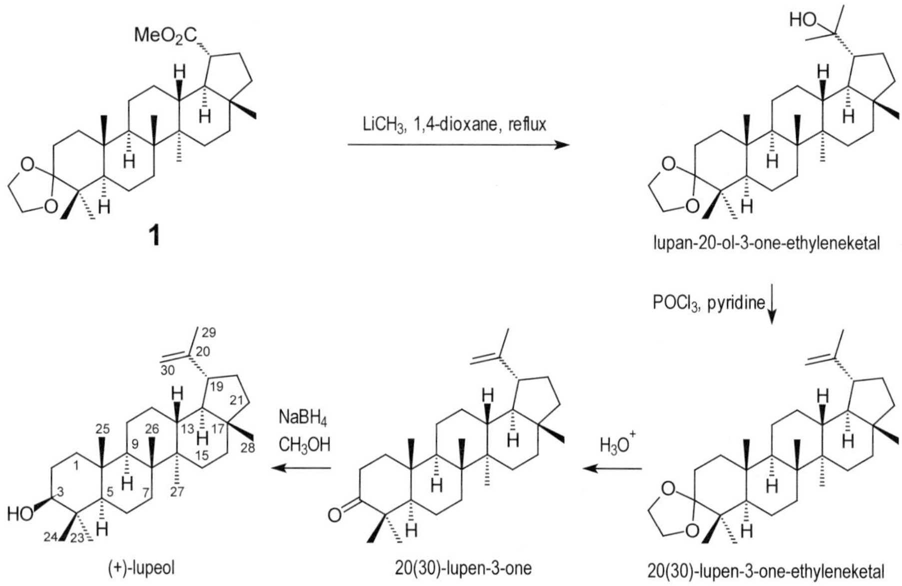

Retrosynthetic disconnection of the ketal ester **1** appropriately begins at the cyclopentane ring *E*, which is expected to close by an intramolecular nucleophilic substitution of OR⁻ (tosylate) by the carbanion α to the methoxycarbonyl function in **2**. The primary alcohol as the precursor of the methylether **2** turns out to be the product of esterification and reduction of the aldehyde carboxylic acid which is formed by ozonolysis of the enol **3** as a tautomer of the ketone **4**.

9.4 Triterpenes

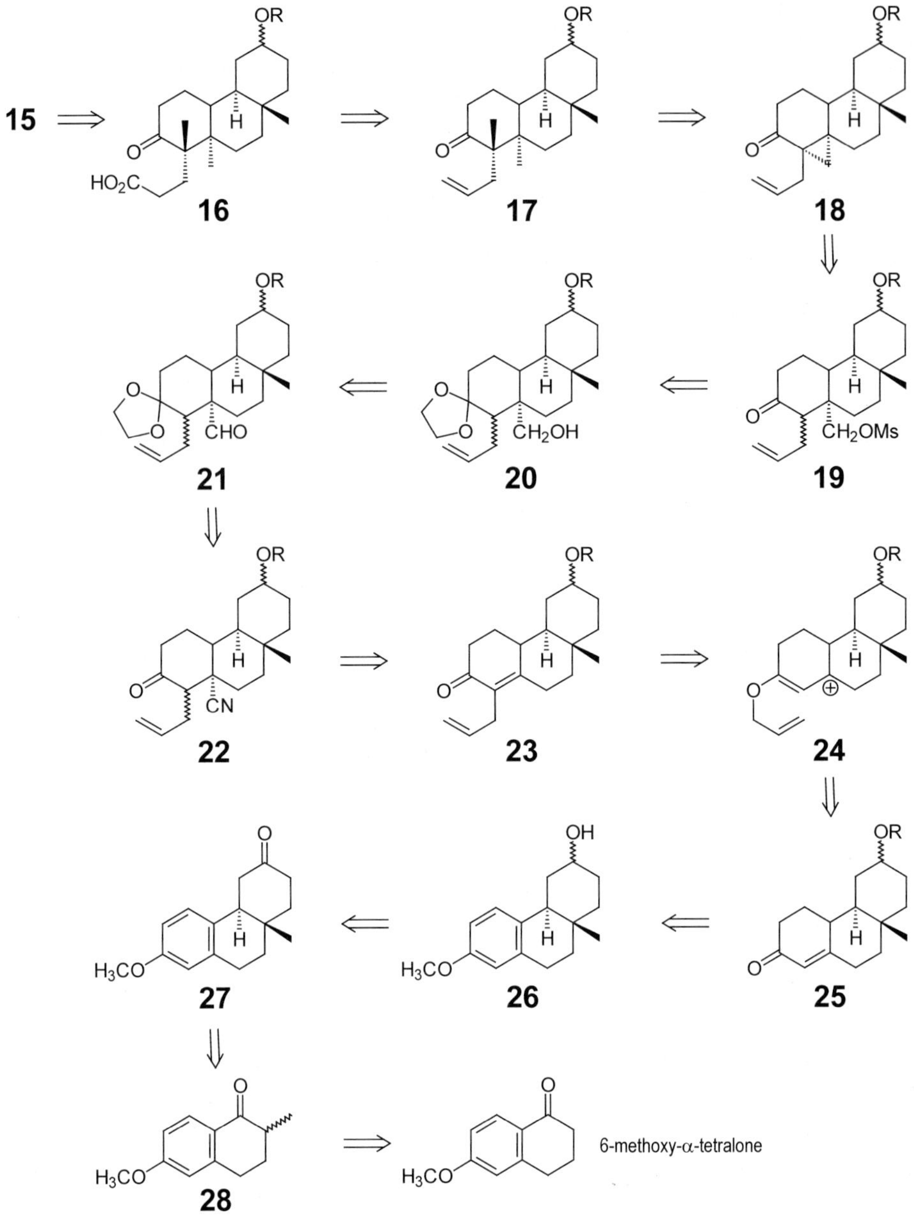

154   9 Selected Syntheses of Terpenes

6-methoxy-α-tetralone

Ketone **4** clearly results from oxidation of the secondary alcohol **5** which is, of course, a ketal derivative of the hydroxy-α,α-dimethylketone **6**. The latter arises from methylation of the mono-α-methylketone **7** obtained by hydrogenation of the CC double bond in the enone **8** which, on its part, originates from an intramolecular KNOEVENAGEL alkenylation of the diketone **9**. **9** is expected to be synthesized from the δ-keto acid **11** *via* the corresponding enol lactone **10** involving a nucleophilic addition of ethylmagnesium halide to the electrophilic lactone carbonyl carbon. The preparation of cyclic enones of type **8** from δ-keto acids and the parent enol lactones by addition of alkylmagnesium halide followed by intramolecular KNOEVENAGEL alkenylation was successfully applied for the synthesis of various steroids [40].

The δ-keto acid **11** is supposed to be formed by hydroboration and oxidation of the terminal CC double bond in **12**, feasible by allylation of the enone **13**. Enone **13**, once again, originates from an intramolecular KNOEVENAGEL alkenylation of the α,δ-diketone **14** prepared by addition of ethylmagnesium halide to the enol lactone **15** of the δ-keto acid **16**. Preceding this keto acid, the allyl compound **17** is formed by nucleophilic opening of the cyclopropane ring with hydride and methylation in α position of the carbonyl in the cyclopropyl ketone **18**.

Cyclopropanation will be achieved by the mesylate **19** of the primary alcohol **20**, obtained by reduction of the aldehyde **21**, provided that the keto function is protected as a ketal before reduction. The aldehyde is introduced *via* the imine by reduction of the β-keto nitrile **22**, formed by nucleophilic addition of hydrogen cyanide to the electrophilic double bond of the enone **23**. The latter originates from a thermal oxy-COPE rearrangement of the intermediate tricyclic allylenolether carbenium ion **24** derived from the enone **25**. This emerges from reduction of the arylmethylether **26** and its precursor **27**. 1,9,10,10a-Tetrahydro-7-methoxy-3(2*H*)-phenanthrone **27** is readily available by anellation of 2-methyl-6-methoxy-α-tetralone **28** which can be synthesized from commercial 6-methoxy-α-tetralone by various procedures; these include α-methylation of the ketone involving the enamine or the enolether; alternatively, the tetralone **28** could be subjected to CLAISEN condensation with a formic acid ester yielding the α-hydroxymethyleneketone so that the methyl group would be introduced by reduction of the hydroxymethylene function.

The synthesis of lupeol [39] starts with the cyclization of 6-methoxy-β-methyl-α-tetralone **28** with 4-*N,N*-dimethylamino-2-butanone methiodide in the presence of potassium *t*-butanolate to 1,9,10,10a-tetrahydro-7-methoxy-3(2*H*)-phenanthrone **27**. Reduction with sodium borohydride and subsequent hydrogenation of the enone CC double bond in the presence of palladium and strontium carbonate as slightly deactivated catalyst gives the octahydrophenanthrol **26**. Partial reduction of the benzenoid ring to the enone is accomplished by lithium in liquid ammonia. The enone is derivatized to the benzoate **25** in order to protect the hydroxy group prior to the subsequent synthetic steps.

Conversion of the benzoate **25** to the allylenolether intermediate succeeds with a mixture of triallyl orthoformate and allyl alcohol in tetrahydrofuran (THF) at room temperature, catalyzed by *p*-toluenesulfonic acid; oxa-COPE rearrangement to the α-allylenone **23** occurs by refluxing in pyridine. Nucleophilic addition of diethylaluminum cyanide to the enone in a mixture of benzene and toluene gives the cyanoketone **22** which, after protection of the keto function as a 1,3-dioxolane, is reduced by lithiumaluminumhydride to the aldehyde **21** involving the intermediate imine. After rebenzoylation required because of hydrolysis of the benzoate during work-up, sodium borohydride is used to reduce the aldehyde **21** to the primary alcohol **20**. The mesylate **19** obtained by mesylation of **20** with methanesulfonyl chloride in pyridine undergoes simultaneous deprotection and cyclopropanation upon addition of diluted hydrochloric acid in THF solution. After rebenzoylation with benzoyl chloride in pyridine, the cyclopropane ring opens to the methyl compound with lithium in liquid ammonia, and the resulting intermediate directly undergoes reductive methylation by iodomethane in hexamethylphosphoric acid triamide α to the carbonyl function in **17**. Hydroboration and oxidation of the terminal vinyl group affords the propionic acid **16** and the corresponding enol lactone **15** crystallizing in the reaction mixture. Addition of ethylmagnesium bromide in diethylether and THF primarily yields the α,δ-diketone **14** (p. 153) which immediately cyclizes to the enone **13** during work-up in aqueous sodium hydroxide and methanol solution.

9.4 Triterpenes

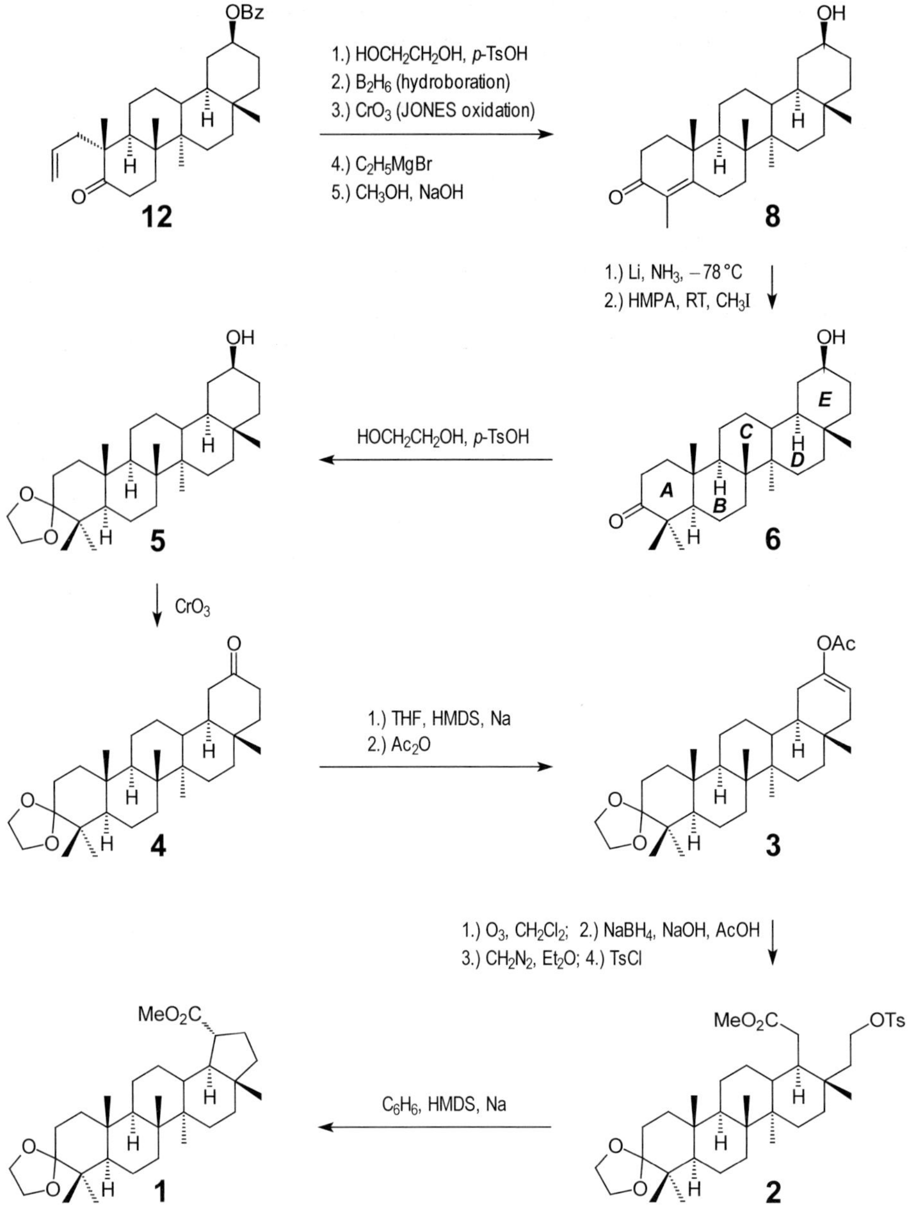

158   9 Selected Syntheses of Terpenes

9.4 Triterpenes

After trapping the enolate of **13** with triallyl orthoformate in the presence of *p*-toluenesulfonic acid at room temperature, subsequent oxy-COPE rearrangement of the resulting allylenolether in refluxing pyridine and benzoyl chloride affords the allylketobenzoate **12**. Ketalization to the 1,3-dioxolane protects the keto function in **12** so that a sequence of hydroboration, JONES oxidation, and 1,2-addition of ethylmagnesium bromide affords the benzoate enone **8** after rebenzoylation, involving the δ-keto acid **11**, its enol lactone **10** (p.153) and the α,δ-diketone **9** as intermediates. The *geminal* methyl group is inserted by reducing the CC double bond of the enone with lithium in liquid ammonia and by reductive methylation of the enolate anion / carbanion of the resulting cycloalkanone **7** (p. 153) with iodomethane. The desired α,α-dimethylketone **6** must be separated chromatographically from some byproducts also formed in the reaction.

In order to contract the cyclohexane ring *E* in the pentacyclic ketoalcohol **6**, the keto function is protected as a 1,3-dioxolane **5** prior to oxidation of the secondary alcohol function of ring *E* to the ketone **4**. Its enolate generated with sodium and hexamethyldisilazane (HMDS) in THF is trapped as enol acetate **3** with acetic anhydride. Ozonolysis of the enol acetate **3** in dichloromethane followed by reduction with sodium borohydride in sodium hydroxide at 0 °C, smooth acidification with diluted aqueous acetic acid, esterification of the resulting carboxylic acid with diazomethane, and tosylation affords the tosylate ester **2**. The latter finally cyclizes in a modified DIECKMANN reaction when heated in benzene solution upon addition of HMDS and sodium to the target **1** as the desired precursor of racemic lupeol.

10 Isolation and Structure Elucidation

10.1 Isolation from Plants

Various volatile mono-, sesqui- and diterpenes are used as flavors and fragrances because of their pleasant taste and odor. Enriched mixtures of terpenes are obtained from the chopped parts of the plants (seed, flowers, fruits, leaves, stems, roots and rhizomes) on a larger scale by steam distillation or by extraction in the ethereal oils, where ethereal means only volatile. Owing to their odor some ethereal oils serve as valuable raw materials used in perfumery; others are used to spice foods because of their taste, or serve as phytomedicines because of the pharmacological activity of their constituents. Various nonvolatile higher terpenes isolated from plants also play an important role as pharmaceuticals or as emulsifying agents in the pharmaceutical and nutrition industries.

Qualitative and quantitative analysis of ethereal oils is usually achieved by gas-chromatography (GC), and the combination of this method of separation with mass spectrometry (GC-MS) for identification. Pure terpenes are obtained on a larger scale from ethereal oils by distillation; chromatographic methods of separation predominantly in the gas (GC) or liquid phase (LC) permit the isolation of small amounts with high purities.

In order to isolate low-volatile sesqui-, di-, sester- and triterpenes with polar groups from plants, fungi, and other organisms, the natural material is dried, chopped or ground and then extracted with inert solvents at the lowest possible temperature in order to prevent the formation of artifacts. Petroleum ether is a suitable solvent for the extraction of less-polar terpenes. Polar terpenes including saponins are extracted with water, ethanol, or methanol. The extract is evaporated to dryness in vacuum or freeze-dried and then fractionated by column chromatography. Well-resolved spots in thin-layer chromatography (TLC) localize separable constituents in the crude fractions which are usually further purified by column chromatography. For elution of the constituents, petroleum ether or cyclohexane with increasing concentrations of more polar solvents such as dichloromethane, chloroform, methanol or ethanol are applied according to the experience based on preceding TLC analysis. Final purifications of the constituents for spectroscopic identification or structure elucidation and pharmacological screenings are frequently performed using liquid chromatography with medium- or high-pressure (MPLC or HPLC).

10.2 Spectroscopic Methods of Structure Elucidation

In exceptional cases, terpenes crystallize after chromatographic purification, thus enabling determination of their three-dimensional structure in the solid state by X-ray crystallography. For the most part, an amorphous or oily material rather than a single crystal is obtained after chromatographic separation, and high-resolution NMR has been identified as the most efficient method to elucidate the three-dimensional structure of molecules [41-44] in solution, requiring sample quantities of less than one mg.

Other spectroscopic methods such as UV- and visible light absorption- and IR-spectroscopy [41] predominantly permit the identification of known terpenes, e.g., by computer-assisted spectral comparison on the basis of digitized spectroscopic data files or spectra catalogues. In the case of unknown terpenes, chromophores such as carbonyl groups (with an absorption maximum at about 280 nm) and their structural environment (conjugation) can be detected by UV spectroscopy. Nearly all functional groups of a terpene are identified in the IR spectrum by means of characteristic vibration frequencies; OH single bonds, for example, vibrate with 3600 cm^{-1}, carbonyl double bonds with 1700 cm^{-1}, and carbon-oxygen single bonds with 1200 cm^{-1} detected as absorption bands in the IR spectra.

Mass spectrometry (MS) [46] detects the molecular mass of a compound with a precision of 10^{-4} mass units. Owing to the isotope mass defect of elements the molecular formula of a terpene can be determined by high-resolution mass spectrometry of the molecular ion. For example, the molecular formula $C_{17}H_{22}O_4$ of acanthifolin from *Senecio acanthifolius* (Asteraceae) [47], is calculated from the molecular mass of 290.1525 determined by high-resolution mass spectrometry of the molecular ion. Additionally, partial structures of molecules can be derived from the masses of the ions arising from fragmentation of the molecular ion and detected in the mass spectrum.

10.3 Structure Elucidation of a Sesquiterpene
Elucidation of Acanthifolin by NMR

10.3.1 Double Bond Equivalents

High-resolution molecular mass analysis provides the molecular formula $C_{17}H_{22}O_4$ of acanthifolin, which corresponds to seven double bond equivalents. The proton broadband decoupled ^{13}C NMR spectrum (Fig. 6a) displays all 17 carbon atoms of the molecule, including a carboxy group ($\delta_C = 170.5$) and four additional C atoms in the sp² range of ^{13}C chemical shifts ($\delta_C = 145.8, 138.1, 119.8$ and 116.2), indicating two CC double bonds. Only three of all seven double bonds of the molecule are detected by NMR. To conclude, acanthifolin incorporates a tetracyclic ring system [47].

10.3.2 Functional Groups and Partial Structures detected by ^{13}C NMR

The coupled ^{13}C NMR spectrum (Fig. 6b) and the DEPT subspectra (Fig. 8c, d) for unequivocal detection of CH multiplicities (C, CH, CH$_2$, CH$_3$) show that acanthifolin contains six non-protonated (C$_6$), four CH (C$_4H_4$), three CH_2 (C$_3H_6$) and four CH_3 carbon atoms (C$_4H_{12}$). These fragments (C$_6$ + C$_4H_4$ + C$_3H_6$ + C$_4H_{12}$) sum up to the CH partial elementary composition C$_{17}H_{22}$ (Table 9, p. 164) in accordance with the molecular formula C$_{17}H_{22}O_4$ determined by mass spectrometry. In conclusion, no OH group is present in the molecule.

In the ^{13}C chemical shift range of alkoxy groups with $\delta_C > 60$ (Fig. 6), the doublet signal with $\delta_C = 61.1$ attracts attention because of its very large CH coupling constant ($J_{CH} = 177$ Hz); its value identifies the CH bond of an oxirane ring. This coupling occurs only once, so that the oxirane ring turns out to be trisubstituted with the second, non-protonated ring carbon detected at $\delta_C = 63.1$ [47].

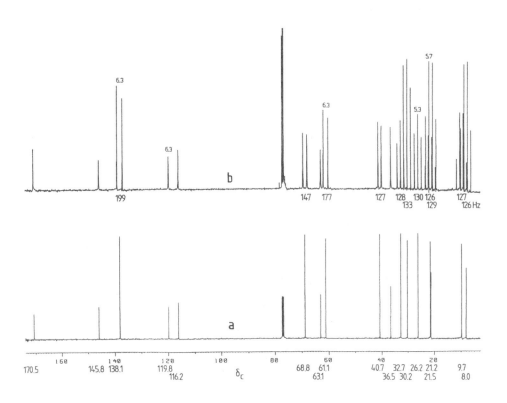

Figure 6. ^{13}C NMR spectra (100 MHz) of acanthifolin in deuterochloroform, CDCl$_3$, a) with, b) without proton broadband decoupling [47].

10.3 Structure Elucidation of a Sesquiterpene

In the ^{13}C chemical shift range of CC double bonds, the doublet signal at δ_C = 138.1 with its outstandingly large CH coupling constant (J_{CH} = 199 Hz; Fig. 6b) reveals the α-CH fragment of an enol ether. This result, and a total of four signals for carbons contributing to CC double bonds between δ_C = 116 and 146 including one CH and three non-protonated carbons, reasonably indicate a trisubstituted furan ring.

The functional groups and partial structures found so far (carboxy group, furan and oxirane ring) sum up to five of all seven double bond equivalents. Thus, the remaining two double bond equivalents reveal two additional rings.

10.3.3 Skeletal Structure (Connectivities of Atoms)

Two-dimensional NMR correlation experiments [45] are the actual methods to determine the skeletal structure (constitution, atom connectivities) of organic compounds [48]. Cross signals of the frequently used homonuclear HH COSY experiment (Fig. 7) detect the proton connectivities (*geminal, vicinal* and longer-range proton-proton relationships). Analysis of the ^1H NMR spectrum (Fig. 7, above the HH COSY plot) between δ_H = 2 and 3 seems difficult at first glance because of signal overcrowding. A two-dimensional CH correlation resolves these signals in the second ^{13}C dimension, performed either by the CH COSY with ^{13}C detection (Fig. 8) or by the more sensitively ^1H detected and therefore much less time-consuming HMQC experiment, also referred to as inverse CH COSY. ^1H and ^{13}C chemical shifts of the CH bonds of a molecule are detected by the coordinates of the cross signals in the contour plots of these experiments. Reading Fig. 8, the proton with δ_H = 7.04 is attached to the previously mentioned carbon with δ_C = 138.1 in α position of the furan ring, and the protons with δ_H = 2.04 and 2.17 forming an AB spin system are attached to the carbon with δ_C = 26.2; this identifies a methylene group with non-equivalent protons. Table 9 summarizes all CH bonds assigned by the CH-correlation (Fig. 8) [47].

With the CH_2 fragment (protons with δ_H = 2.04 and 2.17 attached to the carbon with δ_C = 26.2) assigned by Figure 8 and Table 9 as mentioned above, the cross signals of the HH-COSY (Fig. 7) reveal the inserted partial structure **1a**.

Additional partial structures **1b-f** (Table 10) are detected by two-dimensional CH-correlation experiments using pulse sequences adjusted to the much smaller CH coupling constants of C and H nuclei separated by two, three, or more bonds. Such experiments are known as the CH COLOC with ^{13}C detection (Fig. 9) or as the more sensitively ^1H detected and therefore less time-consuming HC HMBC. Contour plots of these experiments (Fig. 9) permit the localization of carbon atoms two or three bonds apart from a certain proton. Thus, the methyl protons with δ_H = 2.02 in Fig. 9 display cross signals with the carbon nuclei at δ_C = 170.5 and δ_C = 68.8.

To conclude, this methyl group belongs to an acetoxy function attached to the carbon with $\delta_C = 68.8$ (Table 10, partial structure **1d**).

Table 9. CH connectivities (δ_C, δ_H) from Fig. 8, CH multiplicities from Fig. 6 and 8, as well as CH coupling constants J_{CH} from Fig. 6.

δ_C	CH_n	δ_H	J_{CH} [Hz]
170.5	C	----	----
145.8	C	----	----
138.1	CH	7.04	199.5
119.8	C	----	----
116.2	C	----	----
68.8	CH	5.13	147.3
63.1	C	----	----
61.1	CH	3.05	177.0
40.7	CH	1.72	129.5
36.5	C	----	----
32.7	CH_2	2.13, 2.84	126.0
30.2	CH_2	2.19, 3.06	130.5
26.2	CH_2	2.04, 2.17	129.0
21.5	CH_3	1.15	126.0
21.2	CH_3	2.02	129.5
9.7	CH_3	1.04	126.0
8.0	CH_3	1.89	127.0
$C_{17}H_{22}$			

Table 10. Partial structures derved from the CH correlation in Fig. 9: carbon atoms two, three or four bonds apart from the protons with δ_H = 1.15 (**1b**), 1.89 (**1c**), 2.02 (**1d**), 2.84 (**1e**), and 3.10 (**1f**), assembling the skeletal structure (constitution) **1** of acanthifolin.

10.3 Structure Elucidation of a Sesquiterpene

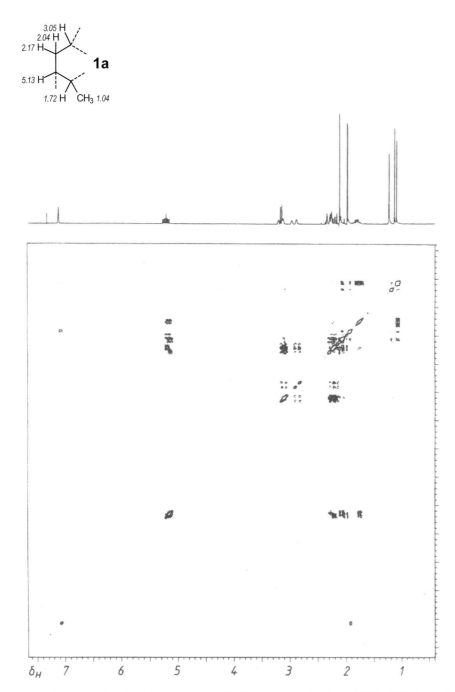

Figure 7. Two-dimensional HH shift correlation (HH-COSY) of acanthifolin (200 MHz, CDCl₃) revealing *geminal* and longer-range HH connectivities.

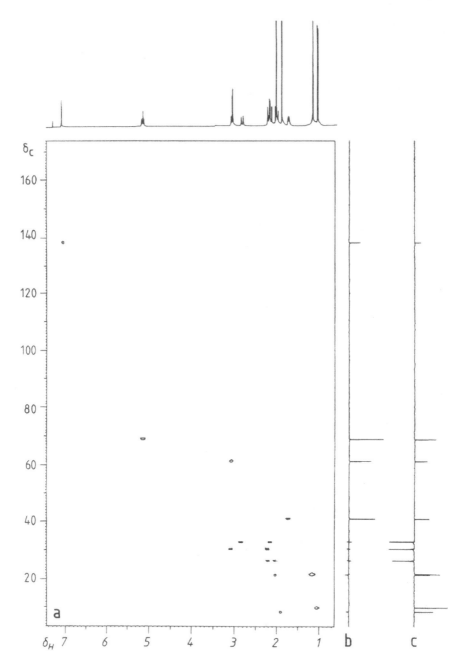

Figure 8. Two-dimensional CH correlation (CH-COSY) **a** assigning all CH bonds of acanthifolin in Table 9 (^{13}C: 100 MHz, ^{1}H: 400 MHz, CDCl$_3$) with DEPT-subspectra **b** and **c** for analysis of CH multiplicities [47].

10.3 Structure Elucidation of a Sesquiterpene

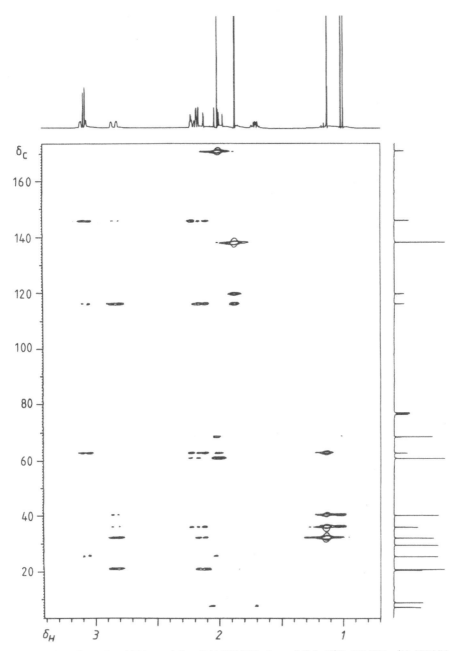

Figure 9. Two-dimensional CH correlation (CH-COLOC) of acanthifolin (^{13}C: 100 MHz, ^{1}H: 400 MHz, CDCl$_3$) detecting two-bond, three-bond and longer-range CH-connectivities [47] giving the partial structures **1b-f** in Table 10.

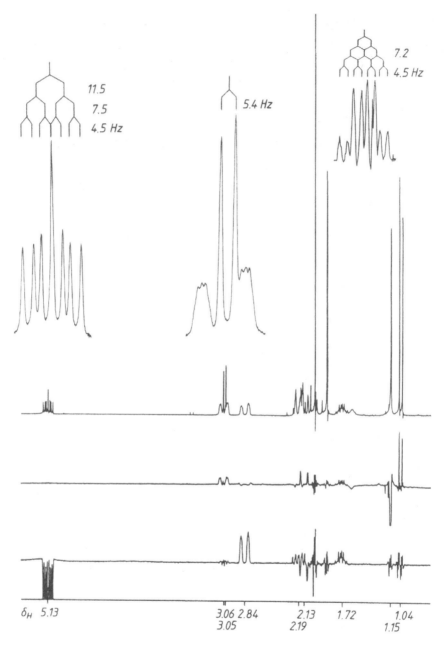

Figure 10. ^1H NMR spectra of acanthifolin (200 MHz, CDCl$_3$), with expanded partial spectra of HH multiplets at δ_H = 1.72, 3.05 and 5.13 and ^1H NOE difference spectra to derive the relative configuration from $^3J_{HH}$ coupling constants and to localize closely spaced protons [47].

10.3.4 Relative Configuration

Coupling constants of protons separated by three bonds ($^3J_{HH}$, *vicinal* couplings) correlate with the dihedral angle enclosed by the C*H*-bonds involved [48], thus reflecting their geometry in the molecule. Therefore, these coupling constants are frequently evaluated for the determination of the relative configuration. In first-order spectra, these data are obtained by measuring the signal distances (Hz) within the multiplet. In the ^1H NMR spectrum of acanthifolin (Fig. 10), some multiplets clearly show the geometry of substituents attached to ring *A* of this molecule.

The ^1H signal of the protons at $\delta_H = 1.72$ splits into a quartet ($^3J_{HH} = 7.2$ Hz) due to coupling with the *vicinal* methyl protons. The *vicinal* ring proton at $\delta_H = 5.13$ causes an additional doublet splitting with a coupling constant of $^3J_{HH} = 4.5$ Hz which corresponds to a dihedral angle of 60° enclosed by the C*H*-bonds involved, thus revealing an *axial* methyl group ($\delta_H = 1.04$) and an *equatorial* acetoxy function. The proton at the oxirane ring ($\delta_H = 3.05$) displays a coupling of $^3J_{HH} = 5.4$ Hz with one of the *vicinal* methylene protons ($\delta_H = 2.22$) corresponding to a dihedral angle of the C*H*-bonds between 30 and 40°. The coupling to the other methylene proton ($\delta_H = 2.04$) is too small to be resolved in the spectrum, revealing the C*H*-bonds involved to enclose a dihedral angle near 90°. To conclude, oxirane ring, acetoxy function and methyl group are all *cis* to each other.

The relative configuration of both adjacent methyl groups and the *cis* or *trans* fusion of the rings *A* and *B* still remain open. This problem cannot be solved by means of the *HH* coupling constants because no protons are attached to the bridgehead carbon atoms.

In such cases, nuclear OVERHAUSER effects (NOEs) in proton NMR are useful. NOEs are changes of the intensities of signals induced by irradiation of a certain proton with a frequency adjusted to its chemical shift (precession frequency). Nuclear OVERHAUSER enhancements increase with decreasing radial distance of the protons independent of the number of bonds separating them. They are detectable by various one- and two-dimensional procedures, for example by NOE difference spectroscopy [45,48]. In an NOE difference spectrum, the irradiated proton appears as a very strong signal with negative amplitude, while closely spaced protons are detected by significant positive signals; more distant protons exhibit weaker signals in the dispersion mode. NOE difference spectroscopy finally solves the relative configuration of acanthifolin.

Significant NOE signal enhancements are observed for the methyl proton signal at $\delta_H = 1.04$ upon irradiation of the methyl protons at $\delta_H = 1.15$ (Fig. 10). These methyl groups are *cis* to each other as a result; in *trans* configuration, the distance of the methyl protons would be too large for a significant NOE. Irradiation of the proton at $\delta_H = 5.13$ in an additional experiment induces a strong signal enhancement for the proton at $\delta_H = 2.84$. To conclude, both protons approach each other, and this requires a *cis* fusion of the rings *A* and *B*, as drawn in the projection and stereo formula of acanthifolin **1**.

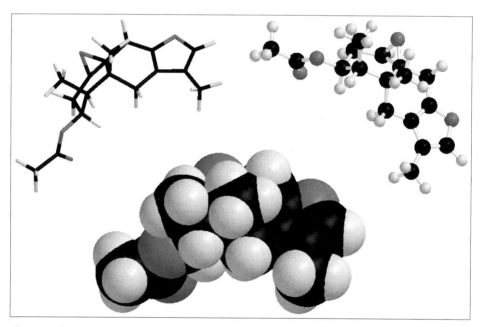

1: projection formula

1: stereo formula

Based on this elucidation, an energy-minimized molecular model can be calculated. Fig. 11 displays the tube, ball-and-stick and the space-filling model of acanthifolin obtained as a result of this calculation by means of commercial software in order to visualize the angle relations and shape of the molecule.

Figure 11. Molecular modeling of acanthifolin based on the skeletal structure and relative configuration determined by NMR; **left top**: tube model; **right top**: ball-and-stick model; **bottom**: space-filling model (H: white; C: black; O: gray).

10.3 Structure Elucidation of a Sesquiterpene

10.3.5 Absolute Configuration

NMR spectra of enantiomers are identical, so that the question of which one of the enantiomers of acanthifolin [$(1R,3R,4S,5S,10R)$ or $(1S,3S,4R,5R,10S)$] really exists in the investigated sample remains open. In fact, the absolute configuration of the majority of all furanoeremophilanes is reported to be $(1S,4R,5R,10S)$ in the literature [2]. For reference, 1,10-epoxyfuranoeremophilane isolated from *Senecio glastifolius* (Asteraceae) exhibits a specific rotation of $[\alpha]_D^{24} = -24.3°$ (c = 0.85 in chloroform). A similar value, $[\alpha]_D^{24} = -15.0°$ (c = 0.8 in chloroform), with the same sign is obtained for acanthifolin from the same plant family (*Senecio acanthifolius*). Therefore, until proven otherwise, (−)-acanthifolin is thought to have the $(1S,3S,4R,5R,10S)$-configuration reported for the majority of authentic furanoeremophilanes as depicted in the formulae.

(−)-acanthifolin

(−)-1,10-epoxy-furanoeremophilane

Chemically non-equivalent diastereotopic protons or ^{13}C nuclei with different chemical shifts in the proximity of asymmetric carbon atoms permit an empirical determination of the absolute configuration of a specific asymmetric carbon [48]. Examples include the diastereomers of 3β,19α-dihydroxy-12-ursen-28-oic acid, which differ only by the absolute configuration at C-20 (20R and 20S, respectively), and this is very clearly detected by the ^{13}C chemical shifts of carbon nuclei close to the stereocenter C-20.

^{13}C chemical shifts δ_C

(20R)-(+)-3β,19α-dihydroxy-12-ursen-28-oic acid

(20S)-(+)-3β,19α-dihydroxy-12-ursen-28-oic acid

Empirical chiroptical methods such as circular dichroism refined for special classes of compounds such as terpene and steroid ketones have been widely applied to investigate the absolute configuration [49,50].

Chemical correlation referring to authentic reference compounds with known absolute configuration, however, is the general method used to determine the absolute configuration. This is exemplified for the cases of (–)-α-*trans*-bergamotene **1** occurring in various plants and (–)-α-*trans*-bergamotenone **4** derived from **1** [51], which is a minor constituent of sandalwood oil with a pleasant milky odor of walnut. The absolute configurations of these levorotatory sesquiterpenes with prenylpinane as the parent hydrocarbon, (1*S*,5*S*) as drawn or (1*R*,5*R*), was unknown.

The problem was expected to be solved by chemical degradation to authentic *cis*-pinane, the levorotatory enantiomer of which has the absolute configuration (1*S*,2*R*,5*S*). Provided that the degradation exactly yields this enantiomer, as identified by value and sign of its specific rotation, then the asymmetric carbon centers of the sesquiterpenes **1** and **4** certainly possess the absolute configurations (1*S*,5*S*).

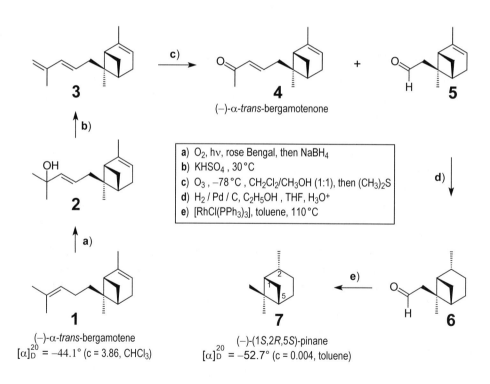

In order to perform the chemical degradation as outlined by the scheme (p. 172), (−)-α-*trans*-bergamotene **1** is subjected to photooxidation followed by reduction with sodium borohydride, yielding the allylalcohol **2**, which undergoes dehydration to the triene **3** in the presence of potassium hydrogen sulfate as acidic catalyst. Reductive ozonolysis with dimethylsulfide as reducing agent splits the CC double bond to a ketone and an aldehyde; chromatographic separation of the reaction mixture yields (−)-α-*trans*-bergamotenone **4** and (−)-9-formylpinene **5**. The latter is hydrogenated catalytically to (−)-9-formylpinane **6**. Deformylation to *cis*-pinane **7** is achieved by the WILKINSON catalyst. The pinane **7** obtained in this manner is levorotatory, thus having the absolute configuration (1S,2R,5S). To conclude, the absolute configurations of (−)-α-*trans*-bergamotene **1** and (−)-α-*trans*-bergamotenone **4** are indeed (1S,5S) [51].

10.4 Determination of the Crystal Structure

Provided that suitable crystals can be cultivated from a solid terpene, these can be used to determine the three-dimensional molecular structure within the crystal by means of X-ray diffraction [52].

X-ray diffraction caused by a crystal generates a diffraction pattern which characterizes the crystal, including the molecular structure embedded therein. The data set of reflexes is used to calculate the relative atom coordinates of the molecule. Various algorithms and refinements are available for this calculation. As a result, the *three-dimensional structure of the molecule within the crystal*, including its *atomic distances* (bond lengths) and *bond angles* are obtained with or without calculation of the vibrational ellipsoids of the atoms. Fig. 12 [53] displays all information delivered by the crystal structure of an interesting heptacyclic triterpene artifact obtained by acidic hydrolysis of a crude saponin extracted from *Panax notoginseng*. For a clear presentation of the molecular geometry, not only the atomic distances (Fig. 12a) and bond angles (Fig. 12b) but also *stereo pictures* (Fig. 12c) are calculated from the data set of reflexes. These stereo pictures can be viewed with special glasses [54] so that both partial pictures melt together, giving a three-dimensional presentation of the molecule. Molecular modeling based on crystal structure is also possible, as shown in Fig. 13, which clearly illustrates the skeleton (tube model) and the shape of the molecule (space-filling model).

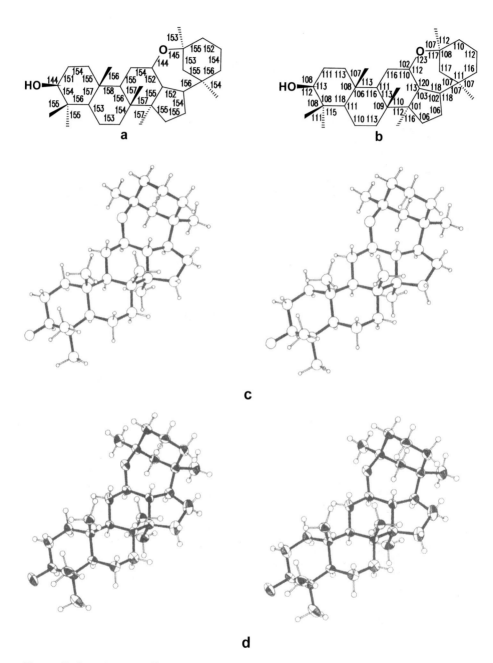

Figure 12. Crystal structure [53] of a saponin hydrolyzate obtained from *Panax notoginseng*; **a)** atomic distances in pm; **b)** bond angles in ° (grad); **c,d)** stereo pictures without (**c**) and with (**d**) vibrational ellipsoids of atoms for viewing the three-dimensional picture of molecular structure with stereo glasses [54].

10.4 Determination of the Crystal Structure

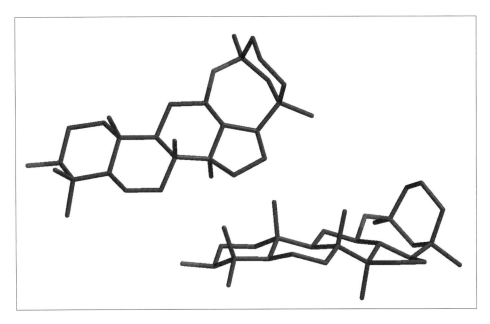

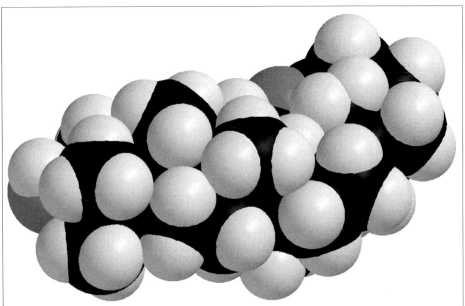

Figure 13. Energy-minimized molecular model of the saponin hydrolyzate from *Panax notoginseng*; **top**: two views of the tube model without hydrogen atoms for clarity; **bottom**: space-filling model giving the shape (H: white; C: black; O: gray).

10.5 Molecular Structure and Odor of Terpenes

A compound smells if it is sufficiently volatile. This applies predominantly not only to monoterpenes, but also to various sesqui- and diterpenes, as fragrances reach the appropriate receptors of the epithelium of the olfactory organ in the upper part of the nose. A molecule induces a specific sense of smell in the nose provided that its shape exactly matches a complementary cavity of the receptor, much as a key fits into a lock. Therefore, according to the stereochemical theory of odor developed by AMOORE [55,56], which only crudely interprets the complicated process of sensory perception, the smell of a compound correlates with the shape of its molecule.

Energy-minimized molecular modeling based on the stereostructure found by X-ray crystallography or spectroscopic elucidation provides the best means of obtaining the shape of molecules by using space-filling models, as demonstrated for three monoterpenes in Fig. 14.

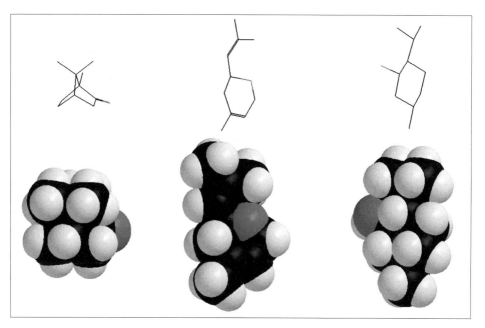

Figure 14. Energy-minimized Dreiding and space-filling models of camphor (left), rose oxide (center) and menthol (right).

Spherical molecules stimulate the typical pleasant but strong odor of camphor, provided that they match into a bowl-shaped receptor cavity of about 900 pm length, 750 pm width, and 400 pm depth (1 pm = 10^{-12} m). Elongated molecules with a monocyclic ring and a side chain such as rose oxide, as well as the ionones

10.5 Molecular Structure and Odor of Terpenes

and irones, comparable with the shape of a paper dragon, induce a flowery odor [55,56], suitable for a receptor cavity of 1650 pm length providing space for the "dragon head" with about 900 pm width and a "dragon tail" with 400 pm width and 700 pm depth. Elongated, drop-like molecules fitting receptor cavities of 1300 pm length, 650 pm width and 400 to 600 pm depth, stimulate the refreshing minty odor such as that from (−)-menthol [18,55,56,58].

Moreover, the olfactory organ is able to detect *functional groups*, structural characteristics such as *skeletal structure* (constitution), *relative* and *absolute configuration*. Therefore, the sense of smell is indeed regio-, stereo-, and enantioselective, as demonstrated by some typical examples.

The specific odor of functional groups is well known from non-terpenoid organic compounds containing an aldehyde or nitro group attached to a benzenoid ring such as benzaldehyde and nitrobenzene (pleasant smells like marzipan and bitter almonds), or volatile aliphatic amines (fishy odor). In terpenes, exchange of the hydroxy function for a sulfhydryl group clearly demonstrates how functional groups influence the sense of smell. Thus, (S)-(−)-α-terpineol smells like the flowers of lilac, while the sulfur analog (S)-(−)-p-menth-1-en-8-thiol emits the intense fragrance of grapefruit juice [18,57]. Minty-smelling *trans*-8-hydroxy-*p*-menth-3-one differs remarkably from its sufur analog *trans-p*-menth-3-one-8-thiol, which has a strong smell and taste of blackcurrant ("cassis"). It is obtained from the oil of bucco, a steam distillate of the leaves of South African bucco trees *Barosma betulina* (Rutaceae) [57]. Notably, it is not *trans-p*-menth-3-one-8-thiol but the hemiterpene 4-methoxy-2-methyl-2-butanthiol (section 2.1) which is known as the shaping fragrance of blackcurrant.

(S)-(−)-
α-terpineol

(S)-(−)-
p-menth-1-ene-8-thiol

trans-p-menth-
8-ol-3-one

(+)-trans-p-menth-
3-one-8-thiol

Regioselectivity of the sense of smell is demonstrated by the homodrimane derivative ambrinal, a pleasant woody-smelling ambergris fragrance used in perfumery, and its odorless isomer differing only in the position of its CC double bond [18]. β-Ionone, the flowery smelling fragrance of violets belonging to the megastigmanes, and the regioisomeric rose ketone containing a doubly conjugated carbonyl function with more of a pleasant fruity odor and a touch of camphor, represent additional examples.

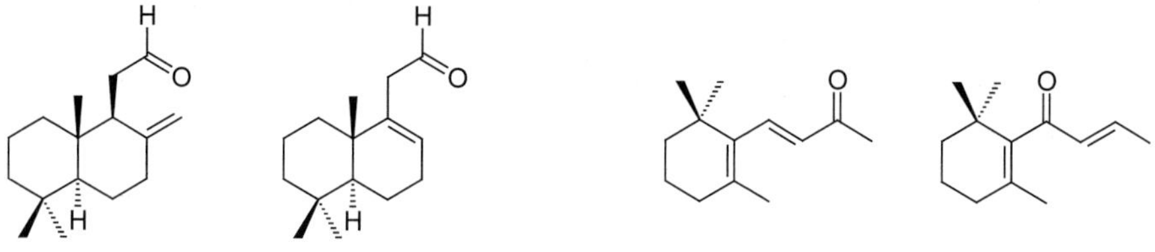

ambrinal   isoambrinal   β-ionone   rose ketone

The strikingly different odors of *cis-trans*-isomers and diastereomers is demonstrated by various other pairs of terpenes. For example, the synthetic fragrance (*E*)-8-methyl-α-ionone has a lovely flowery smell like the blossoms of violets, whereas the (*Z*)-isomer emits a woody tobacco-like odor. Likewise, the strongly woody smelling sesquiterpene (−)-*epi*-γ-eudesmol from Algerian oil of geranium contrasts with the almost odorless diastereomer (+)-γ-eudesmol [18], which is wide-spread among ethereal oils.

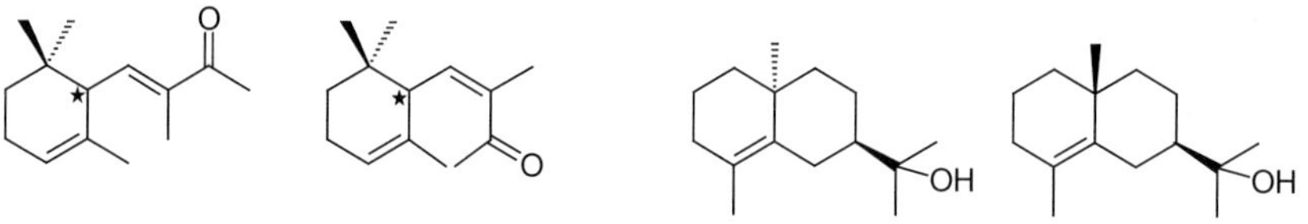

(*E*)-8-methyl-α-ionone   (*Z*)-8-methyl-α-ionone   (−)-10-*epi*-γ-eudesmol   (+)-γ-eudesmol

Many pairs of enantiomers are reported to exemplify the enantioselectivity of the sense of smell. (*S*)-(+)-Linalool from the oil of Coriander has a flowery smell with a touch of citrus; this contrasts with the (*R*)-(−)-enantiomer from the oils of Rose, Neroli and Lavender, with a woody lavender odor [18]. (1*R*,3*R*,4*S*)-(−)-Menthol (shown as molecular model in Fig. 14) smells and tastes sweet and minty, cooling and refreshing, and therefore is widely used in perfumery and confectionery. In contrast, the (1*S*,3*S*,4*R*)-(+)-enantiomer radiates a more herby, weaker minty and less refreshing odor [18].

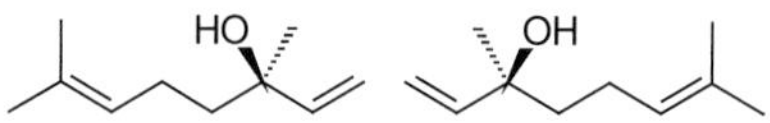

(*R*)-(−)-linalool   (*S*)-(+)-linalool

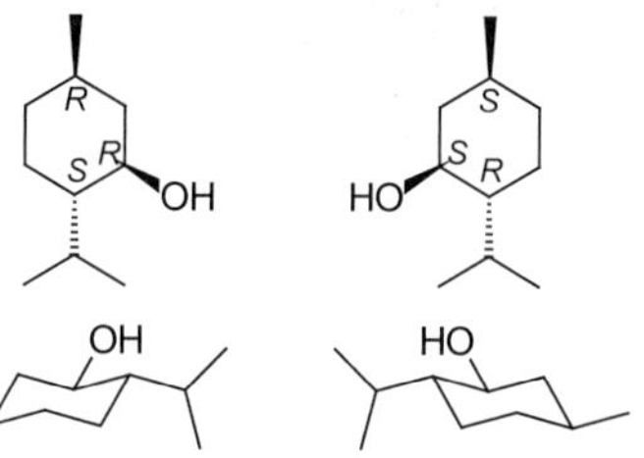

(−)-menthol   (+)-menthol

10.5 Molecular Structure and Odor of Terpenes

(*S*)-(+)-Carvone produces the typical odor and taste of caraway, whereas its (*R*)-(−)-enantiomer in the oil of spearmint from *Mentha spicata* (Labiatae), in contrast, smells like peppermint (p. 18). (−)-Patchoulol (also referred to as patchoulialcohol; p. 40) smells intensely woody and earthy with a touch of camphor, similar to the natural oil of Patchouli used in perfumery, while the weak odor of the synthetically produced (+)-enantiomer is quite untypical.

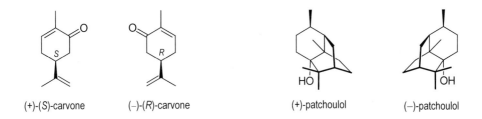

(+)-(*S*)-carvone   (−)-(*R*)-carvone   (+)-patchoulol   (−)-patchoulol

More sensitive human noses can even distinguish between enantiomeric terpene hydrocarbons: (*R*)-(+)-Limonene, the major constituent of the ethereal oil of mandarin peels, is recognized as a characteristic clean orange fragrance, while the (*S*)-(−)-enantiomer from the oil of fir-cones radiates a less pleasant orange odor with a touch of turpentine. Fig. 15 illustrates the sensitivity of the human olfactory organ by means of the limonene enantiomers between it may differentiate.

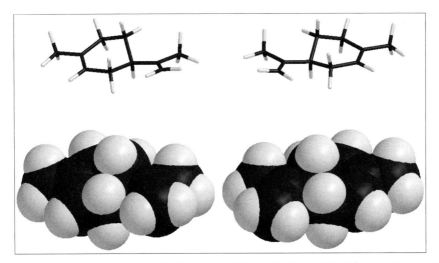

Figure 15. Energy-minimized tube (top) and space-filling molecular models of the enantiomers of limonene, (*R*)-(+)- on the left, (*S*)-(−)- on the right.

Bibliography

Isoprene Rule

[1] L. Ruzicka, *Proc. Chem. Soc. (London) 1959*, 341.

Review of Terpenes until 1990

[2] J.D. Connolly, R.A. Hill (Hrsg.), *Dictionary of Terpenoids*, Vol. 1: *Mono- and Sesquiterpenoids*, Vol. 2: *Di- and higher Terpenoids*, Vol. 3: *Indexes*, Chapman & Hall, London, New York, Tokyo, Melbourne, Madras, 1991.

This dictionary cites the primary literature until 1990 for all terpenes covered in the text.

Monographs and Reviews since 1960

[3] A.A. Newman (Hrsg.), *Chemistry of Terpenes and Terpenoids*, Academic Press, London, New York, 1972.

[4] T.W. Goodwin, *Aspects of Terpenoid Chemistry and Biochemistry*, Academic Press, London, New York, 1971.

[5] W. Templeton, *An Introduction to the Chemistry of Terpenoids and Steroids*, Butterworth, London, 1969.

[6] J.P. Pridham, *Terpenoids in Plants*, Academic Press, London, New York, 1967.

[7] A.R. Pinder, *The Chemistry of Terpenes*, Chapman & Hall, London, New York, 1960.

[8] G. Rücker, Sesquiterpene, *Angew. Chem. 85* (1973) 895; *Angew. Chem. Int. Ed. Engl. 12* (1973) 793.

Biosynthesis of Terpenes

[9] J.W. Porter et al., *Biosynthesis of Isoprenoid Compounds*, J. Wiley & Sons, New York, 1981.

[10] T.W. Goodwin, Biogenesis of Terpenes and Steroids, in: M.F. Ansell (Ed.), *Rodd's Chemistry of Carbon Compounds IIc Supplement*, Elsevier, Amsterdam, 1974.

[11] F. Lynen: Der Weg von der "aktivierten Essigsäure" zu den Terpenen und Fettsäuren (NOBEL lecture), *Angew. Chem. 77* (1965) 929; F. Lynen, U. Henning, *Angew. Chem. 72* (1960) 820.

[12] M. Rohmer, M. Seemann, S. Horbach, S. Bringer-Meyer, H. Sahm, *J. Am. Chem. Soc. 118* (1996) 2564.

Review of Steroids until 1990

[13] R.A. Hill (Ed.), *Dictionary of Steroids*, Chapman & Hall, London, New York, Tokyo, Melbourne, Madras, 1991.

Terpenes as Pheromones

[14] J.M. Brand, J.C. Young, R.M. Silverstein, Insect Pheromones: A Critical Review of Recent Advances in their Chemistry, Biology and Application, *Fortschr. Chem. Org. Naturst. 37* (1979).

[15] J.G. MacConnell, R.M. Silverstein, Neue Ergebnisse der Chemie von Insektenpheromonen, *Angew. Chem. 85* (1973) 647; *Angew. Chem. Int. Ed. Engl. 12* (1973), 644.

[16] M. Jacobsen, *Insect Sex Pheromones*, Academic Press, London, New York, 1972.

[17] M. Beroza, *Chemicals Controlling Insect Behaviour*, Academic Press, London, New York, 1970.

Terpenes as Fragrants and Flavors

[18] G. Ohloff, *Riechstoffe und Geruchsinn – Die molekulare Welt der Düfte*, Springer, Berlin, Heidelberg, New York, 1990; *Düfte – Signale der Gefühlswelt*, Verlag Helvetica Chimica Acta, Zürich, Wiley-VCH, Weinheim, 2004

Terpenes as Phytopharmacia

[19] *The Merck Index*, 13th edn, Merck & Co., Inc., Whitehouse Station, NJ, USA, 2001.

[20] G. Rücker, Artemisinin, *Pharm. Unserer Zeit 23* (1994) 223.

[21] H. Kolodzej, Sesquiterpenlactone - Biologische Aktivitäten, *Dtsch. Apotheker Ztg. 133* (1993) 1795.

[22] C.H. Brieskorn, Triterpenoide, physiologische Funktionen und therapeutische Eigenschaften, *Pharm. Unserer Zeit 16* (1987) 161.

Selected Total Syntheses of Terpenes

Acylic Mono- and Sesquiterpenes

[23] W. Hoffmann, Industrielle Synthesen terpenoider Riechstoffe, *Chemiker Ztg. 97* (1973) 23.

(R)-(+)-Citronellal and (–)-Menthol

[24] Y. Nakatani, K. Kawashima, *Synthesis 1978*, 147; K. Takabe, T. Katagiri, J. Tanaka, T. Fujita, S. Watanabe, K. Suga, *Org. Synth. 67* (1989) 44; B.B. Snider, *Acc. Chem. Res. 13* (1980) 426.

Rose oxide

[25] G. Ohloff, E. Klein, G.O. Schenk, *Angew. Chem. 73* (1961) 578; G. Ohloff, *Fortschr. Chem. Forsch. 12/2* (1969) 185.

trans-Chrysanthemic acid methyl ester

[26] J. Martel, C. Huynh, *Bull. Soc. Chim. Fr. 1967*, 985; P.F. Schatz, *J. Chem. Educ. 55* (1978) 468.

α-Terpineol

[27] T. Inukai, M. Kasai, *J. Org. Chem. 30* (1965) 3567.

Campholenealdehyde

[28] B. Arbuzow, *Ber. Dtsch. Chem. Ges. 68* (1935) 1430.

(–)-Ipsdienol and (R)-(–)-Linalool from α-Pinene and Derivatives

[29] G. Ohloff, W. Giersch, *Helv. Chim. Acta 60* (1977) 1496; G. Ohloff, E. Klein, *Tetrahedron 29* (1973) 1559.

Hexahydrocannabinol

[30] Z.G. Lu, N. Sato, S. Inoue, K. Sato, *Chem. Lett. 1992*, 1237.

Sesquiterpenes

β-Selinene (stereoselective)

[31] B.D. MacKenzie, M.M. Angelo, J. Wolinski, *J. Org. Chem. 44* (1979) 4042.

Isocomene

[32] M.C. Pirrung, *J. Am. Chem. Soc. 103* (1981), 82; *101* (1979) 7130.

Cedrene (racemic)

[33] E.J. Corey, R.D. Balanson, *Tetrahedron Lett. 1973*, 3153.

Periplanone B

[34] S.L. Schreiber, C. Santini, *Tetrahedron Lett. 22* (1981) 4651; S.L. Schreiber, C. Santini, *J. Am. Chem. Soc. 106* (1984) 4038.; S.L. Schreiber, *Nature (London) 227* (1985) 857.

Diterpenes

Vitamin A

35 H. Pommer, *Angew. Chem. 72* (1960) 811, 911.

β-Carotene from Vitamin-A-Aldehyde by McMurry Coupling

36 D. Lenoir, *Synthesis 1989*, 883

Cafestol (racemic)

37 E.J. Corey, G. Wess, Y.B. Xiang, A.K. Singh, *J. Am. Chem. Soc. 109* (1987) 4717.

Taxol

38 K.C. Nicolaou, R.K. Guy, *Angew. Chem. 107* (1995) 2247; *Angew. Chem. Int. Ed. Engl. 34* (1995) 2079; K.C. Nicolaou, Z. Yang, J.J. Liu, H. Ueno, P.G. Nantermet, R.K. Guy, C.F. Claiborne, J. Renaud, E.A. Couladouros, K. Paulvannan, E.J. Sorensen, *Nature (London) 367* (1994) 630.

Triterpenes, Lupeol and Precursor

39 G. Stork, S. Uyeo, T. Wakamatsu, P. Grieco, J. Labowitz, *J. Am. Chem. Soc. 93* (1971) 4945.

40 G. Stork, H.J.E. Loewenthal, P.C. Mukharji, *J. Am. Chem. Soc. 78* (1956) 501.

Structure Elucidation

Introductions to Spectroscopic Methods of Structure Elucidation

41 B. Hesse, H. Meier, B. Zeeh, *Spectroscopic Methods in Organic Chemistry*, 7th edn, Georg Thieme, Stuttgart, 2005.

42 D. Joulain, W. A. König, *The Atlas of Spectral Data of Sesquiterpene Hydrocarbons*, E. B. Verlag, Hamburg, 1999 (collection of spectroscopic data).

NMR Spectroscopy

43 H. Friebolin, *Basic One- and Two-Dimensional NMR Spectroscopy*, 4th edn, Wiley-VCH, Weinheim, 2004.

44 H. Günther, *NMR-Spektroskopie*, 3rd edn, Georg Thieme, Stuttgart, 1992.

45 S. Braun, H.-O. Kalinowski, S. Berger, *200 and More Basic NMR Experiments*, A Practical Course, 3rd edn, Wiley-VCH, Weinheim, 2004.

Mass Spectrometry

46 H. Budzikiewicz, *Massenspektrometrie, eine Einführung*, 4th edn, Wiley-VCH, Weinheim (see also [41,42]), 1999.

Exemplary Structure Elucidations of Terpenes by NMR

47 M. Garrido, S. Sepúlveda-Boza, R. Hartmann, E. Breitmaier, *Chemiker- Ztg. 113* (1989) 201; *111* (1987) 301.

48 E. Breitmaier, *Vom NMR-Spektrum zur Strukturformel organischer Verbindungen*, 3rd edn, Wiley-VCH, Weinheim, 2005; *Structure Elucidation by NMR in Organic Chemistry, A Practical Guide*, 3rd edn, Wiley, Chichester, 2002.

Chiroptical Methods for Determination of Absolute Configuration

49 G. Snatzke, *Angew. Chem. 80* (1968) 15; *Angew. Chem. Int. Ed. Engl. 7* (1968) 14.

50 N. Harada, K. Nakanishi, *Circular Dichroic Spectroscopy*, University Science Books, New York, 1983.

Determination of Absolute Configuration by Chemical Correlation

51 C. Chapuis, M. Barthe, B.L. Muller, K.H. Schulte-Elte, *Helv. Chim. Acta 81* (1998) 153.

X-Ray Crystal Structure Determination

52 W. Massa, *Kristallstrukturbestimmung*, 2nd edn, B.G. Teubner, Stuttgart, 1996.

Crystal Structure of a Triterpene

53 W. Junxiang, C. Liangyu, W. Jufen, W. Chen, E. Friedrichs, H. Puff, E. Breitmaier, *Planta Med. 1984*, 47.

54 F. Vögtle, *Stereochemie in Stereobildern* (including a stereoscope designed by the author), VCH, Weinheim, 1982.

Structure and Odor

55 H. Boelens, *Cosmetics & Perfumes 89* (1974) 452.

56 J.E. Amoore, Specific Anosmia, *Nature (London) 214* (1967) 1095.

57 P. Kraft, J.A. Baygrowicz, C. Denis, G. Fráter, *Angew. Chem. 112* (2000) 3106; *Angew. Chem. Int. Ed. Engl. 39* (2000) 2980.

58 P. Kraft, K.A.D. Swift: *Perspectives in Flavor and Fragrance Chemistry*, Wiley-VCH, Weinheim, 2005.

Survey of Important Parent Skeletons of Terpenes

Nomenclature and positional numbering according to Ref. [2].

Parent skeletons of Monoterpenes
Hemiterpenes, acyclic, mono- and bicyclic Monoterpenes

For illustration of the isoprene rule, isoprene units of acyclic and monocyclic monoterpenes are printed in boldface.

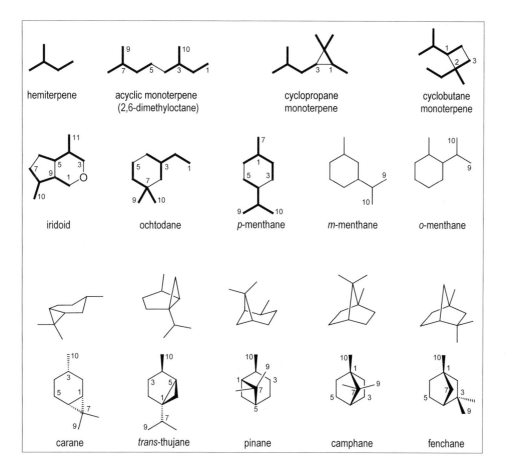

Parent skeletons of Sesquiterpenes I
Monocyclic Sesquiterpenes derived from Farnesane

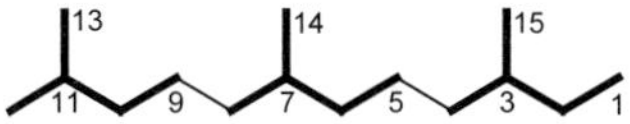

Farnesane
(2,6,10-trimethylundecane)

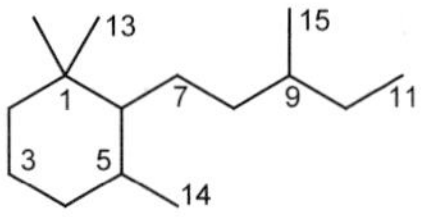

cyclofarnesane

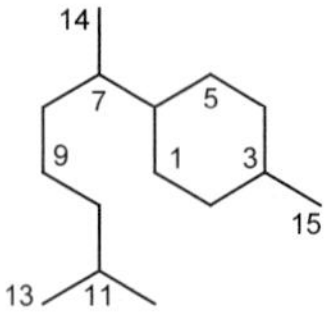

bisabolane

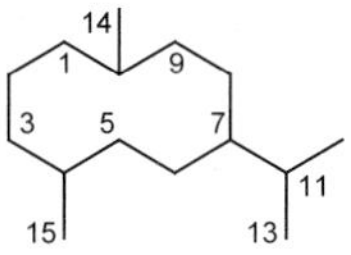

germacrane

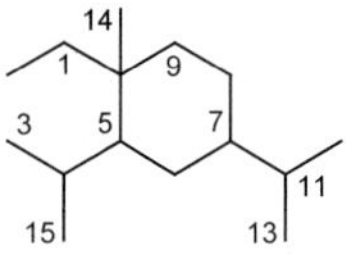

elemane

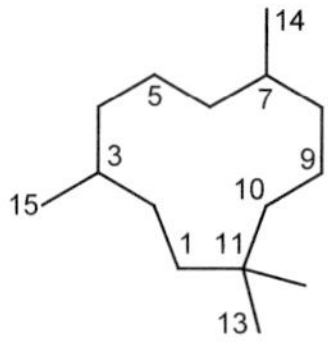

humulane

Parent skeletons of Sesquiterpenes II
Bi- and Tricyclic Sesquiterpenes derived from Farnesane I

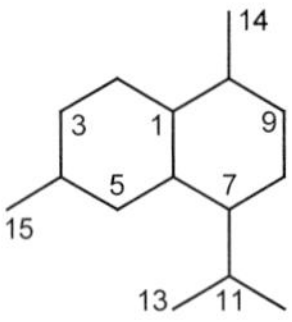

cadinane

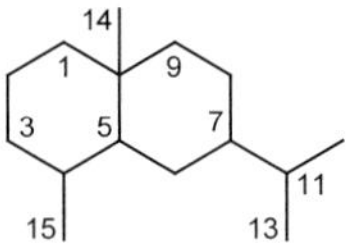

eudesmane

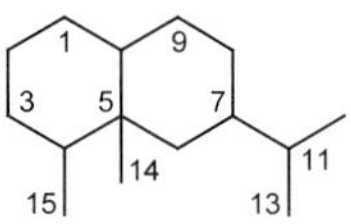

eremophilane

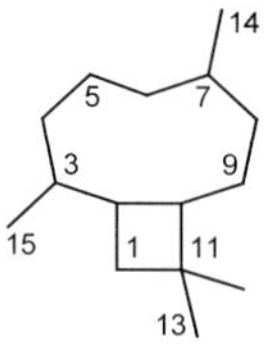

caryophyllane

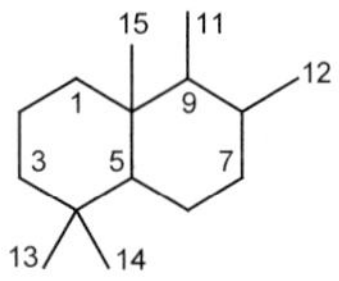

drimane

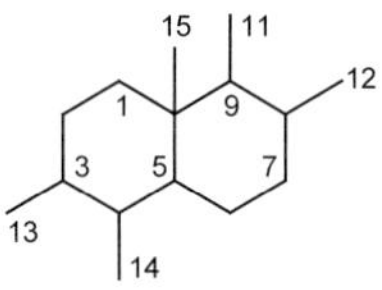

colorane

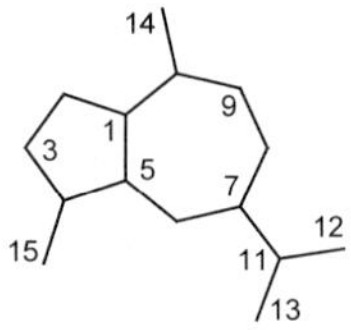

guaiane

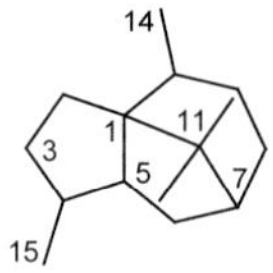

patchoulane

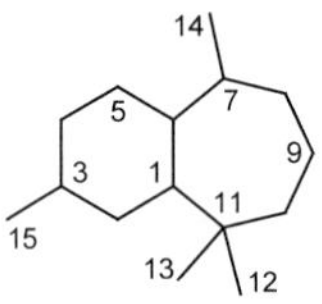

himachalane

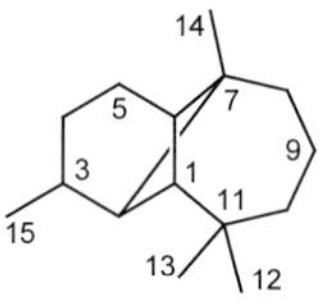

longipinane

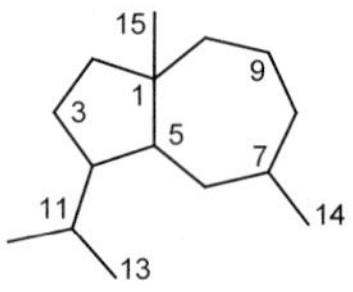

isodaucane

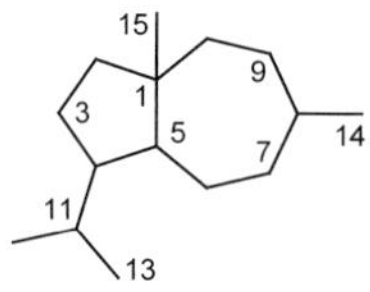

daucane

Bi- and Tricyclic Sesquiterpenes derived from Farnesane II

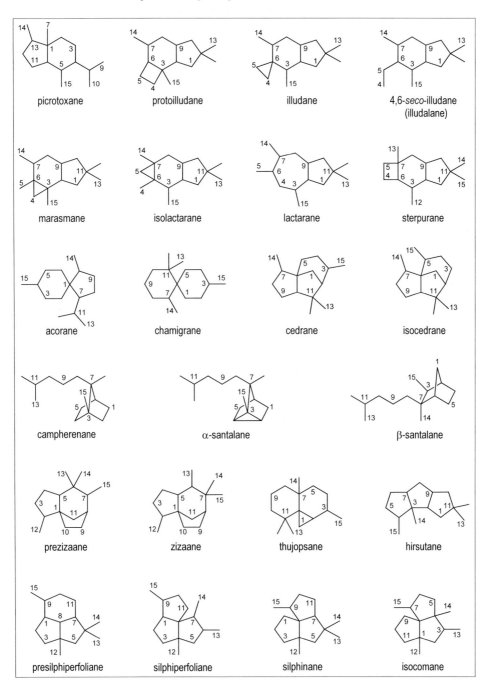

Parent skeletons of Diterpenes I
Mono- and Bicyclic Diterpenes derived from Phytane

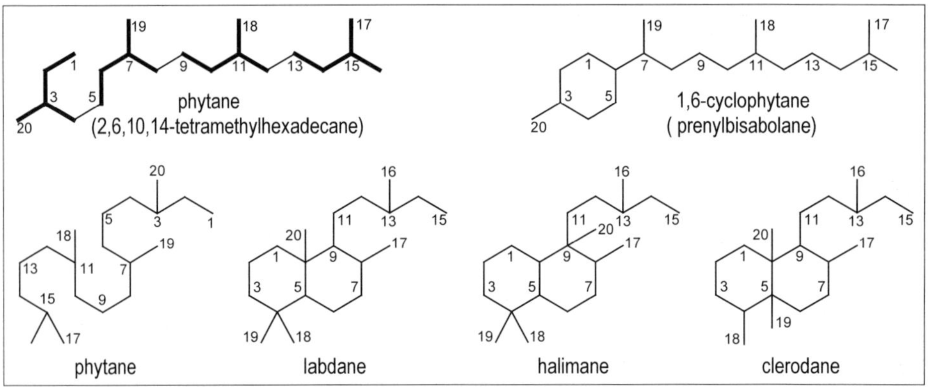

phytane
(2,6,10,14-tetramethylhexadecane)

1,6-cyclophytane
(prenylbisabolane)

phytane

labdane

halimane

clerodane

Parent skeletons of Diterpenes II
Tricyclic Diterpenes derived from Pimarane

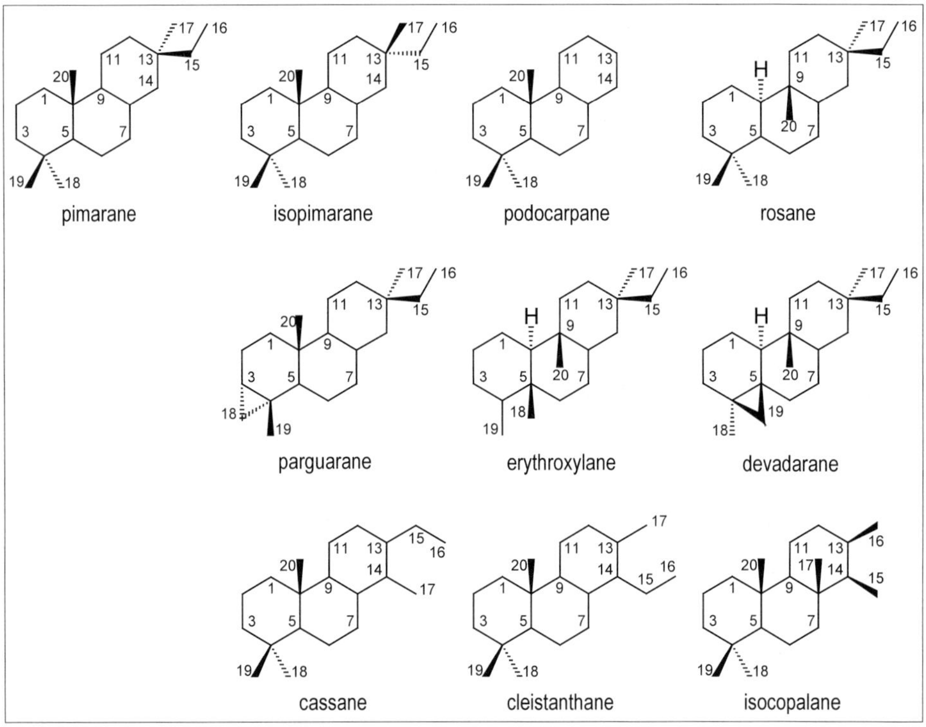

pimarane

isopimarane

podocarpane

rosane

parguarane

erythroxylane

devadarane

cassane

cleistanthane

isocopalane

Parent skeletons of Diterpenes III
Tricyclic Diterpenes derived from Abietane

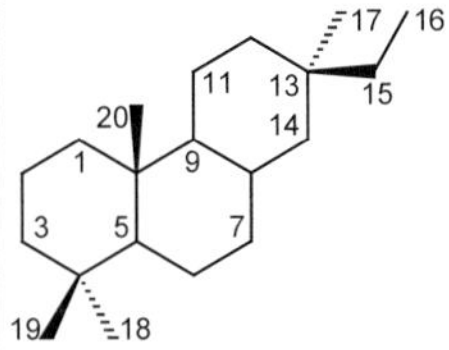

pimarane

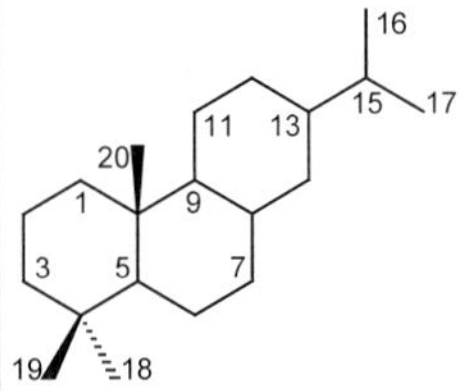

abietane

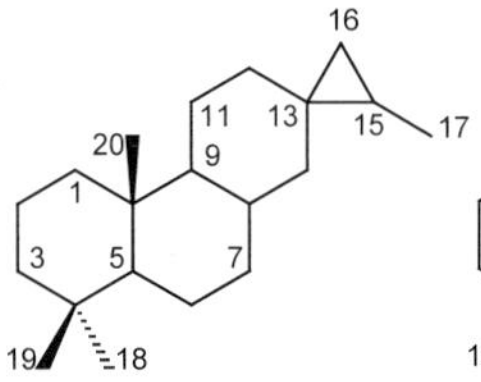

13,16-cycloabietane

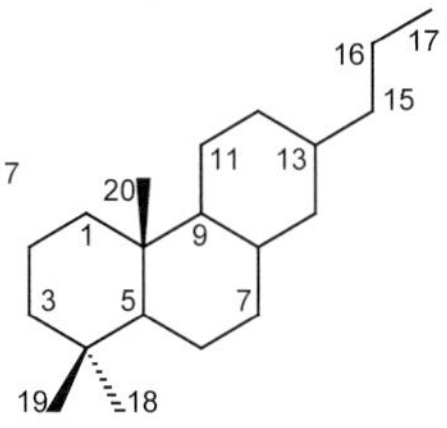

17(15-16)-Abeo-abietane

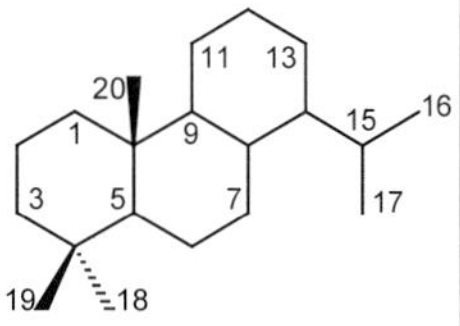

totarane

Parent Skeletons of Diterpenes IV
Tetracyclic Diterpenes derived from Pimarane

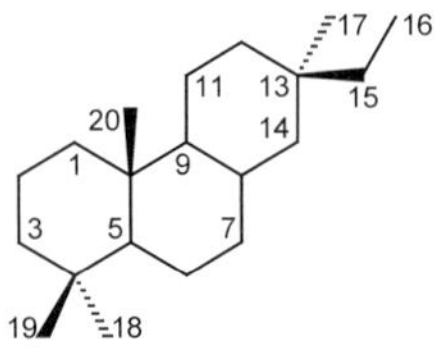

pimarane

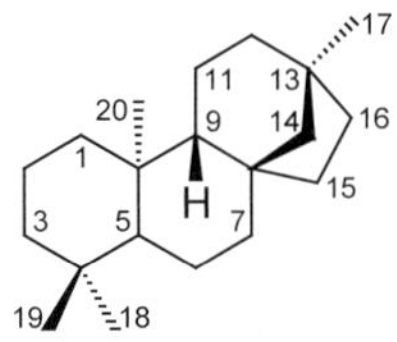

beyerane

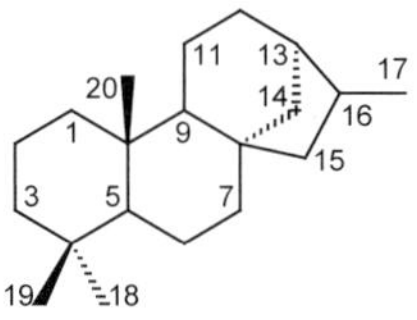

kaurane

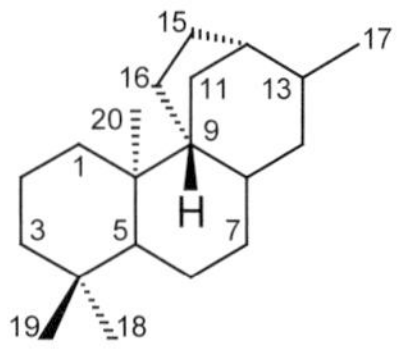

villanovane

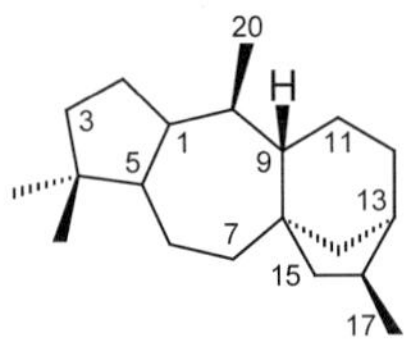

grayanotoxane

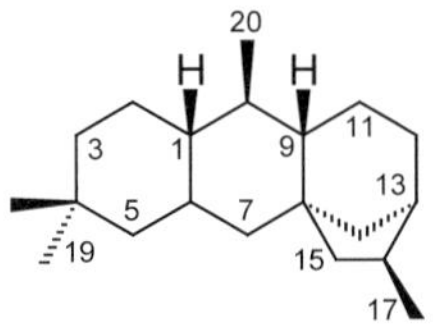

leucothole

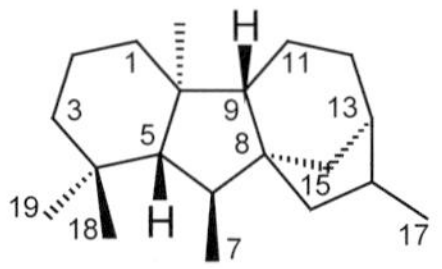

giberellane

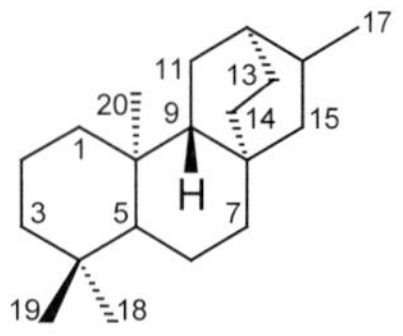

atisane

Parent skeletons of Diterpenes V
Bi-, Tri- and Tetracyclic Diterpenes derived from Cembrane

Bold-face bonds indicate ring fusions and methyl shifts.

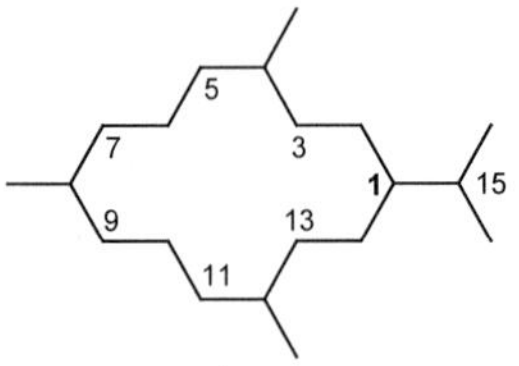

cembrane

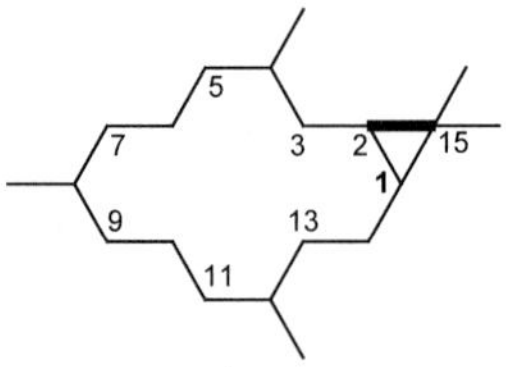

casbane

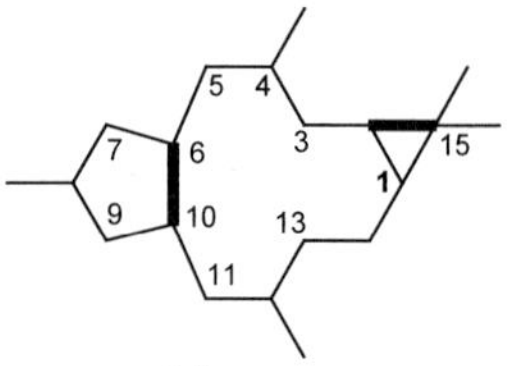

lathyrane

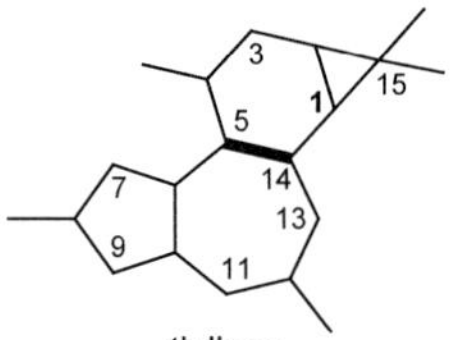

tigliane

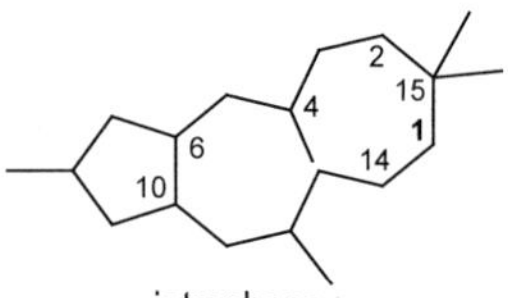

jatrophane

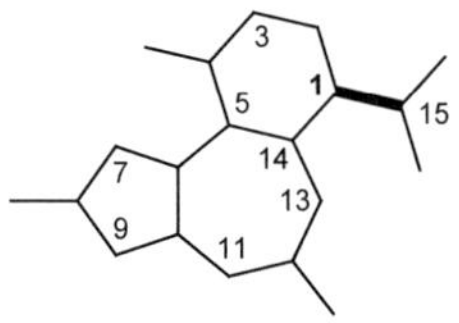

rhamnofolane

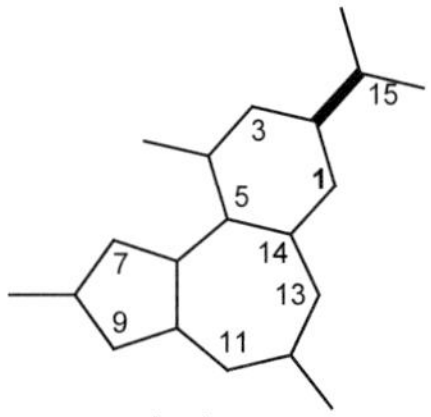

daphnane

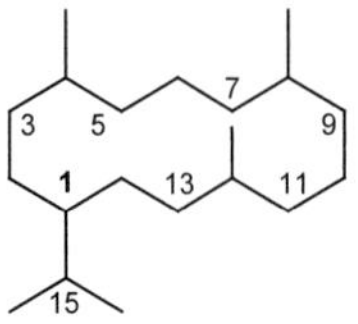

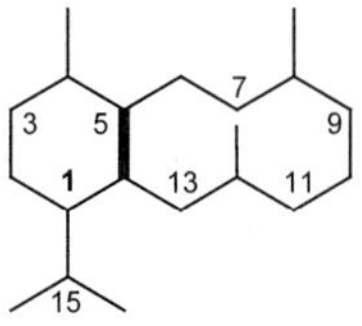

eunicellane

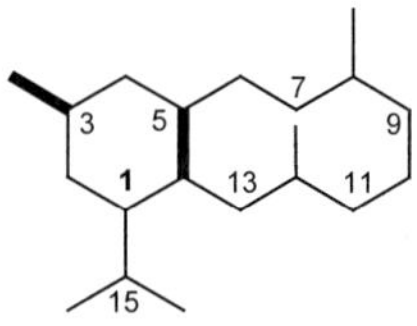

asbestinane

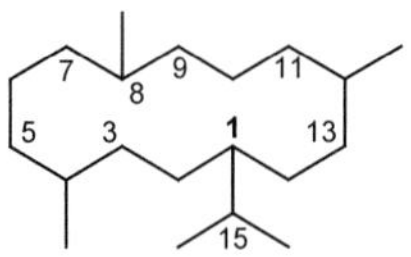

cembrane

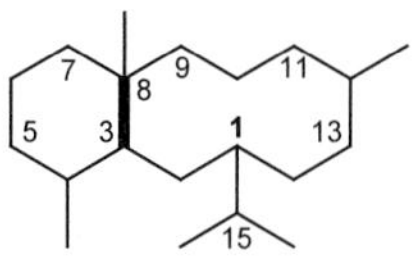

briarane

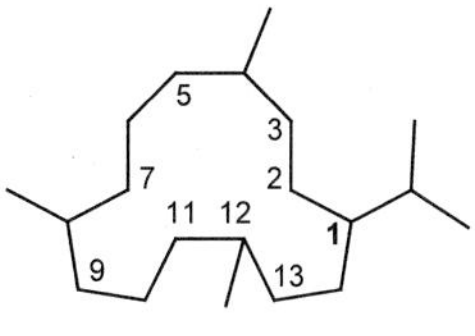

Survey of Important Parent Skeletons of Terpenes

cembrane

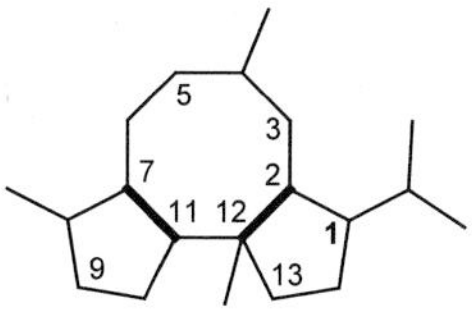

basmane

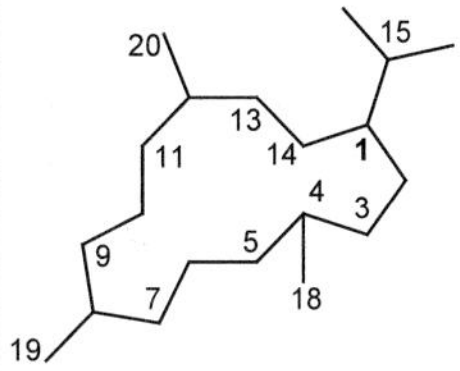

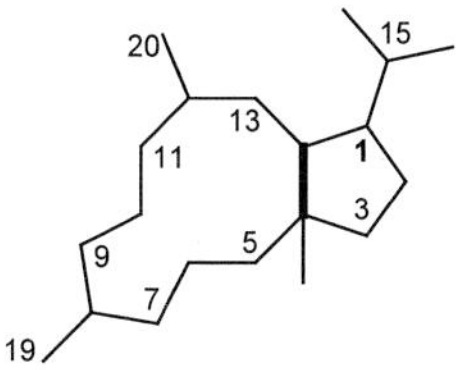

4,14-cyclocembrane

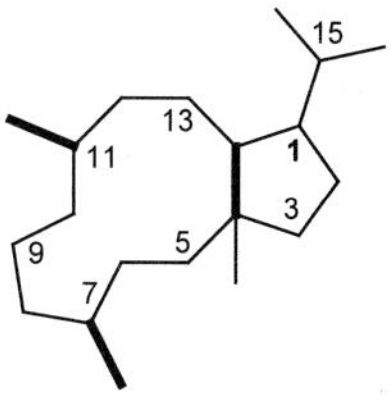

dolabellane

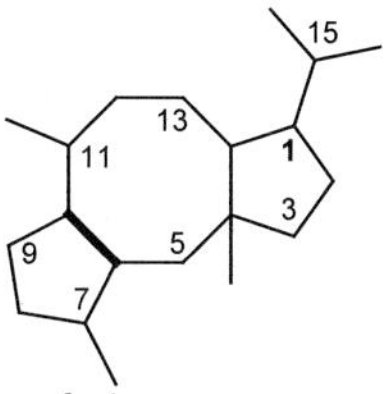

fusicoccane

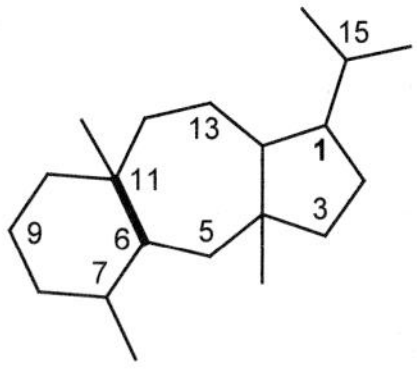

dolastane

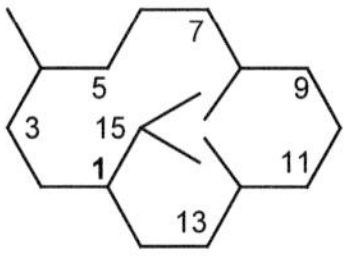

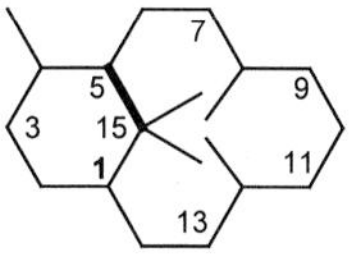

verticillane

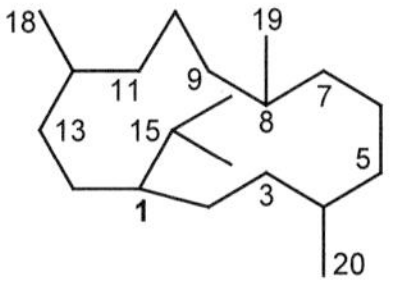

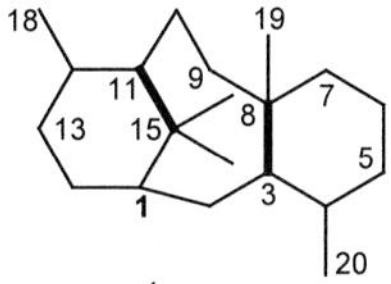

taxane

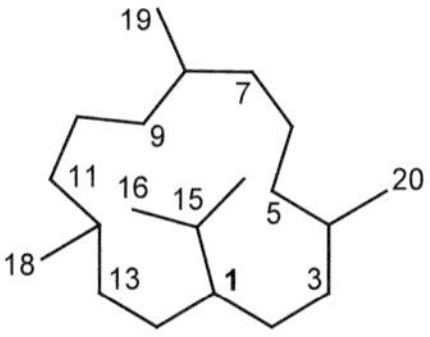

cembrane

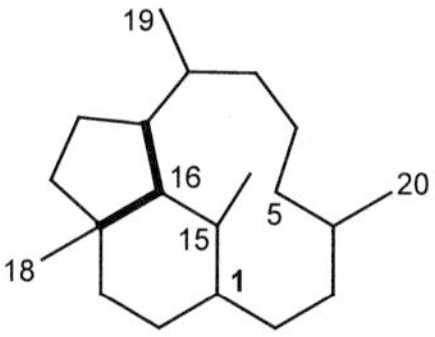

trinervitane

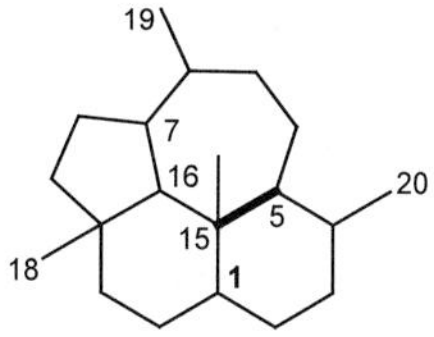

kempane

Parent Skeletons of Diterpenes VI
Prenylsesquiterpenes

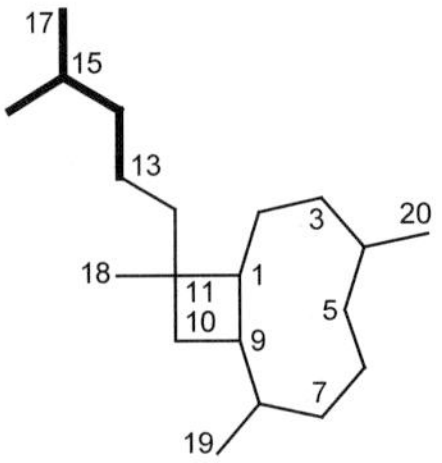

prenylcaryophyllane
= xeniaphyllane

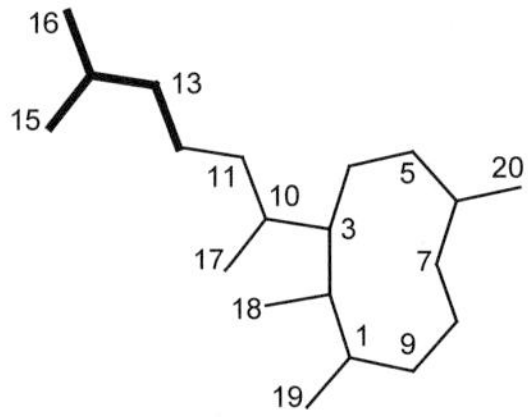

opened prenylcaryophyllane
= xenican

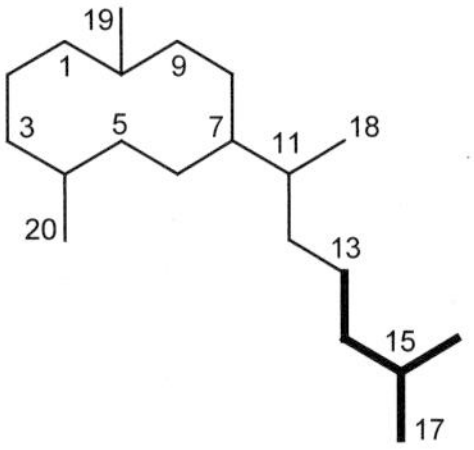

prenylgermacrane

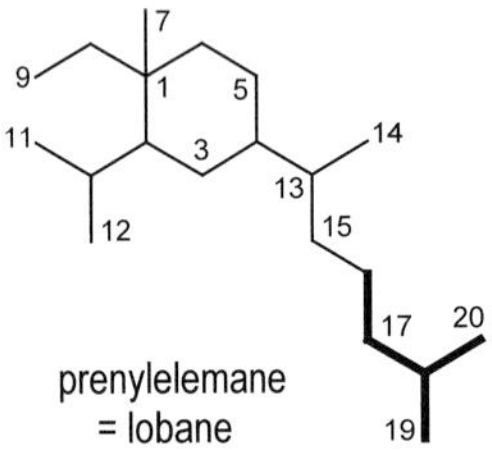

prenylelemane
= lobane

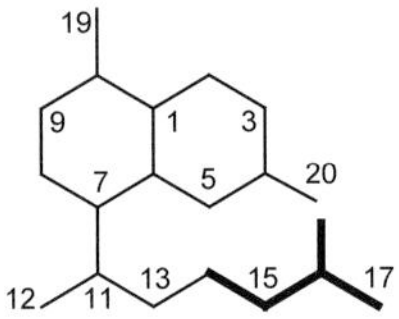

prenylcadinane
= biflorane

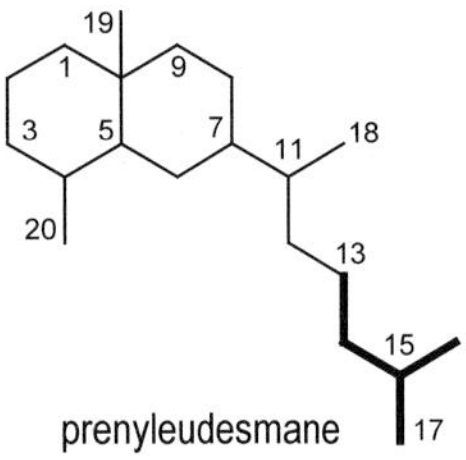

prenyleudesmane

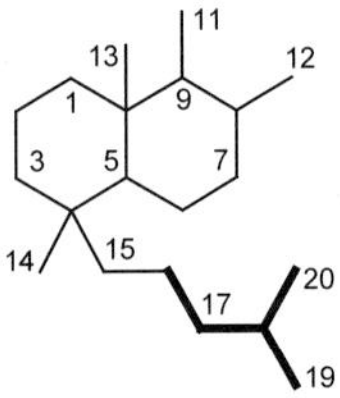

prenyldrimane
= sacculatane

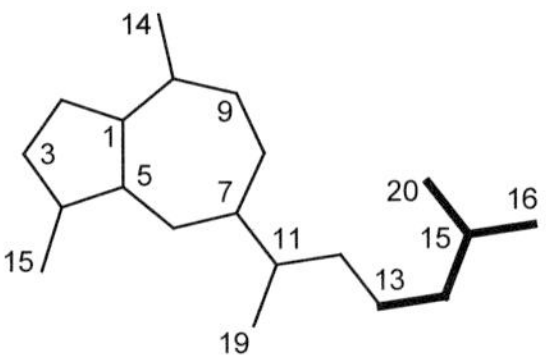

prenylguajane

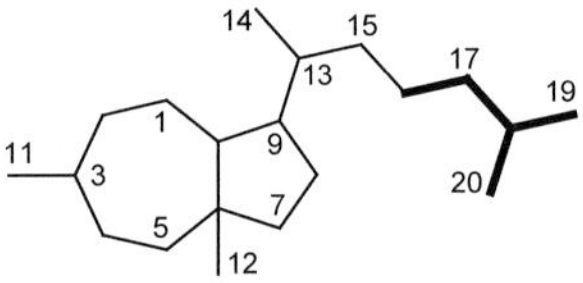

prenyldaucane
= sphenolobane

Survey of Important Parent Skeletons of Terpenes

Parent skeletons of Sesterterpenes
Diprenylsesquiterpenes, Prenylditerpenes, Scalaranes

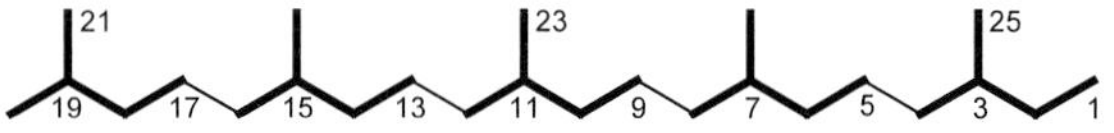

2,6,10,14,18-pentamethylicosane

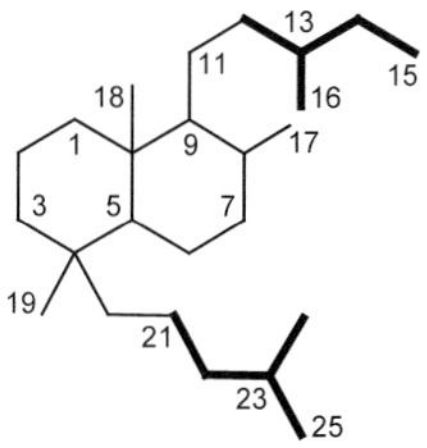

diprenyldrimane
(diprenylsesquiterpene)

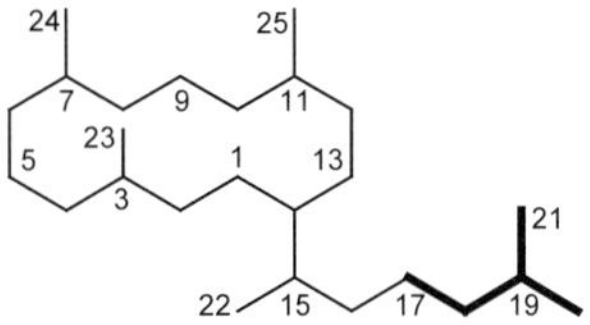

cericerane
(prenylcembrane)

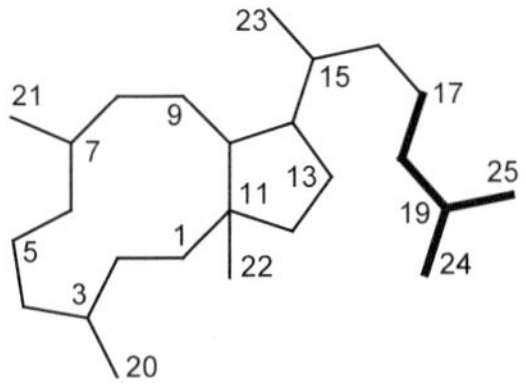

prenyldolabellane

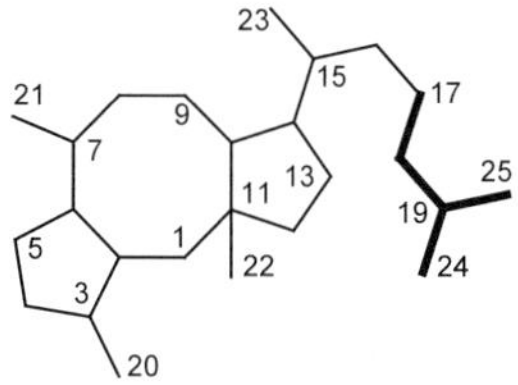

ophiobolane
(prenylfusicoccane)

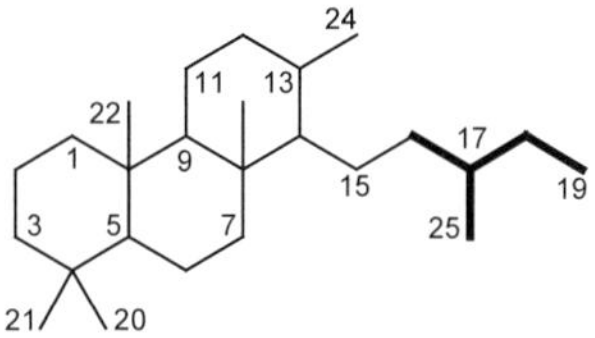

cheilanthane
(prenylisocopalane)

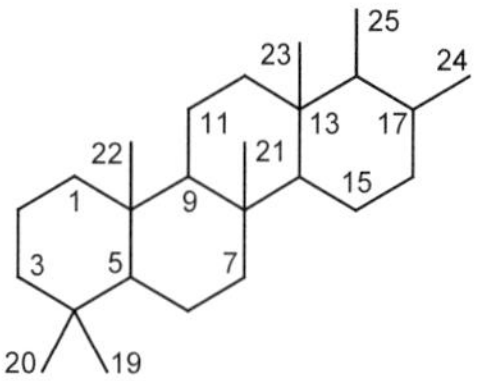

scalarane

Parent skeletons of Triterpenes I
Tetracyclic Triterpenes with Gonane Core Skeleton

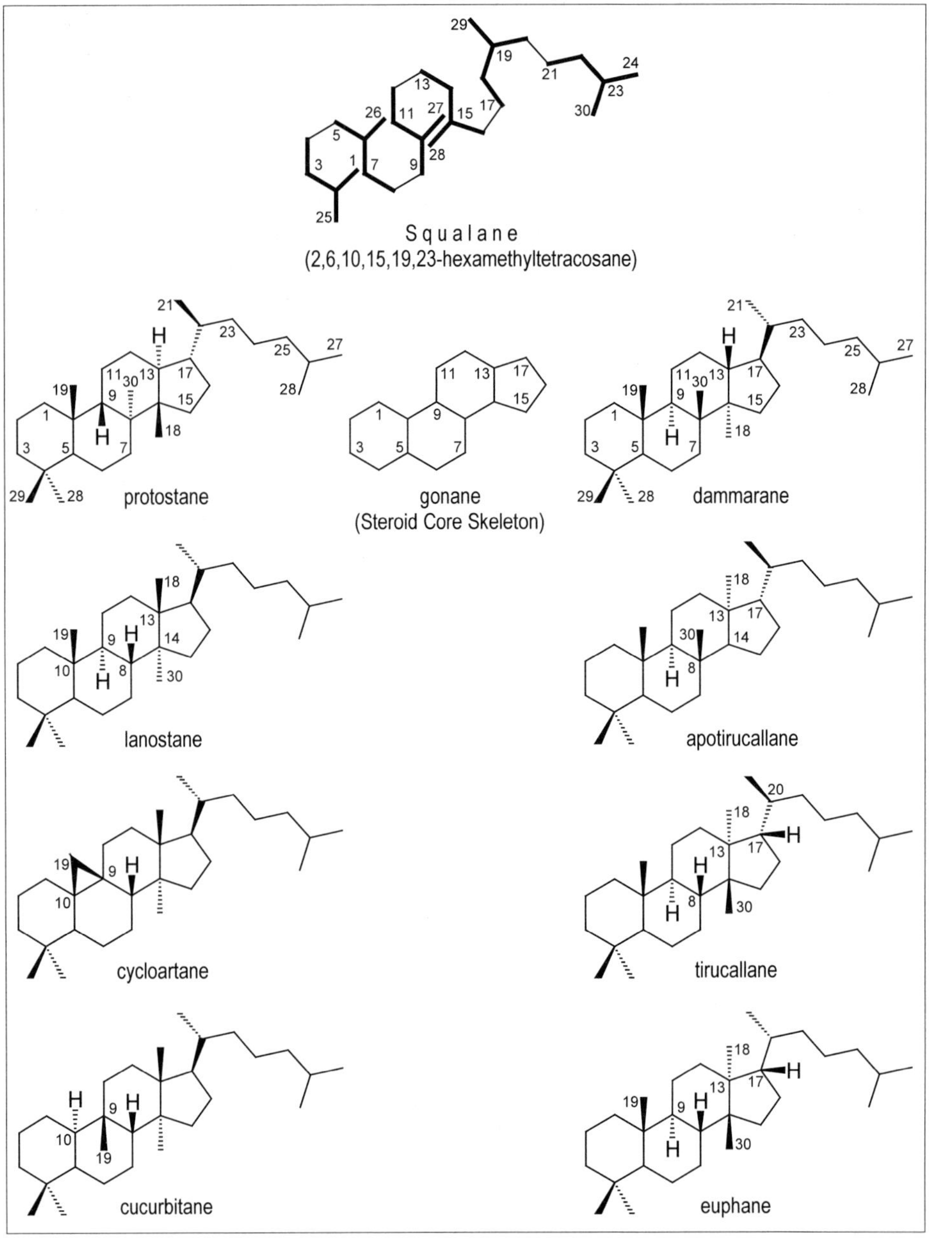

Squalane
(2,6,10,15,19,23-hexamethyltetracosane)

protostane

gonane
(Steroid Core Skeleton)

dammarane

lanostane

apotirucallane

cycloartane

tirucallane

cucurbitane

euphane

Parent skeletons of Triterpenes II
Pentacyclic Triterpenes derived from Baccharane

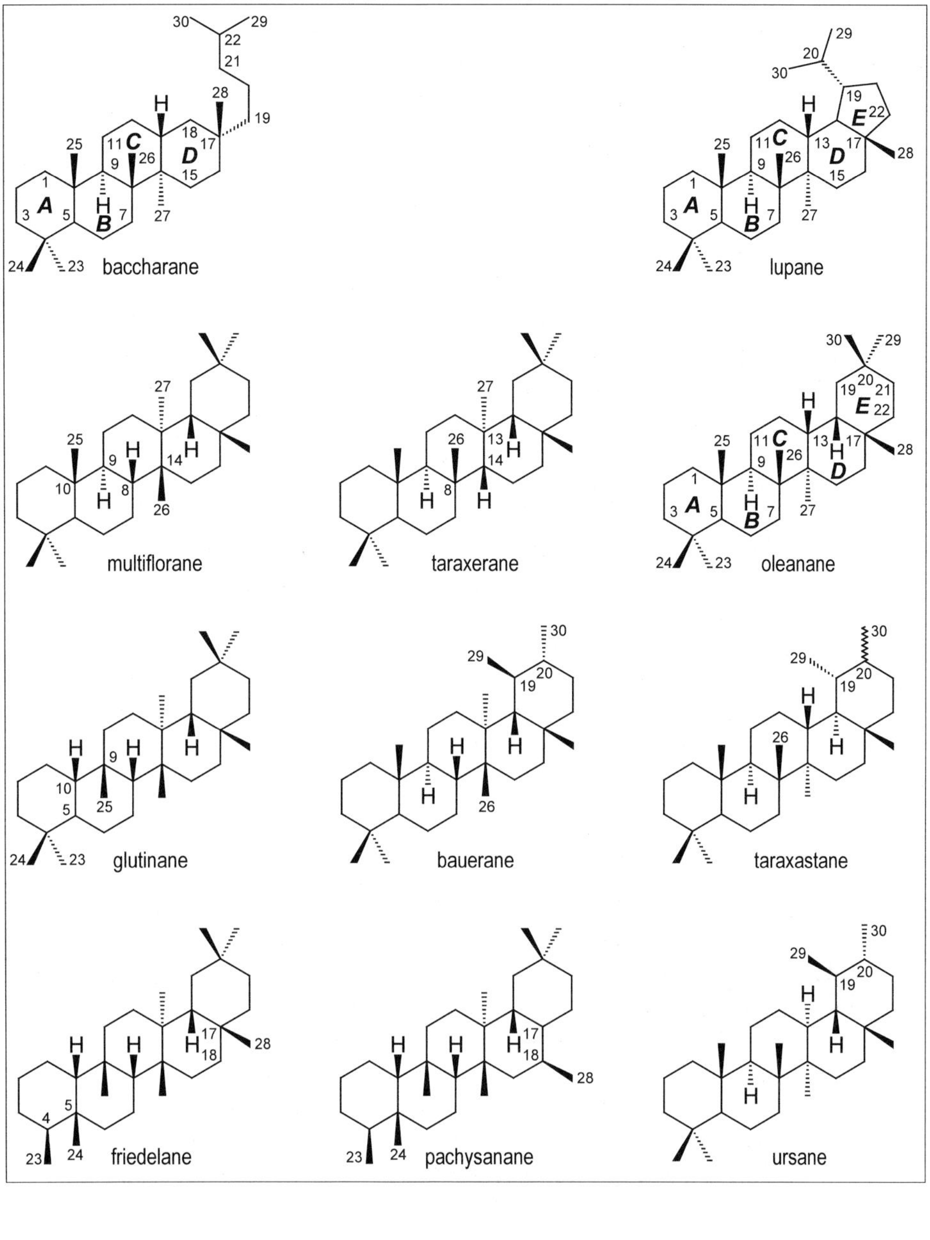

baccharane

lupane

multiflorane

taraxerane

oleanane

glutinane

bauerane

taraxastane

friedelane

pachysanane

ursane

Parent skeletons of Triterpenes III
Pentacyclic Triterpenes derived from Hopane

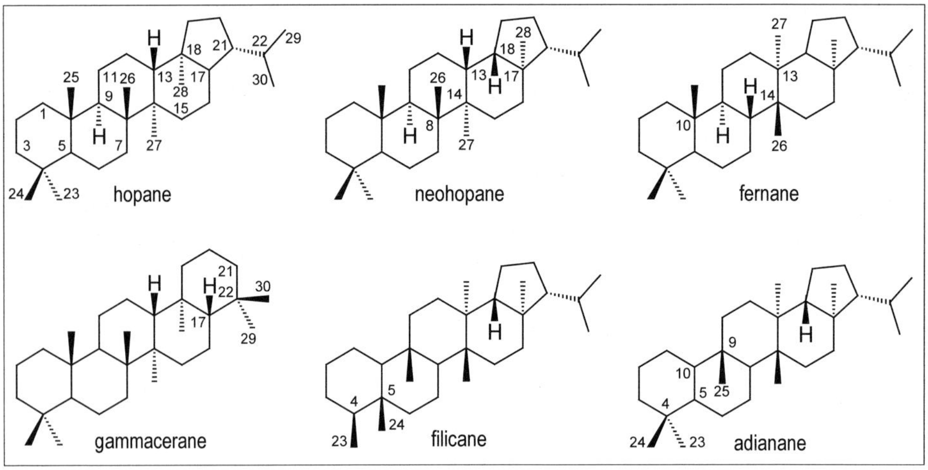

hopane

neohopane

fernane

gammacerane

filicane

adianane

Parent skeletons of Triterpenes IV
Stictanes, Arborinanes, Onoceranes, Serratanes

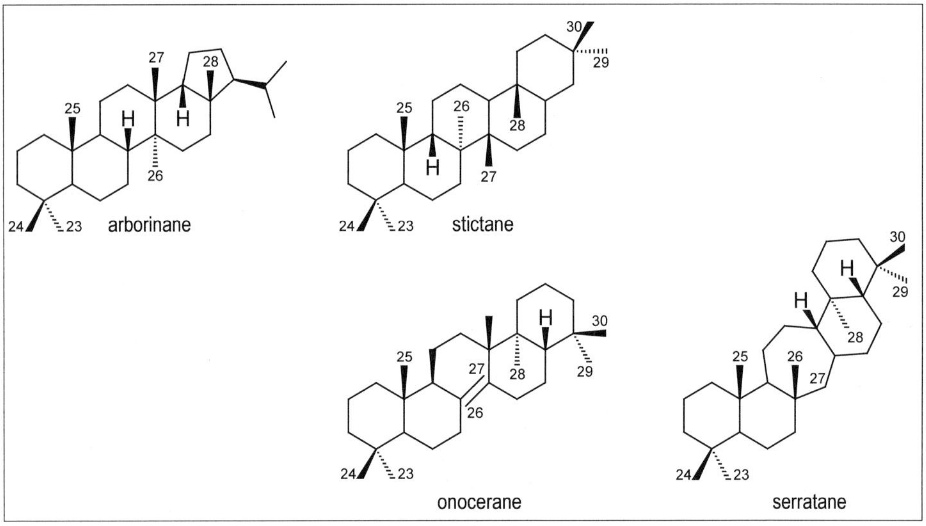

arborinane

stictane

onocerane

serratane

Subject Index

A
abienol, (+)- 54
Abies (Pinaceae)
  alba 16
  balsamea 16
  mariesii 92
  sibirica 34, 62
abieslactone, (−)- 92
abietadiene, 8,13-, (−) 62
abietanes 61
  abeo- 61
abietatriene, 8,11,13-
  (−)- 62
abietenol, (−)- 62
abietic acid, (−)- 62
abortives 39
abscisic acid 25
absinthin 38
absinthium 38
Acacia farnensiana
  (Mimosaceae) 24
acanthifolin
  structure elucidation 161
acetaldehyde
  activated 5
acetic acid
  activated 3
  dihydroxy-γ-ionylidene
    (+)- 25
acetoacetyl-CoA 3
acetyl-CoA 3
Achillea (Asteraceae)
  filipendulina 21
  millefolium 39
achillicin 39
Achronychia baueri 99
Aconitum heterophyllum
  (Ranunculaceae) 66
acoradiene, 3,7(11)-
  (+)- 45
acoranes 45
acorene, 4-
  one, 3-, (−)- 45

Acorus calamus
  (Araceae) 45
addictives 23
adiananes 102
adianene, 5-
  (−)- 104
  ol, 3β-, (+)- 104
agelasidines 53
agelasines 53
aglyca 90, 98
Ailanthus glandulosa
  (Simarubaceae) 91
Ajuga reptans
  (Labiatae) 56
ajugareptansone A
  (−)- 56
alder, black 99
aldol reaction 128
algae
  Aplysia species 80
  Botryococcus braunii 87
  Dictyota
    acutiloba 79
    cervicornis 75
    dichotoma 75, 78, 79
    linearis 75
    species 74, 78, 80
  Dilophus ligulatus 78
  Eunicella stricta 73
  Laurencia
    obtusa 46, 59
    okamurai 87
  Lobophytum species 79
  Pachydictyon
    coriaceum 78
alkaloids
  diterpenoid 59, 66, 76
  hemiterpenoid
    lysergic acid 10
  monoterpenoid
    indole 15
allyl rearrangement 119,
  123, 138, 139
Alnus glutinosa
  (Betulaceae) 99

alnusenol, (+)- 99
ambergris, grey 37, 107
ambrein, (−)- 107
ambrinal 177
ambrinol, α-, δ- 36
Ambrosia maritima
  (Asteraceae) 39
ambrosic acid 39
ambrosin 39
American rosin 57
amijol 75
amorphadiene, 4,11-
  (−)- 34
amorphanes 34
analeptics 22
analgesics 22, 23, 83, 94
  dental 28
Anamirta cocculus
  (Menispermaceae) 42
ananas acid, (+)- 94
Ananas comosus
  (Thymeliaceae) 94
androstanes
  biogenetic origin 9
anesthetics 22, 91
  local 59
Anethum graveolens
  (Umbelliferae) 18
angelic acid 10
antibacterials 14, 25, 34,
  44, 53, 56, 65, 82, 84,
  89, 97
antibiotics 44, 49
anticonvulsants 53
antifeedants 1, 14, 44, 50,
  56, 58, 80, 91
antifungals 71
antihelmintics 18, 19, 30
antihypertonics 15, 59, 94
antiinfectives 48
antiinflammatories 65, 74,
  83, 85, 101, 118

antileukemics 61, 72, 73, 76, 100
antimalarials 36
antimycotics 44
antineoplastics 39, 49, 60, 65, 71, 73, 76
antioxidants 62, 118
antipyretics 39, 42
antirheumatics 22, 23, 39, 94, 118
antirrhinoside 14
Antirrhinum species
   (Scrophulariaceae) 14
antiseptics 19
antiulceratives 52
antivirals 61, 74, 81, 94, 97
ants
  Dendrolasius
    fulginosus 24
  Iridomyrmex humilis 14
  Lasius fulginosus 11
Apium graeveolens
  (Umbelliferae) 29
apocarotene
  β-
    al, 8′- 111
  β,ψ-
    oic acid, 4′- 112
apocarotenoids 111
apotirucallanes 88, 91
Aralia racemosa
  (Araliaceae) 58
arborinanes 105
arborinene, 9(11)-
  ol, 3α- and 3β-
    (+)- 106
arborinols 106
Archaebacteriae
  Sulfolobus
    solfataricus 117
Arctostaphylos uva-ursi
  (Ericaceae) 100
arglabin, (+)- 38
aristolanes 33
aristolene
  1(10)- 33
    al, 12- 33

9-
  one, 8- 33
Aristolochia
  (Aristolochiaceae)
    debilis 33
    indica 33
    triangularis 64
armillarin, (+)- 44
Arnica (Asteraceae)
  montana 39, 100
  parryi 50
arnicenone, (+)- 50
arnidenediol 100
aromadendranes 39
aromadendrene, 1(10)-
  ol, 7-, (+)- 39
artabsin, (−)- 38
artemether 36
Artemisia (Asteraceae)
  absinthum 38
  annua 34, 36
  fragrans 13
  glabella 38
  ludiviciana 13
  pauciflora 30
  species 18
  taurica 30
artemisic acid 34
artemisinine 36
asbestinanes 70, 73
asbestinine 2, (−)- 73
ascaridole 18
Asperula odorata
  Rubiaceae) 14
asperuloside 14
astaxanthin 111
astellolide A 36
astringents 30
Atis plant 66
atisane(s) 63, 66
  one, 3-
    dihydroxy-, 16α,17-, (−)- 66
atisene
  13-
    diol
      16β,17-, (+)- 66

16-
  (−)- 66
  oic acid, 19-
    dihydroxy-, 15,20-
    olide, 19,20-
      (−)- 66
atisine, (−)- 66
ATP 3
auricularic acid, (+)- 60
autumnal coloring
  of leaves 111
axerophthene 53
Azadirachta indica
  (Meliaceae) 58, 91
azadirone, (+)- 91
azulenes, terpenoid 37

B

baccatin
  deacetyl-, 10-, (−)- 76
  III
    retrosynthetic
      disconnection 145
    synthesis 150
bacchar-21-ene
  hydroxy-, 3β- 95
baccharadiene
  12,21-, (+)- 97
baccharanes 95, 97
bacteria
  Flavobacterium
    dehydrogenatus 116
  halophilic 116
bacteriohopane
  tetrol, 32,33,34,35- 102
bacterioruberin 116
Bakanae desease 67
balm fir 16
balsamum
  tolutanum 38, 39
bark beetle 12, 20
Barosma betulina
  (Rutaceae) 177
basmene, 4-, one, 6-
  epoxy-, 7,8-, (+)- 71
baueranes 95, 99
bauerene, 7-, ol, 3β-
  (−)- 99
bearberry 100

Subject Index

benzopyrans
   monoterpenoid 22
bergamotene
   trans-, α-, (−)-
      configuration
         absolute 172
berkheyradulene 51
Betula (Betulaceae)
   alba 97
   verrucosa 115
betulaprenol
   -9, -11, -12 115
betulin, (+)- 97
betulinic acid, (+)- 97
beyeranes 63
beyerene, 15-
   (+)- and (−)- 64
   oic acid, 19-
      7-hydroxy-, (+)- 64
   one, 3-, (+) 64
   triol, 3,17,19-
      cinnamoyl, 17-O-
         (+)- 64
beyerol, (+)- 64
bicycloelemene, (−)- 27
bifloranes 79
bifloratriene
   4,10(19),15-, (−)- 79
bilobalide A 81
bio-hopanes 102
birch 115
   white 97
bisabolanes 25
   bicyclic 26
bisabolene, (+)-β- 26
bisabolol
   (+)-α- and (+)-β- 26
bitter
   herb 104
   substances 15, 38, 50,
      62, 80, 81, 91, 94
   wood 91
blackcurrant 10, 18, 177
blennin D, (+)- 44
blood plasma 111
boll weevil
   Anthonomus grandis 13
Bombus terrestris 24

bonds
   CH
      assignment
         by NMR 163
borneol, endo-, exo-
   enantiomers 21
bornyl cation 124
Boronia megastigma
   (Rutaceae) 114
Boswellia (Burseraceae)
   carterii 101
   serrata 20, 70, 98, 101
boswellic acid
   α- 98
   α-, β-, and keto- 101
botryococcenes 87
brianthein W and X 74
briaranes 70, 74
Briareum (Araceae)
   asbestinum 73
   polyanthes 74
bugwort 93
bulgaranes 34
bulgarene
   β₁- and ε- 34
bumble bee 24
but-2-en-1-ol, 3-methyl 10
butane
   2-methyl, isoprene unit 3
butanethiol, 2-
   methoxy-, 4-
      methyl-, 2- 10
butene
   1-
      ol, 4-
         carboxy-, 2- 10
   2-
      oic acid
         methyl-, 3- 10
   3-
      ol, 2-
         methyl-, 3-
            (S)-, (−)- 10
butter 52

C

cadalene 34
   ol, 3- 34
   quinone, 2,3- 34

cadinadiene
   3,9-(−)- and 4,9-(−)- 34
cadinane(s) 34
   seco- 36
cadinene, α- and β- 34
cafestol, (−)- 65
   retrosynthetic
      disconnection 141
   synthesis 141
calamenene(s) 34
   (7S,10S)-, (−)- 34
   diol, 3,8-, (+)- 34
Calendula officinalis
   (Asteraceae) 100
Camellia sinensis
   (Theaceae) 92, 114
camomile 39
camphanes 19, 21
camphene
   (+)- and (−)- 22
campherenanes 47
campherenol, (−)- 47
campherenone, (−)- 48
campholenaldehyde, (−)-
   synthesis 126
camphor
   Borneo 21
   bromo-, 3- 126
      sulfonic acid, 3- 126
   carboxylic acid, 3- 126
   enantiomers 21
   hydroxymethylene, 3-
      metal chelates 126
   ic acid
      cis- and trans- 124
   isonitroso-, 3- 126
   Japan 21
   quinone 126
   racemic
      industrial
         synthesis 124
   sulfonic acid, 10- 124
camphor tree 21, 48
Cananga odorata
   (Annonaceae) 10, 34
Canarum luzonicum
   (Burseraceae) 27
cannabidiol, (−)- 23
cannabinoids 22

cannabinol 23
  hexahydro-
    synthesis 128
  tetrahydro-
    Δ^8- and Δ^9- 23
Cannabis sativa var. *indica*
  (Cannabaceae) 18, 22, 29
caoutchouc 115
capsanthin 111
Capsicum annuum
  (Solanaceae) 111
capsorubin 111
caranes 19
caraway 22
carbanions 122, 152
carbenes 141
carbenium ions 6, 87,
  129, 141
carbonium ions 6, 124
cardiotonics 59, 117
carene, enantiomers 20
carminatives 28, 30
carnosol 62
carnosolic acid, (+)- 62
carotenal, β- 111
carotene
  β,β- 109, 111
    coloring agent 111
    synthesis 140
  β,ε- (α-), (+)- 111
  β,ψ- (γ-) 109
  ψ,ψ- 109
carotenoids 109
  apo- 111
  bioynthesis 6
  megastigmanes 113
  metabolites 113
CARROLL
  decarboxylation 119, 139
carrots 43, 45, 109, 114
Carum carvi
  (Umbelliferae) 18, 22
carvacrol 19
carvenone, (−)- 18
carveol, (−)- 17
carvone
  enantiomers 18
  odor 179

caryophylladiene
  3(15),7-, ol, 6-, (−)- 29
caryophyllanes 28
caryophyllene
  3(15)-
    epoxy-, 6,7-, (−)- 29
  6-
    al, 15-, (+)- 29
  β-, (−)- 28
Caryophylli flos
  (Caryophyllaceae) 28
casbanes 68, 71
casbene 71
cassaic acid 59
cassaidine 59
cassain 59
cassamine 59
cassanes 59
cassis 177
Castanopsis species
  (Fagaceae) 99
catalysts
  LINDLAR 119
  WILKINSON 173
cedar: 22
cedranes 46
cedrene
  retrosynthetic
    disconnection 132
  synthesis 133
cedrene, α- and β- 46
cedrol
  (+)- 46
  synthesis 134
Cedrus deodara
  (Pinaceae) 41
cembranes 68, 70
cembratetraene
  3,7,11,15-
    (−)- 70
    biosynthesis 7
cembratriene
  2,7,11-
    diol, 4β,6α-, (+)- 70
  3,7,11-
    ol, 1β-, (−)- 70
cerebroprotective
  substances 81
cericeranes 83
ceriferol I, (−)- 83

Chamaecyparis
  (Cupressaceae)
  lawsoniana 22
  nootkatensis 20, 26,
    32, 45
  taiwanensis 46
chamigradiene
  3,7(14)-
    (−)- 46
chamigranes 45
chamigrene, 7-
  one, 9-
    bromo-, 3β-
      chloro-, 2α- 46
chaminic acid, (+)- 20
cheilanthanes 84
cheilanthatriol, (+)- 84
Chenopodium
  ambrosioides
    (Chenopodiaceae) 18
chlorophyll 52
cholestanes
  biogenetic origin 9
chromatographic
  purification
    of terpenes 160
chromophores
  identification by
    UV spectra 161
chromoproteins 111
chrysanthemic acid
  esters 13
  methyl ester
    retrosynthetic
      disconnection 122
    synthesis 122
chrysanthemol, (+)- 13
Chrysanthemum
  (Asteraceae)
    cinerariaefolium 13
    vulgare 41
Cimicifuga species
  (Ranunculaceae) 93
cimicifugenol, (+)- 93
cineol, 1,4-, 1,8- 18
cinerins 13
cinerolone 13
Cinnamomum camphora
  (Lauraceae) 21, 48
circular dichroism 172

Subject Index

Cistus labdaniferus
  (Cistaceae) 54
citral
  (E,Z)-isomers 12
citronellal 12
  (R)-(+)-
    synthesis 121
    aldol reaction 128
citronellol
  (R)-(+)-
    ene reaction 121
  enantiomers 11
Citrus (Rutaceae)
  aurantium 16, 24
    var. bergamia 26
  junos 26
  paradisi 17
  reticulata 16, 17, 24
  sinensis 20, 24
  vulgaris 11
citrus flour beetle
  Plenococcus citri 13
cladielline 73
CLAISEN
  condensation 126, 141, 155
cleistanthadiene
  13(17),15-, (+)- 60
cleistanthanes 59, 60
cleistanthol, (−)- 60
Cleistanthus schlechteri
  (Euphorbiaceae) 60
cleroda-3,13-diene
  olide, 16,15/18,19-
    dihydroxy-, 2,7-
      (−)- 56
clerodanes 56
club-moss spores 106
Cneoreum tricoccon
  (Cneoraceae) 80
cneorubin (+)- U 80
CNS stimulants 42, 81
coating wax 99
  of fruits 24, 100
cocarcinogenic
  substances 72, 80
cockroach, American 27
cod-liver oil 87
coenzyme(s)
  CoQn, Q$_{10}$ 116

Coffea arabica
  (Rubiaceae) 65
coffee
  roasted 65
coleones, A-Z 63
Coleus (Labiatae)
  forskolii 54
  species 63
coloring agents
  for food 111, 112
colors
  of flowers 111
Commiphora
  (Burseraceae)
    abyssinica 26, 30
    molmol 30
compass plant 50
configuration
  absolute
    chemical
      correlation 172
    determination 171
    in diastereomers 171
  relative
    determination
      by NMR 169
conifers 20, 62, 64
connectivities
  CH, HH 163
COPE
  rearrangement 6, 26, 27, 135
  oxa- 119, 136, 155
corals 74
  Clavularia
    inflata 75
    species 74
  Xenia 80
    crassa 78
    macrospiculata 78
    obscuranata 79
coriolin
  A, B, C 49
cork oak and cork 99
corn 109
cortisone substitutes 101
costol, β-(+)- 29
costus acid, β-(+)- 29
counterirritants 28, 72
cowslip 98
cranberries 100

crocetin 112
  γ- 112
crocin
  α-, (+)- and γ- 112
Crocus sativus
  (Iridaceae) 112
Croton (Euphorbiaceae)
  nitens 71
  rhamnifolius 73
  sublyratus 52
  tiglium 72
croton oil 72
crotonitenone, (+)- 71
Crustaceae 111
crustaxanthin 111
cryptopinone 57
crystal structure 173
  three-dimensional 173
cubebanes 40
cubebanol, (−)-4- 40
cucumbers 94
Cucumis (Cucurbitaceae)
  angolensis 94
  sativus 94
cucurbitacin
  B, (+)- and F, (+)- 94
cucurbitanes 89, 94
cuminaldehyde 19
Cunurea spruceana
  (Euphorbiaceae) 60
Cupressus
  (Cupressaceae)
    macrocarpa 64
Curcuma aromatica
  (Zingiberaceae) 21
cycloabietanes, 13,16- 63
cycloaddition
  [2+1]- 133, 141
  [4+2]- 123, 147
cycloartanes 89, 93
cycloartenol, (+)- 93
cyclobranol 93
cyclobutane
  hydroxymethyl-, 1-
    dimethyl-, 2,2-
      propenyl-, 2-
        3- 13
cyclofarnesanes 25

cyclopentene
    acetyl-, 1-
        isopropenyl-, 4- 14
cyclophytanes 52
cycloreversion 127
Cymbopogon (Poaceae)
    flexuosus 12
    martinii var. *motia* 11
    winterianus 12
cymene(s) 18
    m- 18
    p- 18
        ol, 8- 19
Cyperus articulatus
    (Cyperaceae) 21
cytotoxic
    substances 31, 58, 74, 100

D

daisy flower 39
damar tree 89
damascenone 114
damascone
    hydroxy-, 3β- 114
    α- and β- 114
dammar-24-ene
    tetrol, 3β,6α,12β,20(*R*)-
        (+)- 90
    triol, 3β,12β,20(*S*)-
        (−)- 90
dammara-20,24-diene
    diol, 3β,20(*R*)-, (+)- 89
dammaranes 88, 89
    biosynthesis 87
daphnanes 70, 73
Daphne mezereum
    (Thymeliaceae) 73
daphnetoxin, (+)- 73
daucadiene
    4,8-, (+)- 43
daucane(s) 42
    epoxy-, 5,8-, ol, 9-
        (−)- 43
daucene, 8-
    ol, 5-, (+)- 43
Daucus carota
    (Umbelliferae) 43, 45, 109
dendrobin, (−)- 42

Dendrobium nobile
    (Orchidaceae) 42
dendrolasin 24
deoxyphorbol, 12- 72
devadarane-15,16-diol
    (−)- 59
devadaranes 57
diapocarotenoids 112
dictyol
    (+)- A and (+)- B 80
dictyolene 79
dictyotin B 79
DIECKMANN
    reaction 159
DIELS-ALDER
    reaction 123, 147
        hetero-, intramolecular 128
dihedral angle 169
dihydrocarvone, (−)- 18
dilophol 78
diprenyldrimanes 83
disinfectants 19
diterpenes
    acyclic 52
    bicyclic 54
    bicyclophytanes 54
    biosynthesis 8
    cembranes 68
    cyclocembranes 68
    cyclophytanes 52
    monocyclic 53, 68
    phytanes 52
    prenylsesquiterpenes 77
    syntheses 138
    tetracyclophytanes 63
    tricyclic 57
dolabella-2,6-diene
    triol, 6β,10α,18-
        (−)- 74
dolabella-4,8,18-triene
    diol, 3α,16-
        (+)- 74
dolabellanes 70, 74
dolabellene, 4(16)-
    diepoxy-, 7,8-/10,11-
    diol, 17,18- 74
dolastadiene
    1(15),17-
        diol, 3α,4β-, (+)- 75

1(15),8-
    diol, 4α,14α-, (−)- 75
dolastanes 70, 75
dolastatriene, 1(15),7,9-
    ol, 14β-, (−)- 75
dolichols 116
double bond
    equivalents 161
drimane(s)
    5α,8α,9β,10β- 36
    nor-, 11,15- 36
drimene
    7-
        ol, 11-, (−)- 36
        olide, 11,12-, (−)- 36
    8-, one, 7-, (+)- 36
drimenine 36
drimenol 36
Drimys winteri
    (Winteraceae) 36
dysideapalaunic acid
    (+)- 83

E

Ecballium elaterium
    (Cucurbitaceae) 94
edulane
    I and 8α-hydroxy-I 114
egg yolk 111
electron transfer
    agents 117
elemanes 27
elemenone, β- 27
elemi resin, Manila 27
elemol, β- 27
Elettaria (Zingiberaceae)
    cardamomum
        and *major* 18
emulsifiers 92, 98, 100
enantiomeric purity
    determination
        by NMR 126
enantiomers
    different odor 18, 178
ene reaction 121
epoxydictymene, (+)- 75
eremofortin, A and B 32
Eremophila
    serrulata 79

Subject Index

eremophiladien-9-ones
   1(10),11- and 1,11- 31
eremophilanes 31
eremophilen-2,9-dione
   (+)-11- 32
erythrolides 74
Erythrophleum guinese
   (Leguminosae) 59
erythroxylanes 57
erythroxylene, 4(18)-
   diol, 15,16-, (+)- 59
Erythroxylon monogynum
   (Araliaceae) 58, 59, 64,
   66
estranes
   biogenetic origin 9
estrogen substitutes 94
esulone
   A-(+)- and B-(−)- 72
ethereal
   oils 10, 16, 18, 160
   analysis 160
Eucalyptus (Myrtaceae)
   globulus 14, 20
   phellandra 16
eudesmanes 29
eudesmanolides, 12,6-
   oxo-, 3- 30
eudesmene, 11-
   one, 4-, 15-nor-
      prescursor of β-
      selinene 130
eudesmol
   epi-γ-, (−)- 29, 178
   α- (+)-, β-(+)-, γ-(+)- 29
eunicellanes 70, 73
eunicelline 73
euphanes 89, 91
euphol, (+)- 91
Euphorbia
   (Euphorbiaceae)
   acaulis 66
   esula 72
   fidjiana 66
   helioscopia 72
   ingens 71
   jolkini 71
   kamerunica 71
   lathyris 71
   maddeni 72

   regis jubae 92
   resinifera 73
   species 98
   supina 103
   tirucalli 91
euphorbol, (+)- 92
euphornine
   (−)- and A-(−)- 72
expectorants 34, 54

F

faradiol 100
farnesane 24
farnesene, α-, β- 24
farnesol 24
   dihydro-, (S)-2,3- 24
   synthesis 119
farnesyldiphosphate 3
feather wax
   of birds 10
feathers
   of flamingos 111
FEIST-BENARY
   furan synthesis 141
fenchanes 19
fenchene
   α-(+)- and β-(+)- 22
fenchol
   enantiomers 22
fenchones
   enantiomers 22
fernadiene, 7,9(11)-
   (−)- 103
fernanes 102, 103
fernene
   7-
      (+)- 103
      ol, 3β-, (−)- 103
   8-
      (−)- 103
      dihydroxy-, 3β,11β-
      one, 7-, (−)- 103
ferns 44
   Adiantum
      monochlamys 103,
      104
      pedatum 103, 104
   Cheilanthes farinosa 84
   Lemmaphyllum
      microphyllum 97

   Polypodium vulgare 106
fertility vitamin 118
Ferula galbaniflua
   (Umbelliferae) 38
filicanes 102, 104
filicene, 3-
   al, 23-, (−)- 104
fir (*Abies*) 54, 62
flavors
   bitter 38
   black tea 114
   blackcurrant 10
   caraway 18
   cardamom 18
   celery 29
   cinnamon 28
   fruits 24
   grapefruit 17, 32
   industrial syntheses 119
   peppermint 16
   spearmint 18
flesh
   of fruits 112
foam producers 98
Foeniculum vulgare
   (Umbelliferae) 22
forskolin, (−)- 54
fragrances
   ambergris 54
   balsamic 47
   camomile 26
   camphoric 22
   flowers 11, 12, 17, 24, 28,
   113
   industrial syntheses 119
   iris 108
   lemon peel 12
   orange 16, 24
   orchid 12
   peppermint 16, 18
   roses 114
   sandalwood
      synthesis 126
   violet 108, 114
   woody 29, 48
fragranol, (1S,2S) 13
friedelanes 95, 99
friedelanol
   3α- and 3β- 99
friedelanone, 3- 99
friedelin 99

functional groups
    identification by
        IR spectra 161
        NMR 162
fungal metabolites 89
fungi
    Armillariella mellea 44
    Aspergillus
        fumigatus 89
        ustus 84
        variecolor 36
    Cephalosporium
        caerulens 89
    Ceratocystis
        coerulescens 17
    Cercospora cruenta 25
    Clitocybe illudens 44
    Coriolus consors 49
    Fusicoccum amygdali 75
    Fusidium coccineum 89
    Ganoderma lucidum 93
    Gibberella fujikuroii 65, 66
    in cheese 32
    Lactarius
        blennius 44
        deliciosus 37
        vellereus 44
    Marasmius conigenus 44
    Merulius tremellosus 44
    Ophiobolus species 84
    Penicillium roqueforti 32
    phytopathogenic 45
    Stereum
        hirsutum 49
        purpureum 45
    Trichothecium roseum 58
furanoeremophilane(s) 31
    epoxy-, 1,10-
        dione, 6,9-, (−)- 33
    epoxy-, 1-10-, (−)- 33
    one, 9- 33
furanoeremophilene
    one, 3-, (+)- 33
furanoeudesmadiene
    1,3-, (−)- 30
    1,4-, one, 6- 30
furanoeudesmanes 30
fusicoccanes 70, 75
fusicoccin H, (+)- 75
fusicoplagin A, (−)- 75
fusidanes 88, 89

fusidic acid 89

G

galbanum resin 38
Galium odoratum
    (Rubiaceae) 14
gammacerane(s) 102
    (+)- 104
    nor, 30-
        hydroxy-, 22β-
            one, 21- 104
gammacerene, 16-
    ol, 3β-, (+)- 104
ganoderic acids 93
Gardenia species
    (Rubiaceae) 112
Gelonium multiflorum 99
geo-hopanes 102
geranial 12
geraniol 11
    synthesis 119
Geranium macrorhyzum
    (Geraniaceae) 26
geranyl-
    diphosphate 3
    geranyldiphosphate 6
    terpineol, α-, 9- 52
germacradiene
    1(10),4-, ol, 6- 26
germacranes 26
germacratriene
    1(10)-E,4-E,7(11)- 26
germacrenes 26
germacrone 26
gibberellanes 63
gibberellic acid 66
    dihydro- 67
gibberellin
    A_1 (+)-, A_3 (+)-, A_{18} (−)- 66
Ginkgo biloba
    (Ginkgoaceae) 81
ginkgolide
    A, B, C, J, M 81
Ginseng
    saponins
        and sapogenins 90
Ginseng roots 90, 98
ginsenosides 90
glucosides, glycosides 14

glutinanes 95, 99
glutinene, 5-
    ol, 3β-, (+)- 99
Glycine species
    (Leguminosae) 98
glycosides 98
Glycosmis arboreae
    (Rutaceae) 106
gold-bloom 100
golden ragwort 33
golden rod 56
gonane (sterane) 9, 88
grandisol 13
grapefruit 17
grayanatoxadiene
    10(20),15-
        triol, 3β,5β,6β- 67
grayanotoxane(s) 63
    pentol
        5β,6β,10α,14β,16β-
            epoxy-, 2,3- 67
grayanotoxene-, 10(20)-
    pentol
        2β,5β,6β,14β,16α-
            (−)- 67
GRIGNARD
    reaction 130
Grindelia robusta
    (Asteraceae) 54
grindelic acid 54
growth
    regulators 25, 36, 67, 75
guaiac wood 38
guaiadienes 38
guaiane(s) 37
    abeo- 40
    cyclo- 39
guaianolides, 12,6- 38
guaiazulene 37
guaien-11-ol
    1(5)-(−)-, 1(10)-(+)-, and 9-
        (+)- 38
Guajacum
    (Zygophyllaceae)
        officinale 38
        species 37
gum damar 89
gurjun balm 33
gutta-percha 115

Subject Index

H
halides, terpenoid 45
halimadiene
 1(10),13-
  oic acid, 18-
   oxo-, 15-, (+)- 56
 1(10),13(16)-
  oic acid, -18-
   dihydroxy-14,15-
    (+)- 56
halimanes 56
Halimium (Cistaceae)
 umbellatum 56
 viscosum 56, 81
hallucinatories 23
hashisch 23
head-to-tail link
 of terpenes 3
helenalin 39
Helenium autumnale
 (Asteraceae) 39
helicallenal 52
Helichrysum heterolasium
 (Asteraceae) 52
helvolic acid, (−)- 89
hemiterpenes 10
hemp, Indian 22, 29
Hernandia peltata
 (Hernandiaceae) 21
Hevea brasiliensis
 (Euphorbiaceae) 115
hiba oil 49
hibiscones 34
hibiscoquinones 34
Hibiscus elatus
 (Malvaceae) 34
himachalanes 41
himachalene, α- 41
himachalol 41
hirsutanes 49
hirsutene, 4(15)- 49
hirsutic acid 49
hopane(s) 101, 102
 diol, 6α,22-, (+)- 102
hopene, 22(29)-
 diol, 6α,21β-, (+)- 102
humuladiene(s)
 epoxy- 28
 one, 6-, (+)- 28

humulanes 27
humulene, α- and β- 27
humulol, (−)- 28
Humulus lupulus
 (Cannabaceae) 10, 11,
  28, 34
hyacinths 93
Hylotrupes bajulus
 (Cerambycodae) 19
hyrtial 85

I
Iberis species
 (Cruciferae) 94
ichthyotoxic
 substances 31, 42, 74
icosa-2,6-dien-1-ol
 pentamethyl-,
  3,7,11,15,19- 82
Ilex species
 (Aquifoliaceae) 99
ilexol 99
illudalanes 43
illudanes 43
illudin M and S
 (−)- 44
immune stimulants 90
immunosuppressives 83,
 89
impact compounds 17
incense 20, 70, 98, 101
ingol 71
inhibitors
 of cell division 82
 of protein phosphatase 83
 tumor 94
insect repellants 11, 12,
 46, 58
insecticides 13, 14, 74, 91
insects
 Ceroplastes ceriferus 83
ionone
 (R)-α-, β-, (S)-γ- 114
 methyl-, 8-, α-, (E)- 178
 pseudo- 139
 β- 113
  synthesis 139
Ips confusus
 and *I. typograhicus* 20

ipsdienol
 enantiomers
  synthesis 127
ipsdienol, (R)-(−)- 12
ircinin I
 (−)- 82
 dehydro-, 8,9-, (+)- 82
iridal(s) 107
 (+)- 107
iridoids 14
 seco- 14
iridomyrmecin, (+)- 14
iridosides 14
irigermanal, α-
 (+)- 107, 108
iripallidal, (−)- 108
Iris (Liliaceae)
 florentina 108
 germanica 107, 108
 missouriensis 102
 pallida 108
irone(s)
 dihydro- 108
 α-
  cis-(+)- and
   trans-(−)- 108
 α-, β-, γ- 108
 β-, (+)- 108
IR-spectroscopy 161
ishwarane(s) 33
 ol, 1-, (+)- 33
 one, 3- 33
isoabienol 54
isoamijol 75
isoborneol
 exo-
  enantiomers 21
  synthesis 124
isobornyl acetate 124
isocamphanes 19
isocaryophyllene 28
isocedranes 46, 47
isocedrene, 4-
 al, 15-, (+)- 47
 olide, 15,14-, (+)- 47
Isocoma wrigthii
 (Asteraceae) 51
isocomanes 50

isocomene 51
  retrosynthetic
    disconnection 130
  synthesis 131
isocopalanes 59, 60
isocopalene-15,16-dial
  (+)- 60
isodaucanes 42
isodaucene, 7(14)-
  epoxy-, 6,10-, (+)- 43
  one, 10- 43
isodictyohemiacetal 78
isolactaranes 44
isomenthol, (−)- 17
isomenthone, (−)- 18
isonootkatene, (+)- 32
isopentenyldiphosphate
  precursor of terpenes 3
isophorbol 72
isopimaradiene
  7,15- 58
  8(14),15-
    diol, 3,18- 58
  8,15-
    oic acid, 18- 58
    triol, 3,7,19-
      (+)- 58
isopimaranes 57
isopimaric acid, Δ^8- 58
isoprene 10
  rule 2
isoprenoids 2
  megastigmanes
    (C_{13}-) 113
  survey 8
isopulegol, (−)- 17
  synthesis 124
isopulegone, (−)- 18
isospathulenol 39
isothujol, (+)- 20
isothujone 20
isovaleric acid 10
isovelleral, (+)- 44
isoxeniaphyllenol 78
ivaxillaranes 40
ivaxillarin, (−)- 40

J
jasmine 15

Jasminum multiflorum
  (Oleaceae) 15
jasmolactone A, (+)- 15
jasmolins 13
jasmolone 13
Jatropha gossypiifolia
  (Euphorbiaceae) 72
jatrophanes 68
jatrophone, (+)- 72
Java pepper 40
jolkinol, A and B 71
JONES
  oxidation 159
junceic acid, (−)- 56
Jungia species 47
junionone 13
juniper tree 13, 45, 54
Juniperus (Cupressaceae)
  communis 13, 34, 54
  oxycedrus 34
  rigida 45
  sabina 20, 62
  thurifera 58
  virginiana 46, 49

K
kahweol, (−)- 65
kanshone A, (−)- 33
kaurane(s) 63
  (−)- 64
  furano- 65
kaurene, 16-
  (−)- 65
  dihydroxy-, 7,18-
    olide, 19,6β-, (−)- 65
  one, 15-, trihydroxy-,
    1,7,14-
      (−)- 65
kaurene, 9(11)-
  oic acid, 19-
    dihydroxy-, 16,17-
      (+)- 64
kelnac 52
kempadiene, 6,8-
  diol, 2α,13α- 76
kempanes 76
kempene 1 76
kessoglycol, (−)- 38
khusimol 47

KNOEVENAGEL
  alkenylation 126, 128, 155

L
labdadiene
  11,14-
    diol, 8,13- 54
  8(17),13(16)-
    oic acid, 19-
      dihydroxy-
        14,15- 54
labdane(s) 54
  diol, 8α,15- 54
  oic acid, 15-
    hydroxy-, 8β- 54
  triol, 5,8,18- 54
labdanolic acid, (−)- 54
labdanum resin 54
labdene
  13-
    diol, 8,15-, (−)- 54
  8(17)-
    dioic acid, 15,18-
      (+)- 54
lactaranes 44
lactaroviolin 37
lanolin 92
lanost-8-ene-3,24-dione
  nor-, 27-
    epoxy, 17,23-
      hydroxy-, 28-
        (−)- 93
lanosta-8,24-diene-3β-ol,
  (+)- 92
lanostane(s) 89, 92
  nor-, 27- 93
lanosterol 92
lanugone A 63
larch (*Larix*) 54, 62
latex 115
lathyranes 68, 71
lathyrol, hydroxy-, 7β- 71
laurencenone A 46
Lavandula angustifolia
  (Labiatae) 12
lavandulol
  enantiomers 12
laxatives 23, 72
leaf louses 14
lemon juice 22, 48

Subject Index

Leucothoe grayana
  (Ericaceae) 67
leucothol C, (−)- 67
leucothols 63
lichens (Stictaceae) 102
  Lobaria retigera 85, 103
  Sticta pulmonaria 106
Ligustrum species
  (Oleaceae) 15
lilac 17
  Indian 91
limonene
  (*R*)-(+)-
    precursor
      of selinene 129
  enantiomers 16
  biosynthesis 6
  odor 179
linalool
  (*R*)-(−)-
    synthesis 127
  enantiomers 11
  odor 178
  synthesis 119
Lindera strychnifolia
  (Lauraceae) 27
lindestrene, (−)- 30
lion's tooth 98, 100
liver 52
liverworts
  Anastrophyllum
    minutum 81
  Plagiochila
    acanthophylla 75
  Porella
    perrottetiana 80
    platyphylla 50
  Tricheolepsis
    sacculata 80
loba-8,10,13(15)-triene
  epoxy-, 17,18- 78
  ol, 18-, diacetoxy-, 14,17-
    (+)- 79
lobanes 78
lobster 111
loganin, (−)- 14
longicyclene 42
longifolene 42
longifolin 24
longipinanes 41

longipinene
  3- 41
  one, 5- 41
lubricants 92
lupadiene, 12,20(29)-
  triol, 3β,27,28-, (+)- 97
lupanes 95, 97
lupanium ion
  hydroxy-, 3β- 95
lupene, 20(29)-
  diol
    1,11-
      one, 3-, (+)- 97
    3β,11α-
      (+)- 97
lupeol
  (+)- 97
  precursor
    retrosynthetic
      disconnection 152
    synthesis 155
lupin seeds 97
Lupinus luteus
  (Leguminosae) 67, 97
lutein 111
lyclavatol 106
lycopene 109
Lycopersicon esculantum
  (Solanaceae) 109
Lycopodium clavatum
  (Lycopodiaceae) 106, 107

M

maiden's hair
  fern 104
  tree 81
maize 109
manoalide, (−)- 82
mansonone C 34
marasmanes 44
marasmic acid 44
marihuana 23
mass spectrometry (MS)
  high-resolution 161
Matricaria chamomilla
  (Asteraceae) 39
matricin 39
mayurone 49

McMurry
  deoxygenative
    coupling 140, 145
megastigmanes 113
megastigmene, 7-
  epoxy-, 5,6-
    one, 3-, (−)- 114
melianin A 91
menth-1-ene
  epidioxy-, 3,6-, *p*- 18
  thiol, 8-, *p*- 17
Mentha (Labiatae)
  arvensis 16
  gentilis 17
  piperita 16
  pulegium 18
  rotundifolia 17
  spicata 18
  spirata 17
menthane(s)
  benzenoid 18
  o-, *m*-, *p*- 15
menthol
  (1*R*,3*R*,4*S*)-(−)- 16
    synthesis 124
  enantiomers 16
  odor 178
menthone, (−)- 18
merulidial 44
meteorites 52
methylation
  reductive 159
mevalonic acid 3
Michael
  addition 122
Mimosa pudica
  (Mimosaceae) 112
mistletoe 98
modhephene, (−)- 51
molecular mass
  determination by MS 161
molds 36, 89
molecular formula
  determination 161
molecular modeling 170, 173, 176
molecular structure
  elucidation by
    NMR 163
  in the crystal 173

monoterpenes
  acyclic 10
  benzenoid 18
  bicyclic 19
  biosynthesis 3
  cyclobutane 13
  cyclohexane 15
  cyclopentane 14
  cyclopropane 13
  industrial syntheses 119
  monocyclic 13
  p-menthanes 15
Morus nigra
  (Moraceae) 115
mouthwashes 58
multifloranes 95, 99
multiflorene, 7-
  ol, 3β-, (−)- 99
Muscari comosum
  (Liliaceae) 93
muscle relaxants 44
mutageneous
  substances 44
muurolanes 34
muurolene
  ε- and γ- 34
myrcene
  synthesis 121
  β- 11
  ol, 8- 12
Myroxylon balsamum
  (Leguminosae) 38, 39
myrrh 30
myrtenal, (+)- 20
myrtenol, (+)- 20

N
NADPH + H⁺ 3
nardosinanes 33
nardosinone, (−)- 33
Nardostachys
  (Valerianaceae)
  chinensis 33
  jatamansi 31
nasal inhalers 16
natural rubber 115
neem
  oil 58
  tree
    Indian 58, 91

neoabienol 54
Neochamaelea
  pulverulenta
  (Cneoraceae) 80
neohopanes 102
neohopene, 12-
  ol, 3β-, (+)- 102
neoisomenthol, (−)- 17
neomenthol, (+)- 17
Nepeta (Labiatae)
  cataria 14
  hindostana 97
nepetalactones 14
neral 12
Nerium oleander
  (Apocynaceae) 97, 100
nerol 11
  oxide 12
nerolidol
  (S)-(+)- 24
  dehydro- 119
  synthesis 119
neurosporaxanthin 112
neurotoxic substances 67
Nicotiana tabacum
  (Solanaceae) 36, 52, 54,
    70, 114, 115
nimbiol 58
NMR
  carbon-13 162
  coupling constants
    CH 162
    HH 169
  DEPT subspectra 162
  for structure
    elucidation 161
  NOE
    difference
      spectroscopy 169
  proton 163, 169
  shift correlation
    COSY, HH, CH 163
    HMBC, HC- 163
    HMQC, HC 163
nootka cypress 32
nootkatene, (−)- 32
nootkatone 32
norditerpenes 66
nuclear OVERHAUSER
  effect (NOE) 169

O
obtusadiene 46
ocimene, β- 11
Ocimum basilicum
  (Labiatae) 11
octanoic acid
  dimethyl-, 2,6- 10
octanol
  dimethyl-, 3,7-, (R)- 10
odor
  chemoselectivity 177
  diastereoselectivity 178
  enantioselectivity 178
  influence of
    molecular shape 176
oil of
  basil 11
  bay 11
  bergamot 22, 26
  cade 34
  calamus 45
  camomile 24, 37
  Canada balsam 16
  caraway 18
  cardamom 18
  catnip 14
  cedar wood 41, 46
    Texan 49
  celery 29
  chenopodium 18
  cinnamon 28
  citronella 11, 22
    Java 12, 27
  citrus 17, 24, 28
  cloves 28
  coriander 20, 178
  cypress 40
  dill 18, 20
  eucalyptus 16, 18, 20, 28,
    29
  fennel 22
  fir-cones 16
  galbanum 26
  geranium 10, 17, 178
  ginger 25, 28
  grapefruit 32
  guaiac wood 37, 38, 40
  hemp 29
  hops 10, 11, 24, 28, 29,
    34
  incense 20
  iris 108

Subject Index

juniper 20
kalmus 26
lemon grass 12
mandarin peel 16, 17
myrrh 19, 26, 27
neroli 11, 24
orange 24
origanum 19
palmarosa 11
patchouli 38, 40, 179
peppermint 17, 27
    Japanese 17
pettigrain 11
pine needles 16, 34
pine wood 41
rose 11
    Bulgarian 12, 114
sage 29, 39, 43, 62
sandalwood 172
santal 48
savin 20
spearmint 179
spike 11
sweet flag 45
thuja 20
thyme 19, 28
turpentine 15, 20, 22
verbena (Spanish) 20
vetiver 32, 47
violet 108
water fennel 16, 17
oil slate 24, 52, 102, 104
Olea europaea
    (Oleaceae) 15, 98
oleanane(s) 95, 97
    (+)- 97
    triol, $3\beta,11\alpha,13\beta$-
        (+)- 97
oleander 97, 100
oleandrol, (+)- 97
oleanene, 12- 101
oleanolic acid, (+)- 98
oleuropein, (−)- 15
olfactory organ 176
olibanum 70, 101
olive tree 98
    European 15
olivetol 22, 128
onitin 44
onoceradiene
    7,14(27)-
        dione, 3,21- 106

8(26)/14(27)-
    diol, $3\beta,21\alpha$-
        (+)- 106
onoceranes 105, 106
Ononis spinosa
    (Leguminosae) 106
ophiobolanes 84
ophiobolin, A and G 84
orange peels 111
orchids 12, 42
Orthodon angustifolium
    (Labiatae) 19
Oryza sativa
    (Poaceae) 93
Osmanthus fragrans
    (Oleaceae) 114
oxocativic acid, 6-, (+)- 54

P

pachysanadiene, 16,21-
    diol, $3\beta,28$-, (+)- 99
pachysananes 95, 99
Pachysandra terminalis
    (Buxaceae) 99
Palaquium (Sapotaceae)
    gutta and *oblongifolia* 115
palustradiene 62
palustric acid, (+)- 62
Panax Ginseng
    (Araliaceae) 90
panaxosides 90
paprika 111
parguaranes 57
parguerene, 9(11)-
    bromo-, 15-
        triol, $2\alpha,7\alpha,16$-
            (−)- 59
partial structures
    determination
        by NMR 163
Passiflora (Passifloraceae)
    edulis 94, 114
    incarnata 94
passiflorin 94
passion flowers 94, 114
patchoulanes 40
patchoulene
    α- and β- 40
patchoulenone, (−)- 40

patchouli shrub 38
patchoulialcohol
    (−)- 40
    enantiomers
        odor 179
patchoulol 40
    enantiomers
        odor 179
peach 112
Pelargonium
    (Geraniaceae)
        graeveolens 10
        roseum 17
pentulose-5-phosphate
    deoxy-, 1-
        precursor
            of terpenes 5
pepper 26
Perilla citridora
    Labiatae 11
perillaaldehyde, (+)- 17
perillene 11
Periplaneta americana 27
periplanone(s)
    A-D 27
    B
        retrosynthetic
            disconnection 135
        synthesis 136
peroxides
    terpenoid 36, 39
    cyclic 33
perrottetianal A 80
persicachrom 112
petroleum 97, 102
phellandrene, α-, β-
    enantiomers 16
Phellandrium aquaticum
    (Umbelliferae) 16, 17
pheromone(s) 70
    aggregation 12, 20
        synthesis 127
    alarm 16
    defense 11, 24, 76, 79
    marking 24
    sexual 13, 20
        synthesis 135
    traps 1
phorbol 72
Phormium tenax
    (Cucurbitaceae) 94

photocycloaddition
  [2+2]- 130, 136
phytadiene
  (E)- and (Z)-1,3- 52
  1,3(20)- 52
phytane(s)
  (3R,7R,11R)- 52
phytanoic acid
  (3R,7R,11R)-, (−)- 52
phytohormones 1
phytol 52
phytotoxic substances 84
Picea sitchensis
  (Pinaceae) 107
picrasane 91
Picrasma excelsa
  (Simarubaceae) 91
Picris hieracioides
  (Asteraceae) 104
picrotoxanes 42
picrotoxinin, (−)- 42
pimara-8(14),15-diene
  (+)- 58
  diol, 3,18-, (+)- 57
pimaranes 57
pimaric acid, (+)- 57
Pimenta acris
  (Myrtaceae) 11
pinane(s) 19
  (1R,2S,5R)-(+)- 127
  configuration
    absolute 172
  ol, 2-, (1R,2R,5R)-
    (−)- 127
pine (Pinus) 62
pine moth 20
pinene
  α-
    epoxy-, (+)- 126
    protonation 124
  α- and β- 20
  β-
    cycloreversion 121
pines (Pinus) 54
pinguisanes 50
pinguisene, α-, (−)- 50
pinguisone, (+)- 50
pinocarveol, (−)- 20
pinocarvone, (−)- 20

Pinus (Pinaceae)
  contorta 16
  longifolia 20, 42
  pallasiana 62
  palustris 22, 62
  sibirica 26
  silvestris 20, 34, 57, 58, 62
Piper (Piperaceae)
  cubeba 40
  nigrum 26
piperitol, (−)- 17
piperitone, (−)- 18
Pistazia vera
  (Anacardiaceae) 98
plasters 90
plastoquinones (PQ) 117
plau noi 52
plaunotol 52
Plectranthus lanuginosis
  (Labiatae) 63
podocarpanes 57
podocarpic acid 58
podocarpinol 58
Podocarpus
  (Cupressaceae)
    cupressina 58
    totara 58, 63
podototarine, (+)- 63
Pogostemon (Labiatae)
  auricularis 60
  patchouli 38, 40
polyisoprene
  cis- and trans- 115
polyterpenes 115
polyterpenols 116
pomolic acid 100
positive inotropic
  substances 54
potatoes 82, 93
  sweet 24
power root 90
pregnanes
  biogenetic origin 9
prenol 10
prenyl
  aromadendranes 80
  benzothiophene
    quinones 117
  cadinanes 79

caryophyllanes 78
chromanol 118
daucanes 81
drimanes 80
elemanes 78
eudesmanes 79
fusicoccanes 84
germacranes 78
guaianes 80
isocopalanes 84
naphthoquinones
  1,4- 117
quinones 116
sesquiterpenes 77
presilphiperfolanol
  8β-, (−)- 50
presilphiperfolianes 50
prezizaanes 47
prezizaanol, 7-, (−)- 47
prezizaene, (+)- 47
Primula veris
  (Primulaceae) 98
priverogenin B, (−)- 98
proazulenes 37
propellane, [3.3.3] 51
prostratin 72
protoilludanes 43
proton distances
  evaluation
    by NMR 169
protopanaxatriol 90
protostadiene
  17(20)-(Z)-24-
    ol, 3β-, (+)- 89
protostanes 88, 89
  29-nor- 88
  biosynthesis 87
Prunus persica
  (Rosaceae) 112
pseudoguaianes 39
Pteridium aquilinum
  (Polypodiaceae) 44
pulegol, (−)- 17
pulegone, (+)- 18
pumiloxide 54
PUMMERER
  rearrangement
    selena- 137
pumpkins 94
purgatives 23, 72

pyran derivatives
    terpenoid 12
pyrethric acid 13
pyrethrins 13
pyrethrolone 13
pyruvate
    precursor of terpenes 5

Q
qinghao acid 34
qinghaosu 36
    dihydro- 36
Quassia amara
    (Simarubaceae) 91
quassin, (+)- 91
quassinoids 91
Quercus suber
    (Fagaceae) 99
Quillaja saponaria
    (Rosaceae) 98
Quillaja saponin 98
quillajic acid, (+)- 98

R
rayless golden rod 51
reiswigin A 81
resin
    acids 58, 62
    of pine trees 1
resiniferatoxin, (+)- 73
respiratory stimulants 42
retigeranic acids
    A and B 85
retigeric acid A 103
retinal 53
retinol 53
    acetate
        industrial
            synthesis 139
        retrosynthetic
            disconnection 138
retrosynthetic
    disconnection 122, 129, 138
rhamnofola-1,6,14-triene
    dione, 3,13-
        acetoxy-, 20-,
            hydroxy-, 9-
                (−)- 73
rhamnofolanes 70, 73

Rhododendron (Ericaceae)
    adamsii 26
    japonicum 67
    linearifolium 102, 103
    simiarum 104
    species 99
rhodopsin 53
Ribes nigrum
    (Saxifragaceae) 10, 18
rice 93
Ricinus communis
    (Euphorbiacae) 71
Rosa damascena
    (Rosaceae) 11, 114
rosafluin 112
rosanes 57, 58
rose Bengal
    photosensitizer 121
rose flowers 112
rose furan 11
rose oxide
    diastereomers 12
    trans-
        synthesis 121
rose-hips 109
rosein III, (−)- 58
rosemary 62
rosenonolactone 58
rosmanol, (−)- 62
Rosmarinus officinalis
    (Labiatae) 21, 62
rubber 115

S
sabinene, (+)- 20
Saccharum officinarum
    (Poaceae) 99
sacculatadiene
    7,14-
        hydroxy-, 3β-
            olide, 12,11- 80
    7,16-
        dial, 11,12-,
            hydroxy-, 18- 80
    7,17-
        dial, 11,12- 80
sacculatanes 80
sade tree 62
saffron 112
sage 54, 56, 97

Salvia (Labiatae)
    carnosa 62
    deserta 97
    hypoleuca 83
    melissodora 56
    officinalis 21
    schimperi 54
    sclarea 29, 39, 43, 54
    syriaca 83
salvileucolide
    methylester, (+)- 83
salvisyriacolide, (−)- 83
sandalore
    synthesis 126
sandalwood
    odor 37
santalal, (E)-α-, (+)- 48
santalane
    α- and β- 48
santalene, β-(−)- 48
santalol
    (Z)-α-, (+)- 48
    (Z)-β-, (−)- 48
Santalum album
    (Santalaceae) 48
santonane, (+)- 30
santonine, α- and β- 30
saponins
    and sapogenins 98
sarcinaxanthin 116
Saussurea lappa
    (Asteraceae) 30
scalaranes 85
scalarin
    (+)- 85
    deoxy-, (+)- 85
Sciadopitys verticillata
    (Taxodiaceae) 75
Scilla scilloides
    (Liliaceae) 93
sclareol 54
secologanin, (−)- 14
sedatives 94
selinanes 29
selinene
    α- and β- 29
    β-
        retrosynthetic
            disconnection 129
        synthesis 129

Senecio (Asteraceae)
  alpinus 100
  aureus 33
  glastifolius 33
  isatideus 51
  nemorensis 33
  smithii 33
sense of smell 176
  chemoselectivity 177
  diastereoselectivity 178
  enantioselectivity 178
  regioselectivity 177
serradiol 66
serratanes 105, 107
serratene
  13-
    ol, 21β-, methoxy-,
      3α-, (+)- 106
  14-
    (–)- 106
    trihydroxy-
      3α,21β,24-
        one, 16- 107
serratol 70
serrulatanes 79
serrulatanoic acid, 19-
  dihydroxy-, 8,16-, (–)- 79
sesquicarene 26
sesquiphellandrene, β- 25
sesquirosefuran 24
sesquisabinene 26
sesquiterpene(s)
  acyclic 24
  bicyclic 28, 83
  biosynthesis 6
  farnesanes 24
  lactones 38
  monocyclic 25
  polycyclic 28
  structure elucidation
    example 161
  syntheses 119, 129
  tetracyclic 33
  tricyclic 33
sesquiterpenes
  benzenoid 34
sesquithujene 26
sesterterpenes
  acyclic 82
  monocyclic 82

tetracyclic
  scalaranes 85
  tricyclic 84
sex-excitants 27
shampoos 98
SHAPIRO
  coupling 147
SHARPLESS
  epoxidation 149
shift reagents
  chiral for NMR 126
Shorea wiesneri
  (Dipterocarpaceae) 90
shyobunol 26
Sideritis serrata
  (Labiatae) 66
silk worm
  Bombyx mori 97, 115
silphinanes 50
silphinene, 1-
  (–)- 50
  one, 3-, (+)- 50
silphiperfolene, 5-(–)- 50
silphiperfolianes 50
Silphium perfoliatum
  (Asteraceae) 50
SIMMONS-SMITH
  reaction 133
sinensal, α-, β- 24
sinensiachrom 112
sinensiaxanthin 112
skeletal structure
  elucidation
    by NMR 163
skin creams 58
skin irritants 71, 73, 80
snapdragon 14
sneezing powder 98
soap tree 98
Solanum tuberosum
  (Solanaceae) 82, 93
solenolide A, (–)- 74
Solenopodium species 74
Solidago juncea
  (Asteraceae) 56
soyasapogenol, (+)- 98
soybean 98

sperm whale
  Physeter macrocephalus
    37, 107
sphenolobadiene
  13(15),16-
    diol, 5α,18-
      epoxy-, 3α,4α-
        (+)- 81
  3,17-
    ol, 13-, (+)- 81
sphenolobanediones 81
sphenolobanes 81
sponges
  Agelas nakamurai 53, 56
  Cacospongia scalaris 82,
    85
  Dysidea etheria 83
  Epiupolasis reiswigi 81
  Hyrtios erecta 85
  Ircinia oros 82
  Luffariella variabilis 83
  Spongia officinalis 60, 85
spongiadiene, 13(16)14-
  (–)- 61
spongiane(s) 60
  diol, 15,16-, (–)- 60
spongiene, 12-
  one, 16-
    hydroxy-, 11β-
      (+)- 61
spruceanol, (–)- 60
squalane 86
squalene 86, 87
  biosynthesis 6
  diepoxy-, 2,3-/22,23- 105
  epoxy-
    1,10-, (+)- 87
    2,3- 87, 105
      precursor of
        tetracyclic
        triterpenes 9
steroids
  biogenesis 9, 92
sterpurenes 45
Stevia aristata
  (Asteraceae) 64
stictane(s) 105
  diol, 3β,22β-
    (+)- 106
  triol, 2α,3β,22α-
    (+)- 106

Subject Index

stolondiol, (−)- 74
straw flower 52
Strychnos nux vomica
   (Loganiaceae) 15, 93
sugar
   beet 98
   cane 99
sulfolobusquinone 117
Surinam quassia 91
Syzygium (Myrtaceae)
   aromaticum 28
   claviflorum 97

T

Tagetes glandulifera
   (Asteraceae) 12
tagetone, *(E)*- 12
tail-to-tail link
   of farnesane units 86
   of terpenes 3
Tamarindus indica
   (Leguminosae) 114
Tambooti wood 64
Tanacetum parthenium
   (Asteraceae) 39
tanaparthin-α-peroxide 39
Taraxacum officinale
   (Asteraceae) 98, 100
taraxanes 98
taraxastanes 95, 100
taraxastene
   20-
      diol, 3β,16β- 100
      ol, 3β-, (+)- 100
   20(30)-
      diol, 3β,16β-, (+)- 100
taraxasterol, ψ- 100
taraxeranes 95
taraxerene, 14-
   ol, 3β-, (+)- 98
taraxerol, (+)- 98
taurin, (−)- 30
taxanes 70
taxine A, (+)- 76
taxol, (−)- 76
   syntheis 145
Taxus (Taxaceae)
   baccata 76
   brevifolia 76

tea, black 92, 114
termites 16, 70
   Cubitermes umbratus 79
   Nasutitermes kempae 76
   Trinervitermes
      gratiosus 76
terpenes
   basic structure 2
   biosynthesis 3, 5
   chromatographic
      purification 160
   ecological function 1
   molecular structure
      elucidation 160
   parent skeletons 3
   survey 185
   significance 1
terpinene, 16
terpineol, α- 17
   synthesis 123
terpinolene 16
terrestrol 24
tetrahydrocannabinol
   Δ^8- and Δ^9- 23
tetrasulfides
   cyclic, terpenoid 11
tetraterpenes
   apocarotenoids 111
   carotenoids 109
   diapocarotenoids 112
Teucrium fragile
   (Labiatae) 56
teugin, (−)- 56
THCs 23
theaspirane
   A-(+)- and B-(+)- 114
theaspirone 114
thiophene
   methyl-3-pentenyl-, 4-
      3- 11
Thuja (Cupressaceae)
   occidentalis 20, 22
   plicata 22

thujane(s) 19
   ol, 3α-, (−)- 20
   one, 3-, (+)- 20
thujene
   3- 20
   4(10)-, (+)- 20
thujol 20

thujone 20
thujopsanes 49
thujopsene
   3-, (−)- 49
   4-
      one, 3-
         nor-, 15-, (+)- 49
Thujopsis dolabrata
   (Cupressaceae) 49, 64
thymol 19
Thymus vulgaris
   (Labiatae) 19
tiglianes 70, 72
tiglic acid 10
tirucalla-7,24-diene
   ol, 3β-, (−)- 92
tirucallanes 89, 91
tirucallol
   (+)- 91
tobacco 36, 52, 54, 70,
   114, 115
tocopherol, α-, (+)- 118
tocoquinone 118
tomatoes 109, 114
tonics 42, 90, 93
tormesol 81
totaranes 61
totarol, (+)- 63
tranquilizers 14, 23
tree of life 20
tretinoin 53
trinervita-1(15),8-diene
   diol, 2β,3α- 76
   ol, 2β- 76
trinervitanes 70, 76
triquinane
   skeleton 49, 50, 85
trisulfides
   cyclic, terpenoid 11
triterpenes
   acyclic 86
   crystal structure 173
   degradation products 108
   homo- 92
   iridals 107
   molecular modeling 173
   nor-
      di-, 26,27- 106
      tetra- 91

pentacyclic
   baccharane type 95
   hopane type 101
synthesis 152
tetracyclic
   gonane type 88
TROST's reagent 136
tubipofuran, (+)- 30
turpentine 1, 34, 58, 62
Tussilago farfara
  (Asteraceae) 100

U

ubiquinones UQ-n 116
Ulmus (Ulmaceae)
  lactinata 34
  thomasii 34
ursanes 95, 100
ursanoic acid, 28-
  hydroxy-, 3β-, (+)- 100
ursene, 12- 101
  oic acid, 28-
    dihydroxy-, 3β,19α-
      (+)- 100
    (20R)- and
      (20S)- 171
  hydroxy-, 3β-
    (+)- 100
ursolic acid 100
UV- and visible light
  absorption
    spectroscopy 161

V

Vaccinum macrocarpon
  (Ericaceae) 100
valencanes 32
valepotriates 14
valeranes 31
valerenal 41
valerenanes 40
valerenoic acid, (−)- 41
valerenol 40

valerenones 31
valerian 14, 22, 31, 32,
  38, 41
Valeriana officinalis
  (Valerianaceae) 14, 22,
  31, 32, 38, 41
valerianol, (+)- 32
valtrate, (+)- 14
vasodilators 54
vegetable oils 87, 118
Verbenia triphylla
  (Verbenaceae) 20
verbenol, (+)-*trans*- 20
verbenone, (+)- 20
  conv. to ipsdienols 127
vermouth 38, 45
verticillanes 70, 75
verticillol, (+)- 75
Vetiveria (Poaceae)
  zizanoides 32, 47
vetivone, α- 32
villanovane(s) 63
  diol, 13α,19-, (+)- 65
  triol, 3,13,17-
    and 13,17,19- 65
Viola odorata
  (Violaceae) 108, 114
Viscum album
  (Viscaceae) 98
visual process 53
vitamin
  A 53
    activity 111
    alcohol
      industrial
        synthesis 140
      retrosynth. discon-
        nection 138
    aldehyde
      isomerization 111
    precursors 111
  E 118
  K series 117
vulcanization 115

W

WAGNER-MEERWEIN
  rearrangement 6, 95, 105,
    124
wheat germ 118
WILLIAMSON
  ether synthesis 143, 149
WITTIG
  alkenylation 130, 138, 139
woodruff 14
woodworm 19
wool fat 92
wormseed, American 18
wormwood 38

X

xanthophyll 111
xeniaacetal 78
xeniaphyllanes 78
xenicanes 78
xeniolit A 78
X-ray diffraction 173

Y

yarrow 39
yew tree 76
ylang-ylang oil 10, 34

Z

Zea mays
  (Poaceae) 109
Zingiber (Zingiberaceae)
  officinalis 25
  zerumbeticum 28
zingiberene, (−)- 25
zizaane 47
zizaene
  6(13)-
    (+)- 47
  ol, 12-, (+)- 47
zooplankton 52